SELF-ORGANIZATION AND EMERGENCE IN LIFE SCIENCES

SYNTHESE LIBRARY

STUDIES IN EPISTEMOLOGY,

LOGIC, METHODOLOGY, AND PHILOSOPHY OF SCIENCE

VOLUME 331

SELF-ORGANIZATION AND EMERGENCE IN LIFE SCIENCES

Edited by

BERNARD FELTZ
Université Catholique de Louvain, Louvain-la-Neuve, Belgium

MARC CROMMELINCK
Université Catholique de Louvain, Bruxelles, Belgium

and

PHILIPPE GOUJON
Facultés Universitaires Notre-Dame de la Paix, Namur, Belgium

Springer

A C.I.P. Catalogue record for this book is available from the Library of Congress.

ISBN-10 1-4020-3916-6 (HB)
ISBN-13 978-1-4020-3916-4 (HB)
ISBN-10 1-4020-3917-4 (e-book)
ISBN-13 978-1-4020-3917-1 (e-book)

Published by Springer,
P.O. Box 17, 3300 AA Dordrecht, The Netherlands.

www.springer.com

Printed on acid-free paper

Printed in the Netherlands.

First Edition in French: Feltz, B., Crommelinck, M. and Goujon, Ph. (1999). Auto-organisation et émergence dans les sciences de la vie, OUSIA, Bruxelles.

CONTENTS

B. Contemporary Origins

III. Epistemological and Conceptual Approaches
A. Teleology and Intentionality

B. Explanation

LIST OF CONTRIBUTORS

ATLAN Henri, Human Biology Research Center, Hadassah University Hospital, Jerusalem, Israël and Centre hospitalier universitaire Broussais, Hôtel Dieu, Paris, France.

BERSINI Hugues, IRIDIA, Université libre de Bruxelles, Bruxelles, Belgium.

BAUCHAU Vincent, Netherlands Institute of Ecology, Heteren, Netherland.

BOUQUIAUX Laurence, Université de Liège, Belgium.

BRANDON Robert, Philosophy Department and Zoology Department, Duke University, Durham, USA.

COHEN Irun, Cellular Biology Department, Weizmann Institute of Sciences, Rehovot, Israël.

CROMMELINCK Marc, Laboratoire de Neurophysiologie, Université catholique de Louvain, Bruxelles, Belgium.

DUCHESNEAU François, Département de Philosophie, Université de Montréal, Montréal, Canada.

FELTZ Bernard, Institut Supérieur de Philosophie, Université catholique de Louvain, Louvain-la-Neuve, Belgium.

GOUJON Philippe, Facultés Universitaires Notre-Dame de la Paix, Namur, Belgium.

HEUDIN Jean-Claude, Institut d'Électronique fondamentale, Université de Paris XI, Orsay, France.

KOKOSZKA Valérie, Institut Supérieur de Philosophie, Université catholique de Louvain, Louvain-la-Neuve, Belgium.

LEFÈVRE Philippe, Unité d'Automatique, de Dynamique et d'Analyse des Systèmes, Université catholique de Louvain, Louvain-la-Neuve, Belgium et N.I.H., Washington, USA.

LIVET Pierre, Centre des Lettres et Sciences humaines, Université de Provence (Aix-Marseille I), Aix-en-Provence, France.

MAESSCHALCK Marc, Institut Supérieur de Philosophie, Université catholique de Louvain, Louvain-la-Neuve, Belgium.

MEIRE Philippe, Département de Psychologie clinique, Université catholique de Louvain, Louvain-la-Neuve, Belgium.

MENGAL Paul, Université de Paris XII, Faculté de Lettres et de Sciences humaines, Créteil, France.

MISSAL Marcus, Laboratoire de Neurophysiologie, Université catholique de Louvain, Bruxelles, Belgium.

RICHARDSON Robert, Philosophy Department, University of Cincinnati, USA.

THOMAS René, Chimie Physique, Université Libre de Bruxelles, Bruxelles, Belgium.

THOMPSON Paul, Scarborough College, University of Toronto, Toronto, Canada.

TU C, Unité d'Automatique, de Dynamique et d'Analyse des Systèmes, Université catholique de Louvain, Louvain-la-Neuve, Belgium.

VAN DE VIJVER Gertrudis, NFWO, Gent University, Belgium.

VARELA Francisco (†), Laboratoire de Neurosciences cognitives, CNRS, URA 654 et Centre de Recherche en Épistémologie Appliquée (CREA), École Polytechnique, Paris, France.

WEISBUCH Gérard, Laboratoire de Physique Statistique, École Normale Supérieure, Paris, France.

BERNARD FELTZ, MARC CROMMELINCK, PHILIPPE GOUJON

INTRODUCTION

The concept of self-organization takes a growing place in the evolution of contemporary sciences. Coming from the second cybernetics, which developed in USA at the end of the 1950th, this concept had first implications in biological sciences in the context of the Biological Computer Laboratory founded by Von Foerster and in the works of three symposia on the Self-Organizing systems from 1960 to 1962. During the 1970th, this approach was developed especially by the chilian school of biology. Since the 1980th, the Santa Fe Institute gives a new impulse to these perspectives. These works go on linked with the progress in the algorithm's theories, in artificial intelligence and in the analysis of non linear systems, in particular by the Brussels school. They lead, on the beginning of the 1990th, to books whose explicit purpose is a fundamental new approach of the living.

The concept of emergence refers to the coming out of new properties linked to the complexity of an organization. In scientific context, self-organization models have an important place in the formalization of emergence. The order from chaos, presented by Self-Organizing models, is often interpreted in terms of emergence, *id est* the advent of a higher level of organization.

These two concepts can be analysed according to different perspectives. This explains the structure of this book in three parts: scientific, historic and epistemologic. It will be first analysed in what extent the concepts of self-organization and emergence have some impact in experimentations in the different fields of contemporary life sciences. Second, historical origins, distant or more recent, will be envisaged. This concerns remote intuitions of antiquity, the first approach in philosophy of life in the modern period, as the more recent developments of the first and second cybernetics. Finally, in a third part, emergence and self-organization will be epistemologically analysed in relation with the questions of teleology and explanation.

B. Feltz, M. Crommelinck and P. Goujon, (eds.), Self-organization and Emergence in Life Science, 1-3.

* * *

The scientific approach presents two parts. The first one is an introduction to different formalisms of self-organization and emergence. Physicist G. Weisbuch introduces to the dynamic complex systems. V. Bauchau analyses boolean automata networks in biology and H. Bersini presents the problematic of artificial life. The second part analyses experimental biology and medical practice. R. Thomas shows the importance of positive feed back in the cellular differentiation process. Ph. Lefevre and his colleagues develop an example of emergent properties of neuronal networks and F. Varela studies neuronal synchronization in cognitive functions. Ph. Meire analyses the relevance of self-organization concept in psychiatric practice. Finally, H. Atlan shows the fecundity of self-organization perspective in immunology. The dominant image is one of great potentialities with already actual results but specially a great hope of promise.

For historicist, such a fecundity is not surprising. Self-organization and emergence problematic indeed concerns fundamental debate on specificity of living since antiquity to contemporary period. G. Van De Vijver shows that precisely in a detailed analysis of kantian position. More linked to the history of science, the contribution of F. Duchesneau studies the concepts of "formative force" and "essential force" in the epigenesis theories in the 18th century, while P. Mengal shows how, in the 19th century, the concept of emergence oscillates between biology and theology. This historical survey shows that self-organization and emergence, in their philosophical intuitions, lead to a concept of scientific approach of living which takes distance with mechanistic project. On the contrary, analysis of more recent origin of these concepts places us in a radically mechanicist perspective. The first cybernetics is the starting point of a more complex elaboration which tends to integrate the problematic of self-programmation. J.C. Heudin develops such perspectives in relation with artificial life, while P. Livet studies the relations between self-organization and the logic of deconstruction. Historical approach exhibits clearly ambiguities of self-organization and emergence. Distant origin refers to concepts which lead to vitalism, while proximate context places these concepts in a deliberate mechanistic research programme.

This ambiguity is precisely in the core of epistemological analysis of the third part. All the scientists and philosophers of this book keep away from vitalism without renouncing to the question of the specificity of living which presents new formulations. R. Brandon analyses the relation between self-organization and teleology, which is at the core of living, while

M. Maesschalck and V. Kokoszka envisage the relation between self-organization and the phenomenological intentionality. Moreover, epistemological analysis of emergence is linked to the question of explanation which focalises the last contributions. P. Thompson studies the concept of model in Self-Organizing systems. R. Richardson analyses the relation between explanation and causality in these systems. Finally, B. Feltz proposes an articulation between self-organisation and selection in evolutionary theory and analyses the implication of these concepts in the question of emergence.

* * *

We thank the Fonds National de la Recherche Scientifique de la Communauté Française de Belgique, the Institut Supérieur de Philosophie as the Mécénat of the Université catholique de Louvain without whom this book would not have been possible.

I. SCIENTIFIC APPROACH

A. Self-Organization and Biology: General Standpoints

GÉRARD WEISBUCH

THE COMPLEX ADAPTATIVE SYSTEMS APPROACH TO BIOLOGY

The purpose of this contribution is to describe the applications of concepts and methods derived from statistical physics of disordered systems and non-linear dynamics to certain issues in Theoretical biology. In those applications, the central issue is to study functional organization of a multi-component system based on a simplified description of the components. The first section gives a few examples of complex systems taken from physics and biology. We then describe three formalisms commonly used in theoretical biology. The central concepts of this approach, the *attractors* is introduced in the section on networks. Rather than emergence, we further discuss generic organizational properties of networks and give some examples which characterize the difference between organized and chaotic dynamical regimes. Before concluding, we discuss two implementations of memory in models of the brain and of the immune system.

1. FROM STATISTICAL PHYSICS TO COMPLEX SYSTEM

1.1 The Physics Approach to Simplicity and Complexity

Statistical physics has accustomed us to mathematical descriptions of systems with a large number of components. The thermodynamic properties of ideal gases were understood as early as the end of the 19th century, while those of solids were understood at the beginning of the 20th century. In both cases, two important properties make modeling easy:

These are systems in which all of the components are identical.

B. Feltz, M. Crommelinck and P. Goujon, (eds.), Self-organization and Emergence in Life Science, 7-28.

If the interactions between the components are very weak, they can be ignored, as in the case of ideal gases. Otherwise, as in the case of solids, we can use linearization methods to put the problem into a form in which these simplifications can be made.

These early successes compared to the difficulties encountered in the understanding of biological systems would make us consider the above mentioned systems as rather simple.

On the other hand, here are some examples of complex living systems:

The human brain is composed of approximately ten billion cells, called neurons. These cells interact by means of electrico-chemical signals through their synapses. Even though there may not be very many different types of neurons, they differ in the structure of their connections.

The immune system is also composed of approximately ten billion cells, called lymphocytes with a very large number of specificities which interact via molecular recognition, in the same way as recognition of foreign antigens.

Even the metabolism of a single cell is the result of interactions among a large number of genes which results into the cell function.

Although complexity is now a somewhat overused expression, it has a precise meaning within this text: it a complex system is a system composed of a large number of different interacting elements.

In fact, the great majority of natural or artificial systems are of a complex nature, and scientists often choose to work on model systems simplified to a minimum number of components, which allows to observe "pure" effects. This approach is illustrated by a number of Belgian teams (see Nicolis and Thomas). The complex systems approach, on the other hand, is to simplify as much as possible the components of a system, so as to take into account their large number. This idea has emerged from a recent trend in research known by physicists as the physics of disordered systems.

1.2 Disordered Systems

A large class of physical systems, known as multiphase systems, are disordered at the macroscopic level, but some are disordered even at the microscopic level. Glasses, for example, differ from crystals in that interatomic bonds in a glass are not distributed according to symmetries which we observe in crystals. In spite of this disorder, the macroscopic physical properties of a glass of a given composition are generally the same for different samples, as for crystals. In other words, disorder in a system does not lead to impredictable behavior. The simple models used by physicists are based on periodic networks, or grids, and simplified components of two different types are placed on the nodes, such as for example conductors or insulators in the problem known as percolation. These components are randomly distributed, and the interactions are limited to pairs of neighboring nodes. For large enough networks, we perceive that certain interesting properties do not depend on the particular sample created by a random selection, but of the parameters of this selection. In the case of the aforementioned insulator/conductor mixture, the conductivity between the two edges of the sample depends only on the ratio of the number of conductive sites to the number of insulating sites.

These primeval examples show the approach taken by a number of theoretical biologists:

We choose to oversimplify the components of the system whose global behavior we would like to model. The formal genes, neurons and lymphocytes discussed below are cartoon-like simplifications of biological polymers and cells.

Nonetheless, these simplifications enable us to apply rigorous methods and to obtain exact results.

Furthermore this approach of biology is dynamical. We start from a local description of the state changes of the components due to their interactions. We expect the global description of the system from the method, that is to say the long term behavior of the system as a whole. The global behavior can be very complex, and it can be interpreted in terms of emergent properties. Within this notion is the idea that the properties are not *a priori* predictable from the structure of the local interactions, and that they are of biological functional significance.

2. NETWORKS

2.1 Units

2.1.1 Boolean Automata

A simplified automaton is defined by its sets of inputs and outputs and by the *transition function,* which gives the output at time $t+1$ as a function of the inputs and sometimes also the internal state (*i.e.* the output) at time t. In addition, we will limit ourselves to binary automata, that is to say to two states, for example 0 and 1.

Boolean automata operate on binary variables, that is to say variables which take the values 0 or 1. The usual logic functions AND, OR and XOR are examples of transition functions of boolean automata with two inputs. A boolean automaton with k inputs, or of *connectivity k,* is defined by a truth table which gives the output state for each one of the 2^k possible inputs. There are 2^{2^k} different truth tables, and then 2^{2^k} automata.

Let $k = 2$. Here are the truth tables of four boolean logic functions with two inputs:

Table 1.

	AND				OR				XOR				NAND			
Input	00	01	10	11	00	01	10	11	00	01	10	11	00	01	10	11
Output	0	0	0	1	0	1	1	1	0	1	1	0	1	1	1	0

On the input line of the table, we have represented the four possible input states by 00, 01, 10, and 11. The four truth tables correspond to the standard definitions of the following logic functions: AND returns a 1 only if its two inputs are 1; OR returns a 1 only if at least one of its inputs is a 1; XOR is 1 only if exactly one of its inputs is a 1; and NAND is the complement of AND. In logical terms, if A and B are two propositions, the proposition (A AND B) is true only if A and B are true.

We will further discuss the application of boolean units to genetics.

2.1.2 Threshold Automata

The state x_i of the ith threshold automaton is computed according to:

$$h_i = \sum_j J_{ij} x_j \tag{1}$$

$$x_i = 1 \text{ if } h_i > \theta_i \, ; \, x_i = 0 \quad \textit{otherwise}$$

The sum is computed over all of the inputs, subscripted by j. J_{ij} is the weight of the interaction between the ith and jth automata. In other words, the ith automaton has the value 1 if the weighted sum of the states of the input automata $\Sigma J_{ij} x_j$ is greater than or equal to the threshold θ_i and 0 otherwise.

We will further summarize some applications of threshold units to cognition.

2.1.3 Formal Lymphocytes

Not all networks are made of automata. A number of authors studying neural nets used differential equations as units. In immunology, Perelson and Weisbuch (1997), for instance, started from the following model of lymphocytes proliferation. The time evolution of the population x_i of clone i is described by the following differential equation:

$$\frac{dx_i}{dt} = m + x_i(pf(h_i) - d) \tag{2}$$

where m is a source term corresponding to newly generated cells coming into the system from the bone marrow, the function $pf(h_i)$ defines the rate of cell proliferation as a function of the "field" h_i, and d specifies the per capita rate of cell death.

For each clone i, the total amount of stimulation is considered to be a linear combination of the populations of other interacting clones j. This linear combination is called the field, h_i, acting on clone x_i, *i.e.*,

$$h_i = \sum_j J_{ij} x_j \tag{3}$$

where J_{ij} specifies the interaction strength (or affinity) between clones x_i and x_j. The choice of a J matrix defines the topology of the network. Typically J_{ij} values are chosen as 0 and 1.

The most crucial feature of this model is the shape of the activation function $f(h_i)$, which is taken to be a log bell-shaped dose-response function

$$f(h_i) = \frac{h_i}{\theta_1 + h_i}\left(1 - \frac{h_i}{\theta_2 + h_i}\right) = \frac{h_i}{\theta_1 + h_i}\frac{\theta_2}{\theta_2 + h_i} \tag{4}$$

with parameters θ_1 and θ_2 chosen such that $\theta_2 \gg \theta_1$.

Below the maximum of $f(h_i)$, increasing h_i increases $f(h_i)$, we call this the *stimulatory regime*. Above the maximum, increasing h_i decreases $f(h_i)$; we call this the *suppressive regime*. Plotted as a function of log h_i, the graph of $f(h_i)$ is a bell-shaped curve.

2.2 Networks

2.2.1 Structural Properties

A *network* is composed of units interconnected such that the outputs of some are the inputs of others. It is therefore a directed graph, where the nodes are the units and the edges are the connections from the output of one unit to the input of another.

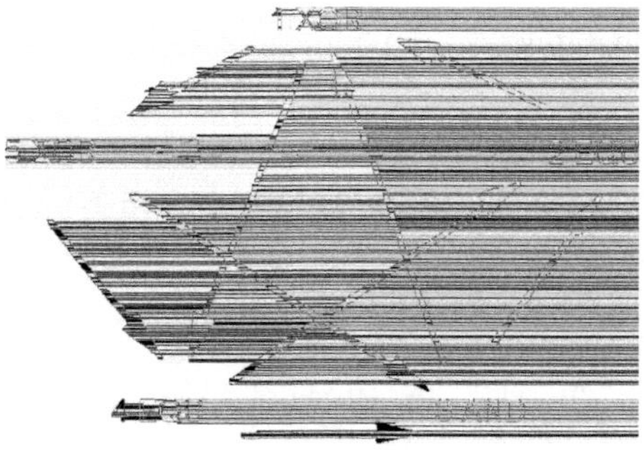

Figure 1. A network of five boolean automata with two inputs. Each automaton has two inputs and transmits its output signal to two other automata. The XOR and AND functions have been previously defined. The EQU(ivalence) function is the complement of the XOR function – it is 0 only if exactly one input is a 1.

Figure 1 represents the graph of the connections of a network of five boolean automata with two inputs.

A network of five boolean automata with two inputs. Each automaton has two inputs and transmits its output signal to two other automata. The XOR and AND functions have been previously defined. The EQU(ivalence) function is the complement of the XOR function — it is 0 only if exactly one input is a 1.

2.2.2 Dynamical Properties

Iteration Mode

Let us discuss here the dynamics of automata networks, since the notion related to attractors are easily defined. Everything discussed here generalizes to continuous dynamics. In fact historically, most notions were first discussed for continuous dynamics.

The dynamics of an automata network are completely defined by its connection graph, the transition functions of the automata, and by the choice of an *iteration mode.* It must be stated whether the automata change their state simultaneously or sequentially, and in what order. In the parallel mode, for instance, all of the automata change their state simultaneously as a function of the states of the input automata in the previous timestep. Conversely, in the case of *sequential iteration,* or iteration in series, only one automaton at a time changes its state. Sequential iteration is therefore defined by the order in which the automata are to be updated. In the discussion that follows, we will talk only of *parallel iteration.*

Iteration Graph

There are 2^N possible configurations for a network of N boolean automata. The network goes from one configuration to the next by applying the state change rule to each automaton. Its dynamics can be represented by a directed graph, the *iteration graph,* where the nodes are the configurations of the network and the directed edges indicate the direction of the transitions of the network from its configuration at time t to a new configuration at time $t+1$.

Figure 2 represents the iteration graph of the previous network for the case of parallel iteration. This graph contains the $2^5 = 32$ possible states. It illustrates the fundamental dynamical characteristics which we will define below.

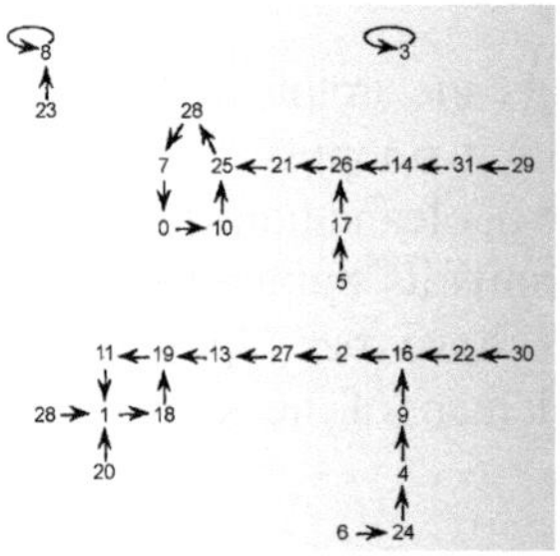

Figure 2. Iteration graph of the network of Figure 1. The numbers from 0 to 31 refer to the decimal representations of the 32 binary configurations of the network. The arrows show the temporal order of the configurations. Note that there are four different basins of attraction. State number 3 is an isolated fixed point. State number 8 is another fixed point. The other, larger, basins are composed of the configurations which converge toward the limit cycles with periods 4 and 5.

Attractors

Since an automata network is a deterministic system, if the network reaches a state for the second time, it will go through the same sequence of states after the second time as it did after the first time. Therefore, the system will go into an infinite loop in state space. These loops are called the *attractors* of the dynamical system, and the time it takes to go around the loop is called the *period* of the attractor. If this period is 1, as is the case for the configuration numbered 8 in the example shown below, the attractor is a *fixed point.* We speak of *limit cycles* if the period is greater than 1. The set of configurations which converge toward an attractor constitutes *a basin of attraction.* The network shown in the example below has four attractors.

Clearly it is only possible to construct a complete iteration graph for small networks. For the large networks we must be content to describe the dynamics of the system by characterizing its attractors.

In this way we can try to determine:

- the number of different attractors,
- their periods,
- the sizes of the basins of attraction (the number of configurations which converge toward each attractor),
- the notion of *distance* is also very important. The *Hamming distance* between any two configurations is the number of automata which are in different states.

3. IN SEARCH OF GENERIC PROPERTIES

In view of all the simplifications that were made to define the units of the model networks, one cannot expect all properties of living systems to be modeled. Only some very general properties, independent of the details of the model will show-up. These are the so-called generic properties of the network. In fact, we are interested not in the particularities of a specific network, but in the orders of magnitude which we expect to observe in studying a set of networks with fixed construction principles. We therefore consider a set containing a large but finite number of networks. We choose some of these networks at random, construct them, and measure their dynamical properties. We then take the average of these properties, and we examine those which are fairly evenly distributed over the set of networks. An example will help to clarify these ideas.

Consider the boolean networks with connectivity $k = 2$, with a random connection structure. The dynamical variable we are interested in is the period, for the set of all initial conditions and networks. Of course, this period varies from one network to the next. We have measured it for 10 randomly chosen initial conditions for 1 000 different networks of 256 randomly connected automata, whose state change functions were generated at random at each node of the network. Figure 3 shows the histogram of the measured periods. This histogram reveals *that the order of magnitude of the period is ten* (this is the generic property), even though the distribution of the periods is quite large.

We can certainly construct special "extreme" networks for which the period cannot be observed before a million iterations. For this, we need only take networks which contain a random mixture of exclusive OR and EQUivalence functions (EQU is the complementary function of XOR; its output is 1 only if its two inputs are equal). But these extreme cases are observed only for a tiny fraction ($1/7^{256}$) of the set under consideration. We consider them to be pathological cases, *i.e.* not representative of the set being studied.

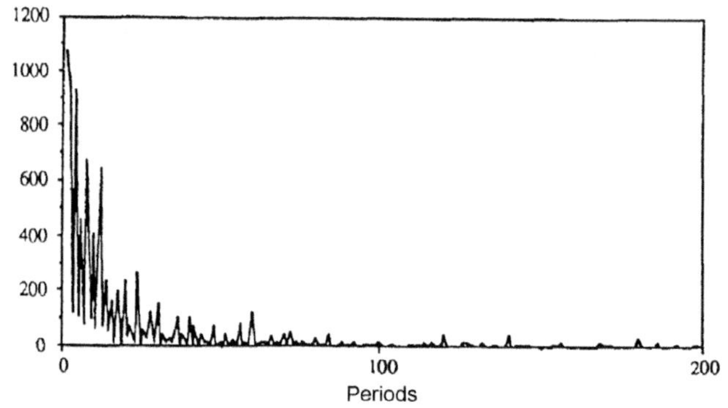

Figure 3. Histogram of the periods for 10 initial conditions of 1 000 random boolean networks of 256 automata.

We then call *generic properties* of a set of networks those properties which are independent of the detailed structure of the network — they are characteristic of almost all of the networks of the set. This notion then applies to randomly constructed networks. The generic properties can be shown not to hold for a few pathological cases which represent a proportion of the set which quickly approaches 0 as the size of the network is increased. In general the generic properties are either:

– qualitative properties with probabilities of being true are close to 1
– semi-qualitative properties, such as the scaling laws which relate the dynamical properties to the number of automata.

The notion of generic properties characteristic of randomly constructed networks is the basis for a number of theoretical biological models. It is similar to the notion of universality classes, developed for phase transitions. Without going into too much detail, we can say that the physical variables involved in phase transitions obey scaling laws which can be independent of the transition under consideration (such as, for example, problems in magnetism, superconductivity, or physical chemistry) and of the details of the mathematical model which was chosen. These laws only depend on the physical dimension of the space in which the transition takes place (for us, this is three-dimensional space) and on the dimension of the order parameter. The set of phase transitions (and their mathematical models) which obey the same scaling laws constitutes a universality class.

In fact, the first attempt to model a biological system by a disordered network of automata by S. Kauffman (1969), a theoretical biologist, predates the interest of physicists in this subject. It is also based on the idea that the properties of disordered systems are representative of the vast majority of systems defined by a common average structure.

3.1 An example: Cell Differentiation and Random Boolean Automata

The apparent paradox of cell differentiation is the following: "Since all cells contain the same genetic information, how can there exist cells of different types within a single multicellular organism?".

Indeed, our body contains cells with very different morphologies and biological functions: neurons, liver cells, red blood cells (...) a total of more than 200 different cell types. Yet the chromosomes, which carry the genetic information, are not different in different cells. Part of the answer is that not all of the proteins coded for by the genome are expressed (synthesized with a non-zero concentration) in a cell of a given type. Hemoglobin is found only in red blood cells, neurotransmitters and their receptors only appear in neurons, etc.

Several mechanisms can interfere with the different stages of gene expression to facilitate or block it. We speak of activation and repression. The best known mechanisms involve the first steps of transcription. In order to transcribe the DNA, a specific protein, DNA polymerase, must be able to bind to a region of the chain, called the promoter region, which precedes the coded part of the macromolecule. Now, this promoter can be partially covered by a control protein, called the repressor; reading the rest of the chain is then impossible. It follows that, depending on the quantity of repressor present, the gene is either expressed or not expressed. The protein which acts as a repressor is also coded for by another gene, which is itself under the control of one or several proteins. It is tempting to model the network of these interdependent interactions by an automata network.

- A gene is then represented by an automaton whose binary state indicates whether or not it is expressed. If the gene is in state 1, it is expressed and the protein is present in large concentrations in the cell. It is therefore liable to control the expression of other genes.
- The action of control proteins on this gene is represented by a boolean function whose inputs are the genes which code for the proteins controlling its expression.
- The genome itself is represented by a network of boolean automata which represents the interactions between the genes.

In such a network, the only configurations which remain after several iteration cycles are the attractors of the dynamics, which are fixed points or limit cycles, at least when the dynamics is not chaotic. These configurations can be interpreted in terms of cell types: a configuration corresponds to the presence of certain proteins, and consequently to the biological function of a cell and its morphology. Consequently, *if* we know the set of control mechanisms of each of the genes of an organism, we can predict the cell types. In fact, this is never the case, even for the simplest organisms. Without knowing the complete diagram of the interactions, S. Kauffman (1969) set out to uncover the generic properties common to all genomes by representing them by random boolean networks. Since there is a finite number of possible boolean laws for an automaton with a given input connectivity k, it is possible to construct a random network with a given connectivity.

S. Kauffman determined the scaling laws relating the average period of the limit cycles and the number of different limit cycles to N, the number of automata in the network. For a connectivity of 2, these two quantities seem to depend on the square root of N (in fact the fluctuations are very large). In fact, these same scaling laws have been observed for the time between cell divisions and for the number of cell types as a function of the number of genes per cell.

It is clear that Kauffman's approximations were extremely crude compared to the biological reality — binary variables representing protein concentrations, boolean (and thus discrete) functions, simultaneity of the transitions of automata, random structures\dots The robustness of the results obtained with respect to the possible modifications of the model (these are random networks) justifies this approach. As for the existence of a large number of attractors, it is certainly not related to the particular specifications of the chosen networks; it is a generic property of complex systems, which appears as soon as frustrations exist in the network of the interactions between the elements.

3.2 Generic properties of Random Boolean Nets

In fact, the results obtained by Kauffman show two distinct dynamical regimes, depending on the connectivity.

For networks of connectivity 2, the average period is proportional to the square root of N, the number of automata. The same is true of the number of attractors. In other words, among the 2^N configurations which are a priori possible for the network, the dynamics selects only a small number of the order of N which are really accessible to the system after the transient period. This selection can be interpreted to be an *organization* property of the network.

As the connectivity is increased, the period increases much faster with the number of automata; as soon as the connectivity reaches 4, the period as well as the number of attractors become exponential in the number of automata. These periods, which are very large as soon as the number of automata is greater than one hundred, are no longer observable, and are reminiscent of the chaotic behavior of continuous aperiodic systems. In contrast with the organized regime, the space of accessible states remains large, even in the limit of long times. Further research has shown that other dynamical properties of these discrete systems resemble those of continuous chaotic systems, and so we will refer to the behavior characterized by long periods as *chaotic*.

3.2.1 Functional Structuring

We have shown that when boolean automata are randomly displayed on a grid their temporal organization in period is related to a spatial organization in isolated islands of oscillating automata as soon as the attractor is reached. In the organized regime, percolating structures of stable units isolate the oscillating islands. In the chaotic regime the inverse is true: few stable units are isolated by a percolating set of oscillating units.

3.2.2 The Phase Transition

The connectivity parameter is an integer. It is interesting to introduce a continuous parameter in order to study the transition between the two regimes: the organized regime for short periods, and the chaotic regime corresponding to long periods. B. Derrida and D. Stauffer suggested the study of square networks of boolean automata with four inputs.

The continuous parameter p is the probability that the output of the automaton is 1 for a given input configuration. In other words, the networks are constructed as follows. We determine the truth table of each automaton by a random choice of outputs, with a probability p of the outputs being 1. If $p = 0$, all of the automata are invariant and all of the outputs are 0; if $p = 1$, all of the automata are invariant and all of the outputs are 1. Of course the interesting values of p are the intermediate values. If $p = 0.5$, the random process described above evenly distributes all of the boolean functions with four inputs; we therefore expect the chaotic behavior predicted by Kauffman. On the other hand, for values of p near zero, we expect a few automata to oscillate between attractive configurations composed mainly of 0's, corresponding to an organized behavior. Somewhere between these extreme behaviors, there must be a change of regimes. The critical value of p is 0.28. For smaller values, we observe small periods proportional to a power of the number of automata in the network. For $p > 0.28$, the period grows exponentially with the number of automata.

3.2.3 Distance

The distance method has recently been found to be one of the most fruitful techniques for determining the dynamics of a network. Recall that the Hamming distance between two configurations is the number of automata in different states. This distance is zero if the two configurations are identical, and equal to the number of automata if the configurations are complementary. We obtain the relative distance by dividing the Hamming distance by the number of automata.

The idea of the distance method is the following: we choose two initial conditions separated by a certain distance, and we follow the evolution in time of this distance. The quantity most often studied is the average of the asymptotic distance, measured in the limit as time goes to infinity. We compute this average over a large number of networks and of initial conditions, for a fixed initial distance. Depending on the initial distance, the two configurations can either evolve toward the same fixed point (in which case the distance goes to zero), or toward two different attractors, or they could even stay a fixed distance apart (in the case of a single periodic attractor), regardless of whether the period is long or short. Again, we observe a difference in the behaviors of the two regimes. On Figure 4, the x-axis is the average of the relative distances between the initial configurations, and the y-axis is the average of the relative distances in the limit as time goes to infinity. In the chaotic regime, we observe that if the initial distance is different from 0, the final distance is greater than 10 %. The final distance seems almost independent of the initial distance. On the other hand, in the organized regime, the final distance is proportional to the initial distance.

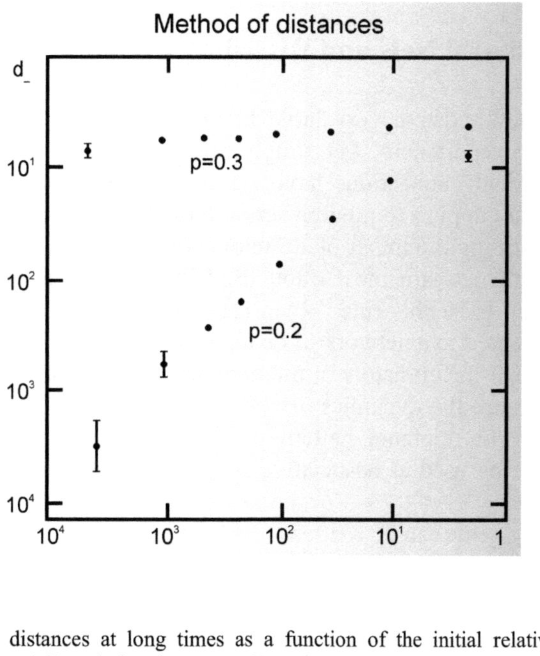

Figure 4. Relative distances at long times as a function of the initial relative distances, in the organized (p = 0.2) and chaotic (p = 0.3) regimes. (From B. Derrida and D. Stauffer (1986) *Europhys. Lett.*, **2**, 739).

3.2.4 Conclusions

This study clearly demonstrates the existence of two types of behaviors, organized and chaotic. Table 2 summarizes the differences in the generic properties of these two regimes.

Table 2. Generic properties of random networks

Properties	Organized regime	Chaotic regime
Period	Short	Long
Scaling law (Periods)	On the square root of N	Exponential in N
Oscillating nodes	Isolated islands	Percolating
Evolution of distances Cellular networks Random connectivity	Proportional to d_0 d_∞ goes to 0	d_∞ finite, independent of d_0 d_∞ stays finite

4. MEMORIES

4.1 Neural Nets and Distributed Memories

There now exists a very large literature on neural nets which we are not going to report here. Let simply summarize the results. Neural nets with symmetrical connections have an exponential number of point attractors. This result applies to random serial iteration, and exponential means that that the logarithm of number of attractors is proportional to the number of units.

Neural nets are most often used in learning tasks. A general learning algorithm is Hebb's rule. When reference patterns (network configurations) are presented to a network to be learned, connections can be constructed that ensure that the attractors of the network dynamics are the reference patterns. Furthermore the dynamics drives the network from initial conditions not to far from the reference patterns to the nearest reference patterns: these nets can then be used as associative memories that can be recalled from partial memories.

Hebb's rule can be written:

$$J_{ij} = \sum_{\mu} S_i^{\mu} S_j^{\mu} \tag{5}$$

where μ refers to the different reference patterns and S_i^{μ} and S_j^{μ} to the states of connected neurones i and j in the corresponding pattern.

Memories are thus distributed in the network as opposed to a memory that would be localized on some part of the net. The memory capacity of a fully connected neural net build according to Hebb's rule scales as the number of units in the net: no more than $0.14N$ patterns, where N is the number of units, can be stored and retrieved in a Hopfiled neural net.

4.2 Immune Nets and Localized Memories

As a memory device, the immune system needs to obey certain constraints: it should be sensitive enough to change attractor under the influence of antigen. It should not be too sensitive and over react when antigen is present at very low doses. The immune system should also discriminate between self-antigens and foreign antigens. Finally, it should be robust — memories of previously presented antigens should not be lost when a new antigen is presented. Thus, in some sense, the system should be able to generate independent responses to many different antigens. This independence property is achieved when attractors are localized, *i.e.,* when the perturbation induced by encounter with antigen remains localized among the clones that are close to those that actually recognize the antigen (see Figure 5).

Our problem is to classify the different attractors of the network and to interpret the transitions from one attractor to another under the influence of antigen perturbation.

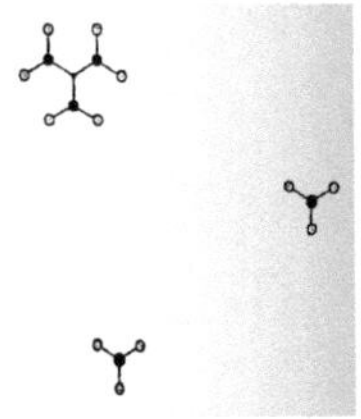

Figure 5. Localized patches of clones perturbed by different antigenic presentations. Two vaccination and one tolerant attractors are represented.

Let us start with the most simple virgin configuration, corresponding to the hypothetical case where no antigen has yet been encountered and all populations are at level m/d, *i.e.* all proliferation functions are 0. After presentation of the first antigen, memorization is obtained if some populations of the network reach stable populations different from m/d. In the case of a localized response, there will be a patch close to the antigen specific clone in which cells are excited out of the virgin state. Each antigen presented to the network will result in a patch of clones that are modified by the presentation. As long as the patches corresponding to different clones do not overlap, the various antigens presented to the network can all be remembered. Once the idea of localized non-interacting attractors is accepted, everything is simplified: instead of solving 10^8 equations, we only have to solve a small set of equations for those neighboring clones with large populations, supposing that those further clones that do not belong to the set have populations m/d. A practical approach to studying localized attractors is to combine computer simulations and analytic checks of the attractors by solving the field equations (see below).

4.2.1 Immunity

Let us examine the case of antigen presented to clone Ab_1, which results in excitation of clones Ab_2, clones Ab_3 remaining close to their virgin level (see Figure 6). We expect that Ab_1 will experience a low field, L, while Ab_2 will experience a large suppressive field, H. From the field equations we can compute the populations x_i. Recall, from

$$h_1 = zx_2 = L = \frac{d\theta_1}{p'} \tag{6}$$

$$h_2 = x_1 + (z-1)\,\frac{m}{d} = H = \frac{p'\theta_2}{d} \tag{7}$$

where $p' = p - d$.

An immune attractor is usually reached for an intermediate initial antigen concentration, and intermediate decay constants. If the initial antigen concentration is too low or if the antigen decays too fast, the immune attractor is not attained and the system returns to the virgin configuration, *i.e.*, Ab_1 and Ab_2 populations increase only transiently and ultimately return to the virgin m/d level. Thus, no memory of antigen encounter is retained.

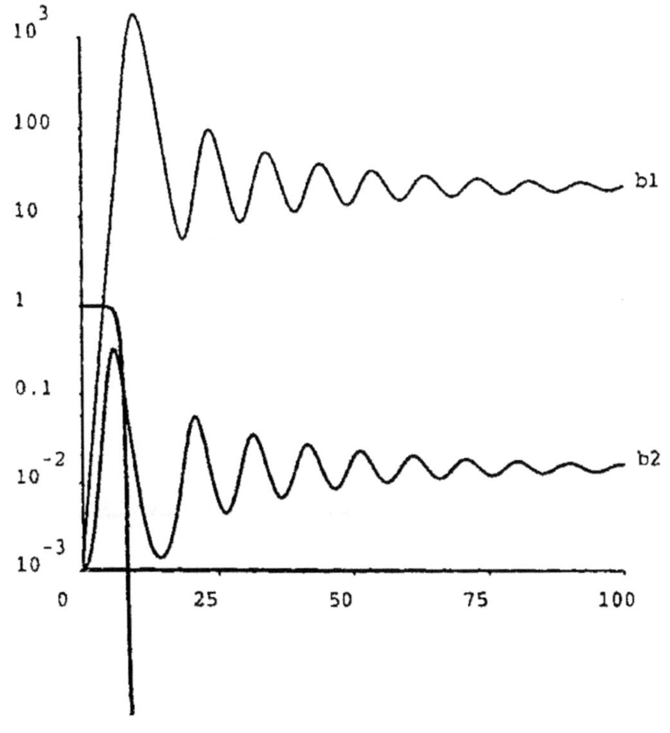

Figure 6. Time plot of an antigen presentation resulting in a vaccination attractor. On the vertical axis are the clone populations on a logarithmic scale. Time in days is on the horizontal axis. In the vaccinated configuration the largest population is localized at the first level. X_1 is high (H) and sustained by an intermediate population (L/z) of X_2. The rest of the clones are virgin (V) (or almost virgin) after the system settles into this attractor. When antigen is presented again, it is eliminated faster than the first time.

4.2.2 Tolerance

Another localized attractor corresponds to tolerance (see Figure 7).

A strong suppressive field acts on Ab_1 due to Ab_2's, the Ab_2's proliferate due to a low field provided by Ab_3's, but Ab_4's remain nearly virgin. The field equations once more allow one to compute the populations:

$$h_2 = x_1 + (z-1)\,x_3 = L = \frac{d\theta_1}{p'} \tag{8}$$

which gives x_3 if one neglects x_1, which is small.

$$h_3 = x_2 + \frac{(z-1)m}{d} = H = \frac{p'\theta_2}{d} \tag{9}$$

and thus for small m/d

$$h_1 = zx_2 \approx zH \tag{10}$$

Substituting h_1 in Eq. (1) gives a very small value for $f(h_1)$, which shows that x_1 is of the order of m/d. The Ab_1 population, experiencing a field several times higher than H, is said to be *oversuppressed.*

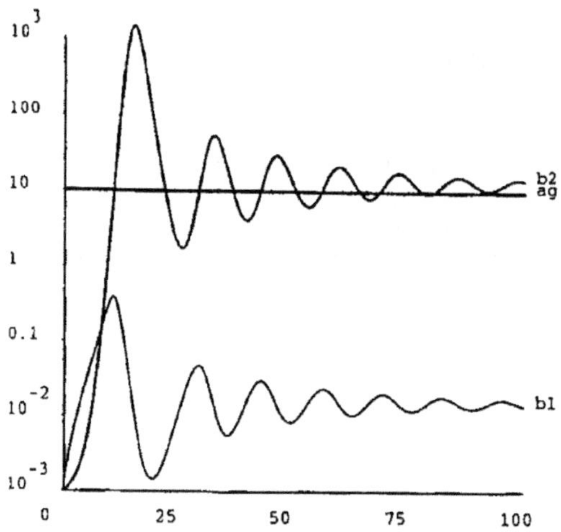

Figure 7. Time plot of an antigen presentation resulting in a tolerant attractor. x_2 is high (H) and sustained by an intermediate population (L/z) of x_3. x_1 is over suppressed by the x_2 and is not able to remove the antigen.

As in the case of the immune attractor, one can study the conditions under which the tolerant attractor is reached when antigen is presented. One finds that tolerance is obtained for large initial antigen concentrations, slow antigen decay rates and large connectivity, z (Neumann and Weisbuch 1993a).

4.2.3 Number of Attractors

Localized attractors can be interpreted in terms of immunity or tolerance. Because these attractors are localized they are somehow independent: starting from a fully virgin configuration, one can imagine successive antigen encounters that leave footprints on the network by creating non-virgin patches, each of these involving a set of p perturbed neighboring clones. An immune patch contains $1 + z$ clones, a tolerant patch $1 + z^2$. Independence of localized attractors implies a maximum number of attractor configurations that scales exponentially with N, the total number of clones. The following simplified argument gives a lower bound. Divide the network into $\frac{N}{(1+z^2)}$ spots. Each spot can be in 3 possible configurations: virgin, immune or tolerant. This gives a number of attractors that scales as $3^{\frac{N}{1+z^2}}$. Few of these attractors are of interest. The relevant question is the following: A living system must face frequent encounters with antigen during its life. Self antigen should elicit a tolerant response; dangerous external antigens should elicit immune responses and subsequent immunity. The nature of the localized response on each individual site of the network is then determined by the fact that the presented antigen should be tolerated or fought against. In this context, we can ask how many different antigens can be presented so that no overlap among different patches occurs?

In the case of random antigen presentation, simple reasoning (Weisbuch 1990, Weisbuch and Oprea 1994) is sufficient to derive the scaling law relating m, the memory capacity (*i.e.* the maximum number of remembered antigens) to N, the total number of clones. Let ns be the number of suppressed clones involved in a patch.

m is given by:
$$m \propto \sqrt{\frac{2N}{n_s}} \qquad (11)$$

and this provides an estimate for the mean memory capacity of the network.

The only assumption to obtain this scaling law is the random character of the network with respect to antigens, *i.e.,* the network is not organized to respond to the set of presented antigens. On the other hand, it can be argued that the clones expressed by mammals have been selected by evolution according to the environment of the immune system, *e.g.,* to be tolerant to self molecules and responsive to frequently encountered parasites and pathogens. If the system were optimized to the antigens in its environment, the network could be filled compactly with non-overlapping patches. The number of antigens (patches) would then scale linearly, *i.e.,*

$$m \,\alpha\, \frac{N}{n_s} \tag{12}$$

Weisbuch and Oprea (1994) discuss more thoroughly the capacity limits of model immune networks with localized responses. They verify by numerical simulations the square root scaling law for the memory capacity. They also examine a number of other features of the network. They show that when the number of presented antigens increases, failures to remove the antigen occur since the relevant clone has been suppressed by a previous antigen presentation. They also show that previous immune or tolerant attractors are rather robust in the sense that destruction of these local attractors by new encounters with antigen is rare, and that the complete reshuffling of the attractors, as in Hopfield nets (Hertz, Krough and Palmer 1990), is never observed.

5. CONCLUSION

This presentation is based on the authors own interests in theoretical biology issues. Still in many other instances concerning the origin of life, evolution of species, co-evolution... emergent organization appears as the dynamical selection of an attractor in a multi-component system.

REFERENCES

Derrida B. (1987). Dynamical Phase Transitions in Random Networks of Automata. In J. Souletie, J. Vannimenus and R. Stora (eds). *Chance and Matter.* North Holland.

Hertz J., Krogh A. and Palmer R.G. (1991). *Introduction to the Theory of Neural Computation.* Redwood City, CA, Addison Wesley.

Kauffman S. (1990). *The Origins of Order: Self-Organization and Selection in Evolution.* Oxford University Press.

Kauffman S. (1969). *J. Theor. Biol.,* **22**, 437-467.

Perelson A.S. and Weisbuch G. (1997). Immunology for Physicists. *Review of Modern Physics,* **69**, 1219-1267.

Weisbuch G. (1990). *Complex Systems Dynamics.* Redwood City, CA, Addison Wesley.

Weisbuch G. and Oprea M. (1994). *Bull. Math. Biol.,* **56**, 899-921.

VINCENT BAUCHAU

EMERGENCE AND REDUCTIONISM: FROM THE GAME OF LIFE TO SCIENCE OF LIFE

"*As a philosopher who looks at this world of ours, with us in it, I indeed despair of any ultimate reduction. But as a methodologist this does not lead me to an antireductionist research program.*" (K. Popper 1974)

"*...clouds on the left and clocks on the right and animals and men somewhere in between.*" (K. Popper 1965)

"*Every philosophy student should be held responsible for an intimate acquaintance with the Game of Life.*" (D. Dennett 1991)

"*My dear old friend, I wish to God there were more automata in the world like you.*" (Charles Darwin 1882, *Letter to T.H. Huxley* (*in* Darwin F. 1887)

1. INTRODUCTION

Reductionism and emergence are two related concepts that are either rejected as meaningless (trivial) or are subject of much philosophical discussions. Not only are these two issues debated by philosophers, it is also a matter of controversy amongst *scientists* (*e.g.* Weinberg against Mayr, *in* Weinberg 1993). How is it that scientists cannot agree on such matters that, somehow, stand at the center of their work? It is irritating that, in spite of all the great successes of Science, we still cannot come to grip with these concepts of emergence and reductionism. The confusion and the recurrent controversy about emergence and reductionism are probably partly due to misunderstandings. One problem is the classification of working scientists in either of two classes, the reductionists on one hand, and those who 'believe' in emergence on the other hand. Actually, many scientists do not really care about this potential classification, while other find themselves to be classified one way by some people, and the other way by other people. Here I want to discuss the idea that all scientists are, in a sense, reductionist,

B. Feltz, M. Crommelinck and P. Goujon, (eds), Self-organization and Emergence in Life Science, 29-40.

but that, at the same time, they cannot ignore emergence (they would not be here without it, in the first place). There is more to gain from seeing these two concepts as compatible and complementary instead of seeing them as contrary. My discussion will largely focus on the relationship between Biology and Physics. To illustrate my point about emergence I will refer to a class of abstract models, the cellular automata, exemplified by the most popular of them, the Game of Life (Gardner 1970). These models will also provide the basis for my suggestion that there are several degrees, and possibly classes, of emergence, the highest degree of emergence being related to universal computation (UC), a property to which more attention should be paid. My discussion of the importance of UC in Life Sciences will be highly speculative but, I hope, suggestive.

2. REDUCTIONISM AND THE UNIVERSE

Decomposing every entity under observation into smaller constituents is one hallmark of reductionism. The idea behind this practice is to look for deeper scientific principles that could all be traced to a small set of fundamental laws, a final theory, which would be expressed, ultimately, in terms of particle Physics (*e.g.* Weinberg 1993).

It is common knowledge that living system display an array of complex structures and processes, like the immune system, the brain, the metabolic network, the development... Complexity is present in the inanimate world as well, for example in the structure of the galaxies. As the reductionist program is to explain the Universe by a small set of simple laws, as opposed to complex laws, it has to face the problem of explaining the existence of complexity in the Universe. If everything ultimately reduces to particles and some fundamental laws, where does the complexity in the Universe come from? In other words, the reductionist program could not be complete without explaining the emergence of complexity.

3. FROM SIMPLE RULES TO COMPLEX DYNAMICS: CELLULAR AUTOMATA

Simple, abstract models known as cellular automata (hereafter CA) are probably the best tools to study the emergence of complexity (*e.g.* Wolfram 1984). CA can be considered as extremely simple universes. 'Matter' consists of an array of cells, in 1 or 2 dimensions (sometimes more). Each cell can only exist in a limited number of states. At each time step, cells can change to another state. The 'laws' of these artificial universes simply state how cell change from state to state over time as a function of their own state and the state of their closest neighbors. These laws (transition rules) are applied recursively over several 'generations', with the help of computers. In CA, time and space are discrete, all actions are local. The number of possible transition rules is generally very large.

CA models date back to some work by John von Neumann in the 1950s: he wanted to show that 'machines' could reproduce, and came up with a design for a CA model that could do so (in a manner that later, when the structure of DNA was unraveled, proved to be very close to what organisms do). This model however was rather complicated — there were 29 possible cell states, and a self-reproducing pattern would occupy 200 000 cells. Around 1970, John Horton Conway, a mathematician from Cambridge University, tried to simplify von Neumann's model as much as possible to get something workable on paper, by hand. After two years of experimenting on a checkerboard drawn on the floor of the departmental tea room, Conway came up with what he called the Game of Life (hereafter LIFE): a 2-dimensional CA with only two states (on/off), and a simple transition rule: a cell stays on if 2 or 3 neighbors are on, else turns off; an off cell turns on when 3 neighbors are on (only the 8 adjacent neighbors count).

With this rule, simple initial patterns might generate very complex evolution. For example, a 5-cell pattern known as the R pentomino takes 1 103 generations before the dynamics stabilized. Other 5-cell patterns however die out very quickly. However, even for patterns as small as 5 (connected) cells, there is no apparent relationship between the pattern and the dynamics that it induces. Although a great number of dedicated people have watched countless simulations of LIFE and analyzed them with sophisticated mathematical tools, no one has come with some rule to relate the shape of a small initial pattern to its long term future. LIFE seems basically unpredictable. Actually, it can be formally shown that LIFE *is* really and fundamentally unpredictable, as we shall see now.

When the simulation is started from a random initial setup, many recognizable patterns appear spontaneously: gliders; still life; oscillators; symmetric patterns... The gliders are small patterns that look like small insects crawling on the screen, until they encounter another pattern and interact with it. What results from the collision of gliders can be extremely variable. When the Game of Life spread into the scientific community, more complex structures were rapidly discovered, most notably the 'glider gun', a cyclic pattern that emits a glider every 30 time steps and can be constructed by the interaction of 13 gliders. By carefully positioning glider guns (and some other simple patterns), it is possible to create continuous flow of gliders that interact in any arbitrary way, more or less like pulses of electricity would interact in wires. With these building blocks, logical gates (AND, OR, NOT) can be build. These gates, in turn, serve as building blocks to set up more complex circuits. All components needed to design a computer could be found in LIFE, and Conway managed to design a computer embedded in LIFE (Berlekamp *et al.* 1982; Poundstone 1985). Moreover, this computer was a *universal* computer, that is one that can perform (although more slowly) any computations that any other computer, be it an IBM, a Cray or a Mac, can do. That is, the Game of Life has the property of Universal Computation (UC), a property described by Turing. He demonstrated that any machine capable of UC can perform *any* algorithm, including some for which it is impossible to know whether they will ever halt or not. Because any arbitrary algorithm can be implemented in LIFE, there is no general procedure to predict the state at time *t* unless by performing a complete simulation of the model for *t* generations (computational irreducibility). In the presence of UC, long-term behavior is unpredictable.

LIFE is so complex that its study is still going on, with new features announced from times to times. Conway says that in a very large array self-reproducing organism should appear, evolve, write PhD theses... This claim may appear silly, but is it more silly than to say that atoms can support life? Although LIFE is a universe with a very simple and perfectly known physics, we have not yet explore all the higher level possibilities. (And this is *not* because of imperfect knowledge about the state of the system, as in chaotic system, because the system can be perfectly known). There are still many questions about LIFE that we cannot answer. It is tempting to compare this with our study of life on Earth: we probably know all the 'rules' (physico-chemistry, DNA molecular machinery, natural selection...) but we still cannot explain many phenomena ranging from diseases, development or the mind. After all, Conway was probably quite right with the name he gave to his CA.

In short, CA illustrates that perfectly known, simple systems may still display unpredictability and irreducibility. In these systems, emergence is the spontaneous appearance of new concepts and new rules, non-deducible from a lower level, be it completely known or not. Obviously, interesting phenomena appear in systems which support (universal) computation. On one hand, these systems are programmable— they can be tuned to do specific task under specific conditions. On the other hand, at least some of their characteristics will forever be beyond our knowledge.

4. CLASSES OF EMERGENCE?

The kind of emergence seen in the Game of Life is very strong (by Turing thesis, the strongest). There are other mechanisms by which simple rules lead to complex dynamics, notably chaos. I suggest that these other kinds could be sorted into classes along a possible gradient. At the lower end, there would be simple aggregates, for example, the biomass, which is a simple sum where connections do not matter. At the highest point of the scale would be the class defined by universal computation. Below this class, chaotic systems would form another class. These are systems where the emergent pattern is an attractor, as in the so-called 'self-organized' systems, like Kauffman's (1993) networks.

This classification is a very tentative proposal. Other classes may be defined, and a finer classification could be devised with the use of other 'measures' of emergence (possibly: number of degree of freedom; type of language in Chomsky hierarchy; importance of connections; relative proportion of intra- vs. inter-levels interactions; independence from and action on lower level; symmetry breaking; knowledge of micro-level, *i.e.* complete, incomplete, not useful, statistical...). But the point I would like to make is that there are several possible mechanisms for emergence. Recognizing which mechanism is at work in a given system would help to understand this system, by allowing the use of tools and concepts devised for other systems in the same class. When a system is recognized as belonging to a given class, this would have immediate consequences. In the case of 'strong emergence' (the class defined by universal computation), one could infer that there is no shorter way (theory) to describe a process than by simulating it completely, *i.e.* that there is no possible reduction above a given level; and that questions about many aspects of the dynamics are undecidable, even with a total knowledge of the initial conditions.

5. UNIVERSAL COMPUTATION IN BIOLOGICAL SYSTEMS?

Thus, universal computation, when present in a system, can have extremely important consequences for the possibility of reducing the system. Therefore it is a crucial, but a difficult and overlooked, question to know how frequent is universal computation in living systems. The necessary building blocks are certainly present at different levels, in DNA, metabolic network, cell interactions, etc. The only problem would be to have the good wiring, which natural selection would provide if it were useful. UC has been formally demonstrated for a few levels: in biochemical networks (Arkin & Ross 1994), DNA (Lipton 1995) and, of course, neural networks[1]. UC can only be demonstrated by (rather painful) construction of a universal computer, as we have seen above for the Game of Life. The number of systems with UC could (but does not need to) be much larger that we think. Also, systems for which UC has been proven may actually perform UC in a completely other way (possibly more 'natural') than the one used for the formal demonstration.

The Game of Life is a very interesting CA but it took two years to a math genius to find the rule. Most CA (*e.g.* with randomly chosen rules) are uninteresting (their dynamics are either very simple or chaotic). Hence the general question: where, in the huge rule space, are the interesting, life-like CA? Langton (1990, 1992) has suggested that complex behavior and UC appear in CA 'at the edge of chaos', that is when their rules closely approach (in some parameter space) the region of chaotic dynamic, away from the ordered regime. He further argued that living and complex systems can only exist at this 'edge of chaos', where information can be stored, changed and transmitted. Followers have applied this idea to a vast range of phenomena, from traffic jams to zooplankton ecology. Langton's idea is controversial. Mitchell et al (1993) for example state that "it is not clear that anything like a drive toward universal-computational capabilities is an important force in the evolution of biological organisms. It seems likely that substantially less computationally-capable properties play a more frequent and robust role". Recently, Lakdawala (1996) has presented theoretical evidence that UC may be a too powerful model to describe 'edge of chaos' systems. This is obviously a field under development.

---

1. Undecidability and computational irreducibility may throw some light on mind problems, especially free will. Penrose (1989) says that, because of Godel theorem, we should abandon the idea that the mind is a machine. I think on the contrary that Godel-like arguments can help to take the idea that the mind might be a machine, because, if it is a machine with UC, then it can have many of the characteristics we have problems to explain. Computational limitations of our algorithmic mind offer a way to conciliate free will and determinism. Free will may come from the balance between self-knowledge of our decision-making processes and lack of it (Hoffstadter 1980, 804; Crick 1994), where this lack of complete knowledge would be guaranteed by computational irreducibility. Free-will would then be a product of the evolution of brains towards the edge of chaos, *i.e.* towards universal computation and unpredictability. Imagine a complex organism developing in LIFE and living its way (actually, its halting problem): should it not feel free? Freedom can be an emergent property in a deterministic system, without any need to resort to quantum indeterminacy. The neuron is a simple, deterministic machine that cannot lie or make an error, but emergence ensures that we (*i.e.* several levels higher) can.

In any case, the degree to which UC permeates living system is important to know, because of the limitations associated with it. For example, if development was computationally irreducible, there would be no hope to find a general procedure to predict, from an arbitrary DNA sequence, the final morphology or behavior of the organism. This would be a limitation to some of the Human Genome objectives. It is also a fact that would shed a new light on evolution: if the phenotype cannot be predicted from the genotype then a trial-and-error process, *i.e.* natural selection, is needed (and Lamarckism cannot possibly work). In short, I suggest that UC is a logical possibility that we should not ignore and that the theory of computation may change our way to look at many biological problems.

6. CAN BIOLOGY BE REDUCED TO PHYSICS?

Physicists often claim that particle physics is the most fundamental science, in the sense that questions about any phenomenon reduce to question about the standard model of particles (*e.g.* Weinberg 1993). Hence, questions about heredity reduce to questions about DNA, questions about DNA reduce to chemistry, questions about chemistry reduce to particle physics. Then follows the related claim that Biology can be reduced to Physics (or similar claims about other sciences from Chemistry to Psychology). In the light of the previous sections, I will show some problems associated with this kind of claim. One problem is the existence of universal laws in Biology that do not belong to Physics (notably, natural selection). Another problem is the relative independence of life from the physical substrate. Let's first look at this latter problem.

We only know one example of life. However theoretical speculations can lead us to see what is needed for life *in general* ('life as-it-could-be', Langton 1989). Living beings can be defined by the possession of those properties needed to ensure evolution by natural selection. The basic ingredients are rather simple: replication, memory. Any physical world that allows its constituents to instantiate these processes could (and probably would) support life. Works in Artificial Life have demonstrated that many important aspects of life could be simulated with a digital computer by ignoring most of the physics of the natural world (Ray 1991, Langton 1989). In a sense, this is not much different from weather prediction not taking into account quantum effects.

The structure of an organism is not determined by the laws of physics and chemistry, it is only constrained by these laws (the structure of a TV set is not determined, but constrained by the laws of electricity). The same point can be made about DNA. The sequence on a string of DNA is not determined by the laws that govern the physical and chemical properties of DNA. If it was so, the string could not contain any information (Polanyi 1968). For DNA to work as carrier of genetic information, it was necessary that this molecule acquire the capability to change its sequence arbitrarily (or nearly so). It is only statistically that DNA contains A, T, G, C in equal proportion. In this respect, DNA is unique among molecules to conserve its physico-chemical properties when its sequence changes. In other words, the genetic information is irreducible to the physico-chemistry of DNA. Strings of DNA can function as symbols, which can be used by Nature (and interpreted by us) without any further regard for its physical or chemical basis. It may be true that quantum mechanics could, in principle, be used to derive many properties of the DNA molecules (molecular weight, denaturation temperature...). In that sense, reduction would have been achieved. However, there is nothing from chemistry or physics that can be used to derive the function of DNA. This function is irreducible.

Similarly, Darwinism is independent of string theory. It is unlikely that any further discovery in particle Physics will make biologists change their mind about evolutionary theory (as Weinberg 1993 admits). Returning to the analogy with the Game of Life, one could imagine that new laws are discovered that would explain the transition rule as deducible from a lower level. This would not change our way to understand the dynamics of LIFE at higher levels. For a biologist, the only thing required from the physical world is to *allow* life. If one think of life as a class of systems (as opposed as a unique, local phenomenon; see Langton), then several physical universes would be compatible with life. Basically, these universes should allow natural selection to work, and many universes could do that. Hence, knowing more about particle physics will not help to understand more about biology (or meteorology, for that matter). Living systems are underdetermined by physical laws. The problem is not that Physics might still be incomplete (nor that Biology might be incomplete, however unlikely).

Let's now look at another problem with the alleged reduction of Biology to Physics: the fact that Biologists have described laws that do not belong to Physics (but which are, of course, compatible with it). At the heart of Biology lies the theory of evolution by natural selection, and natural selection can be seen as a universal law (Reed 1981; Dawkins 1983; Bauchau 1993). It states that a trait distribution will inevitably change from generation to generation whenever the following conditions are met: the trait affects reproduction rate, is (at least partly) heritable, and varies among individuals. Besides species, this law also applies to RNA molecules *in vitro* or to computer programs. Natural selection is both a fundamental law (as fundamental, at least for living organisms, as the standard model of particles) and a law unknown to Physics. It is probably the best example and a very important one, considering the central position of natural selection in Biology, but other potential laws could be listed, although their status is less well established. The existence of these autonomous laws is a consequence of (and evidence for) the irreducibility of living processes to physical laws.

7. LEVELS

One can speak of possible or successful reduction of one level to the level immediately below. On the other hand, when one start to speak of reduction trough more than two levels, or through all (known) levels, the idea becomes either trivial ("everything is composed of particles") or fallacious ("everything is only but a pack of particles"). There is a horizon limit when we navigate through the levels of organization: when more than two levels are crossed, connections fade away (as in a power law?). This rate of dilution may be different for different classes of emergence (being close to zero for simple aggregates).

Dawkins (1986) takes defense of 'hierarchical reductionism', which explains a complex entity in terms of entities only *one* level down the hierarchy. Later, however, Dawkins says that the biologist "can regard is task as done when he has arrived at entities so simple that they can safely be handed over to physicists". However, going down to the lowest possible level is not the best strategy, especially not for an evolutionary biologist. If one asks why peacocks have long tail, it will not help to inquire about the physico-chemistry of feathers; instead one should go higher in the hierarchy, and look at peacocks as individuals competing for access to the females. Considering the *higher* level can be a useful heuristic as much as the one-way reductionism. When looking at a computer screen, it much more useful to see words than pixels (or even letters), even if we lose some precision in the description of the system (there are several ways of drawing a letter on a screen, depending on font, size...). Sometimes one can only understand the parts by looking at the whole. To understand the neuron it helps a lot to know what the brain is for (imagine a Martian scientist confronted to neurons isolated *in vitro*). This is because the working of the brain depends on the neuron machinery as much as the neuron is designed to work in a brain.

Because of the necessity to connect only levels close enough in the hierarchy, it is important to recognize those levels. Sometimes reduction might be incomplete because we have not yet found the intermediate sub-levels (building blocks, schemata, symbols; Hofstadter 1980). Too a detailed knowledge of the lowest level may obscure higher relationships. Would we not know more about the mind if we knew less about neurons and DNA? It is more difficult to describe and understand cathedrals in terms of bricks than in terms of arches (Brexbaum 1995). Similarly, to discuss how to implement a computer in LIFE, it is much easier to speak in terms of logical gates than in terms of cells and their state transitions (Dennett 1991). The low-level physics of the system can be and should be ignored; it becomes irrelevant. (This is also why there is hope to find general laws for complex systems, *i.e.* laws that are independent of the microscopic details; natural selection is an example). Weather forecast is based on concepts like clouds, rain, cold fronts, although we know that it's all gas molecules and we know much more about the physics of gases than the physics of the weather (Hofstadter 1980). In the same way Darwinism and quark theory are independent: you do not need one to build the other (both ways).

Another problem comes from the asymmetry between levels: the lower allows the higher, the higher constraints the lower. We can easily recognize which level is lower or higher. Why is it that Physics could explain Biology, while nobody expects Biology to explain Physics (although, when it comes to explain physicists, I would better bet on Biology)? The reason is a restriction of the objects considered at each level. Biology limits itself to a subset of the objects present in our Universe, while Physics has no such limitation. It is this restriction that allows the enrichment of sciences like Biology. Physicists *have* to see organisms as they see other objects, *i.e.* as clumps of atoms.

8. CONCLUSIONS

The reductionist program has been and is extremely successful and scientists should keep looking for fundamental laws. However, the ability to reduce everything to simple fundamental laws does not imply the ability to start from these laws and reconstruct the universe (Anderson 1972). Applying simple rules recursively to create complex dynamics is possible, as shown, for example, by chaos theory. An older, but less publicized, class of systems where complexity can emerge from simple rules are systems like the Game of Life, which exhibit universal computation. In these systems, although both the rules and the initial states can be perfectly known, the existence of irreducibility and unpredictability can be formally proven. The significance of such strong properties in living systems is still a matter of speculation, but at least it is a logical possibility to which, I believe, more attention should be paid.

Reductionism and emergence do not contradict, they complement each other. The aim of Science is to explain, and explaining a given phenomenon consists as much as the reduction to the micro-level than the prediction of the macro-level. Stating that consciousness is based on brain activity, is that a reductionist or emergentist statement? On one hand, the reductionist approach has been fulfilled by the discovery of the neuron and its working. On the other hand, this only gives us the basic component, which, grouped together in a proper way, support the higher capacity of the brain. Emergence can only be studied properly when the laws of the lower levels are known, *i.e.* when some reduction has been done. On the other hand, the reductionist program has to include the notion of emergence of complexity and, doing so, has to acknowledge potential limitations, as those discussed here. The Theory Of Everything dreamed of by the physicists might be necessary for a complete picture of the Universe, but it might not be sufficient.

REFERENCES

Anderson P. (1972). More is Different. *Science,* **177**, 393.

Arkin A. and Ross J. (1994). Computational Functions in Biochemical Networks. *Biophys Journal,* **67**, 560-578.

Bauchau V. (1993). Universal Darwinism, *Nature,* **361**, 489.

Berlekamp E.R., Conway J.H. and Guy R.K. (1982). *Winning your Ways for Your Mathematical Plays,* vol. 2. Academic Press.

Buxbaum R.E. (1995). Biological Levels. *Nature,* **373**, 567-568.

Crick F. (1994). *The Astonishing Hypothesis. The Scientific Search for the Soul.* New York: Charles Scribner's Sons.

Darwin F. (ed.) (1887). *The Life and Letters of Charles Darwin.* London: John Murray.

Dawkins R. (1983). Universal Darwinism. In Bendall (ed.), *Evolution from Molecules to Men.* Cambridge University Press.

Dawkins R. (1986). *The Blind Watchmaker.* Penguin Books.

Dennett D. (1991). Real Patterns. *J. Philosophy,* **88**, 27-51.

Gardner M. (1970). The Fantastic Combinations of John Conway's New Solitaire Game 'Life'. *Scientific American,* **223**(4), 1970, 120-123.

Hofstadter D.R. (1980). *Godel, Escher, Bach: an Eternal Golden Braid.* Penguin Books.

Kauffman S.A. (1993). *Origins of Order: Self-Organization and Selection in Evolution.* Oxford University Press.

Lakdawala P. (1996). Computational Complexity of Symbolic Dynamics at the Onset of Chaos. *Physical Rev. E,* **53** (5), 4477-4485.

Langton C.G. (1989). Artificial Life. In C.G. Langton (ed.), *Artificial Life.* Addison Wesley.

Langton C.G. (1990). Computation at the Edge of Chaos: Phase Transitions and Emergent Computation. *Physica D,* **42**, 12-37.

Langton C.G. (1992). Life at the Edge of Chaos. In C.G. Langton *et al.* (eds), *Artificial Life II.* Addison-Wesley.

Lipton R.J. (1995). DNA Solution of Hard Computational Problems. *Science,* **268**, 542-545.

Mitchell M., Hraber P. and Crutchfield J.P. (1993). *Revisiting the Edge of Chaos: Evolving Cellular Automata to Perform Computations.* Santa Fe Institute, preprint 93-03-014.

Penrose R. (1989). *The Emperor's New Mind.* Oxford University Press.

Polanyi (1968). Life's Irreducible Structure. *Science,* **160**, 1308-1312.

Popper K. (1974). Scientific Reduction and the Essential Incompleteness of all Science. In F.J. Ayala and T. Dobzhansky (eds), *Studies in the Philosophy of Biology. Reduction and Realed Problems.* MacMillan, 259-284.

Popper (1965, reprinted 1979). *Objective Knowledge.* Oxford: Clarendon Press, 228-229.

Poundstone W. (1985). *The Recursive Universe.* Oxford University Press.

Ray T. (1992). An Approach to the Synthesis of Life. In C.G. Langton *et al.* (eds), *Artificial Life II.* Addison-Wesley.

Reed E.S. (1981), The Lawfulness of Natural Selection. *American Naturalist,* **118**, 61-71.

Weinberg S. (1993). *Dreams of Final Theory.* London: Vintage.

Wolfram S. (1984). Cellular Automata as Models of Complexity. *Nature,* **311**, 419-424.

HUGUES BERSINI

FORMALIZING EMERGENCE: THE NATURAL AFTER-LIFE OF ARTIFICIAL LIFE

1. INTRODUCTION

Originally, the field of Artificial Life was born out of the frustration and isolation felt by some "hackers" keen on cellular automata, game of life, genetic algorithms, L-systems and other computer recreations. Fascinated by this surprising cohabitation of programming simple algorithms and the complex working of these same algorithms (this new perception of complex phenomena as emerging from simple algorithms but iterated, distributed and recursive), convinced of the interest of their works for theoreticians of biology but aware of the lack of dialogue with them, they organized a series of workshops whose desired originality was multidisciplinarity and the coming together of researchers sharing the same will to understand the mechanisms and functions characterizing living organisms. These researchers in computer science, mathematics, physics, biology, robotics, philosophy, now meet every year, alternatively in Europe and the USA.

What is discussed as inherent to all living organisms, and therefore which represents the bulk of the material dealt with during these workshops, are the mechanisms of self-organization or of the "emerging functionalities" opposing a centralized vision of biology, the need to better balance the coupling of the studied objects with their environment opposing a solipsistic methodology still representative of a certain artificial intelligence, the compulsory passage via the mechanisms of learning and adaptation as the most simple and autonomous way to face the complexity typical of the architecture and dynamics of these systems and, finally, the study of this complexity per se. A same motto brings together all these researchers: "some form of complexity can be faced and domesticated very simply by relying on the computer brute force". The mascots that are most representative of artificial life are: robotic insectoids, the game of life and

B. Feltz, M. Crommelinck and P. Goujon, (eds), Self-organization and Emergence in Life Science, 41-60

other cellular automata, genetic algorithms, L-systems and simulations of ecosystems.

These first workshops, due to the originality of the process, created a considerable stir. They undoubtedly seemed to reach their primary target, that is to allow better communication between researchers. Today, however, a certain breathlessness is noticeable which goes not without reminding the same dying down that characterized the cybernetic and systemic trends (Alife fathers) of the forties and fifties. The multidisciplinarity although essential to the inspiration does not survive, in principle, the specialization which arises naturally as a consequence of several years of study dedicated to a same subject and which drive researchers to privilege interlocutors sharing their same narrow and deep interest. Gradually new scientific communities appear with a more focused object of study and which, either free themselves of the mother field (like genetic algorithms or cellular automata) or become connected with existing communities (like robotics, study of ecosystems, study of the origin of life, study of insects societies). As we can notice during these workshops, "life" resists whatever unique and narrow definition. This diversity is the de-stabilizing factor which could cause the burst of artificial life. Besides, the risk is important of a forthcoming divorce, which has already taken place in artificial intelligence, between a so called "strong" science which could fuse with an existing scientific tradition (cognitive science for AI and theoretical biology for artificial life) and its so called "weak" counterpart with a more engineering like aftertaste and leading to technological innovations (expert systems, fuzzy logic and knowledge engineering in AI, neural networks, genetic algorithms and autonomous robotics in artificial life). If the artificial life star turns into a supernovae to finally explode and leaves behind, as relics of its glorious past, one and only one scientific pulsar, more focused, firmly grounded, and, above all, perpetrating as well as possible the original enthusiasm, the best candidate I can see could be a more formalized study of the emergent phenomena.

My contribution to this characterization of emergent phenomena is currently limited to two of them appearing in a large amount of biological networks: the de-stabilizing effect of frustrated connectivity and the tendency to fragment the whole network into small clusters of units showing similar behavior. Among the networks showing these two emergent properties, the attention will be paid to only two of them: Hopfield Neural Networks (HNN) and Idiotypic Immune Networks (INN). Frustrated connectivity is responsible for perturbing the equilibrium dynamics of the network and provoking "wavering" among alternative equilibrium regimes.

When frustrated a homeostatic network exhibits oscillatory behavior while an oscillatory network falls into a new type of chaotic regime which will be designated as frustrated chaos. In HNN, there is a threshold in the degree of connectivity which marks a sharp transition into the dynamics of the network. Below this threshold, *i.e.* in the case of a strongly diluted connectivity, the network clusters itself into small group of oscillatory units. In IIN also, this clustering phenomenon prevails and follows very regular rules for the dimension and the distribution of the clusters. It is clear that these two properties can be regarded as emergent since in order to appear they require a specific collective configuration of the units, and in order to be detected they require a level of observation which transcends each unit taken separately. In this paper, rather than theoretical analysis, results of computer simulation are given and briefly explained to illustrate these common properties.

2. FRUSTRATION AND CLUSTERING IN HOPFIELD NEURAL NETWORKS

2.1 Frustration

Suppose a network of interconnected Boolean units and that this network is further constrained such that two units being connected means that unit 1 in one state can only co-exist with unit 2 in the anti-state. A two unit network can only settle in two possible configurations. Then take a three unit network and connect these units in an open chain fashion: unit 1 is connected only to unit 2 which in turn is connected only to unit 3. Here again two configurations are possible with the 1-2 couple as well as the 2-3 couple containing each two units settled in reverse states. Each couple per se complies with the local effect of the connection: the state/anti-state pairing. The problem gains interest by closing the chain, then getting a odd loop, connecting unit 3 back to unit 1 (see Figure 1). We now have three couples which must each independently complies with the imposed constraint: the state/anti-state pairing. Looking at the Moebius triangle in Figure 2, you'll observe a very similar type of impossible global configuration despite three possible local coupling. As a matter of fact no global configuration turns out to be possible: the couples mutually compete for reaching their state/anti-state configuration. In the modeling of spin glasses, this well-known phenomenon designated by the term frustration[1] is responsible for preventing spin glasses from relaxing to their minimal energy level, but equally for enlarging the set of intermediary solutions among which the network can choose to settle.

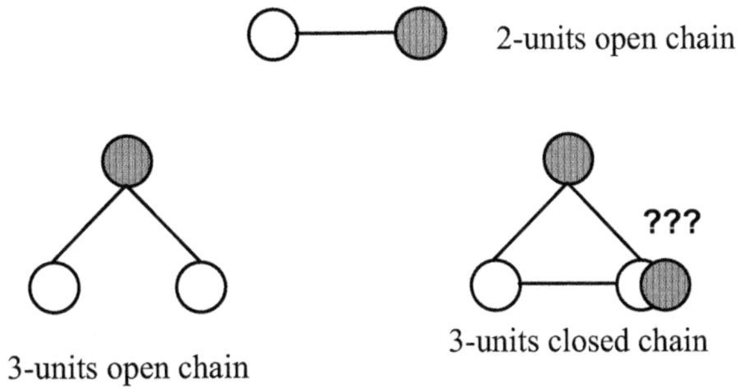

Figure 1. The frustration phenomenon

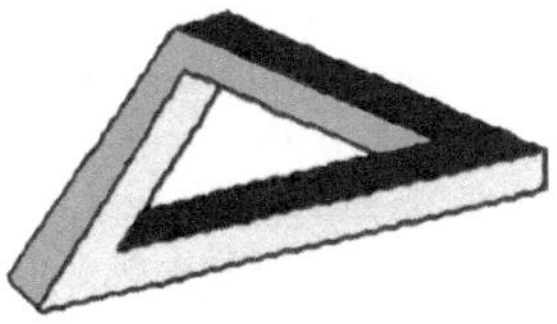

Figure 2. A Moebius triangle is a frustrated figure

In the Boolean network just described and connected in a loop: 1 → 2 → 3 → 1 with all inhibitory connections, it is trivial to see that when updating the units in an asynchronous way such a triangular network will eventually oscillate whereas the presence of an even number of negative interactions would force the network to relax in a stable configuration among others. This is the simplest case of a frustrated network. Thomas relies on such simple structures to study genetic regulatory networks[2]. He has shown how the presence of loop in a network, provided it contains an odd number of inhibitory connections, de-stabilizes this same network by triggering oscillations. He negatively judges this presence since alternatively the

1. Pimm Stuart (1991). *The Balance of Nature.* University of Chicago Press.
2. Thomas R. (1991). Regulatory Networks Seen as Asynchronous Automata: A Logical Description. *J. Theor. Biol.,* **153**, 1-23.

presence of positive loop (frustrated and non frustrated loops sometimes are respectively designated by negative and positive according to the sign of the product of all connections in the loop) is responsible for enlarging the repertoire of possible equilibrium configurations, each possibly expressing a particular cell. When frustrated, the network passes through all the possible configurations in a sequential and recurrent way and cannot stop in any of them. This is the first and simplest illustration of the instability, from fixed point to oscillation, obtained by frustrating a network.

In its original conception, symmetric and without self-connection, Hopfield network dynamics relaxes to fixed points. These fixed points are minima of an energy function which decreases as the network evolves in time. Frustration was first discovered and discussed by Toulouse[3] in the context of spin glasses in which unit settles in one or the other state so as to decrease a similar energy function. Since spin glasses and Hopfield networks share the same energy function, some physicists like Amit[4] or Sherrington[5] have tried to rely on statistical physics results obtained in the field of spin glasses to better characterize the capacities of Hopfield network. In a frustrated network, it is easy to see that there is no configuration which collectively drives the energy to a global minimum. All couples of units are unable to simultaneously settle in their state/anti-state pairing and then the network is driven into one among several intermediary configurations with higher energy. Together frustration raises the minimum energy and increases the degeneracy of the ground state. When Hopfield networks are used as a mechanism for associative memory, it is interesting that there be many available fixed points, namely an energy function with a lot of equivalent degenerate minima, a situation typically arising in frustrated networks. This is one of the few cases where frustration is considered to be beneficial to the network.

More interesting for understanding the de-stabilizing effect of frustration is the study of the same Hopfield network but now allowing for asymmetric connectivity (then loosing the proof of convergence to fixed-point attractors) and taken in its continuous form:

---

3. Toulouse G. (1977). Commun. *Phys.*, **2**, 115.
4. Amit D.J. (1989). Modeling Brain Function: *The World of Attractor Neural Networks.* Cambridge: Cambridge University Press.
5. Sherrington D. (1990). Complexity Due to Disorder and Frustration, *Lectures in the Sciences of Complexity – SFI Studies in the Sciences of Complexity,* Lect. Vol. II, Addison-Wesley: Ed. Erica Jen, 415-455.

$$\frac{da_i}{dt} = -\frac{a_i}{\tau} + \tanh\left(\sum_{j=1}^{N} m_{ij} a_j\right)$$

a_i is the activation of unit i, m_{ij} is the synaptic value connecting i and j, tanh has the classical tangent hyperbolic sigmoid profile and τ is a time constant (simplified to be the same for all neurons and taken to be 100 in our simulations).

Atiya and Baldi[6] have done a detailed analysis on how the units behave when they are interconnected in a loop or a ring, with the asymmetric connection matrix given by (for the 3-neuron version): $\begin{bmatrix} 0 & 0 & m_{13} \\ m_{12} & 0 & 0 \\ 0 & m_{23} & 0 \end{bmatrix}$ that is in presence of an odd inhibitory loops: $m_{12}m_{13}m_{23} < 0$ (they have generalized the study to the presence of any odd number of inhibitory connections). Summarizing their results, their formal analysis confirms a preliminary investigation of Hirsch[7] where it was shown that a necessary condition for the Hopfield network to oscillate is indeed to exhibit frustration in its connectivity. So here the de-stabilizing effect of frustration for odd ring connectivity is clear and can be theoretically justified.

2.2 Clustering

We have performed the same type of *NK* analysis popularized by Kauffman for Boolean Network but now applied to Hopfield Asymmetric Hopfield network given in the previous section. N is the number of binary units and $2K$ the number of units with which any unit is interconnected[8]. K reflects the dilute or not dilute nature of the network. The state transition of any unit is randomly extracted from the 2^{2^K} possible transitions. The dynamics of these Boolean nets is well known and the essential points will be now reminded by borrowing here and there pieces of Kauffman's own literature. Two strongly different regimes have to be stressed: one for very dilute network

6. Atiya A. and Baldi P. (1989). Oscillations and Synchronization in Neural Networks: An Exploration of the Labeling Hypothesis. *International Journal of Neural Systems,* Vol. **1**, 2, 103-124.

7. Hirsch M.W. (1987). Convergence in Neural Networks. *Proc. 1987 Int. Conf. Neural Networks*, San Diego, CA.

8. Kauffman S.A. (1989). Principles of Adaptation in Complex Systems. *Lectures in the Sciences of Complexity - SFI Studies in the Sciences of Complexity.* Addison-Wesley: Ed. D. Stein, 619-712.

($K = 2$) and one for fully connected network ($K = N$). Here again a sharp transition between these two regimes seems to occur at low connectivity. In fully connected network, the regime can be characterized as maximally disordered even chaotic (although this can't be a real chaos due to the finite nature of the network). There are N/e number of cycles. The length of the cycles grows exponentially with K. The network shows extreme sensitivity to initial conditions (a key feature of chaotic regime) because the successor to any state is essentially random and that almost any perturbation that flips one element will sharply change the network subsequent trajectory.

When K drops to 2, so in presence of a very dilute type of connectivity, the properties of the Boolean net change abruptly. The number of cycles is now given by $\sqrt{N}$. The reason why, despite their small number, these cycles keep a short period is due to the fact that the system is partitioned into an unchanging frozen core (this core contains unvarying units) which isolates islands of oscillatory units. This core has several effects: first it blocks the propagation of cyclic behavior favoring then small cycles, secondly it makes each cyclic attractor stable to most minimal perturbations and endows the local network with precious homeostatic quality. In Kauffman[9], it is explained how the properties of the network is highly dependent on the probability of appearance of the frozen core and how for low connectivity a lot of transitory rules are akin to identity rules, while for $K = 4$ and higher the proportion of identity rules falls abruptly. According to Kauffman, random Boolean nets with $K = 2$ provide examples of unexpected and powerful collective spontaneous order.

Let's turn to this same *NK* analysis applied now to asymmetric HNN. Figure 3 shows the synaptic matrix of 10 neurons when $K = 2$. We have done important statistics by randomly generating asymmetric *NK* matrix (m_{ij} could only take values 1 or -1). Our results have shown that, in agreement with Kauffman's results, there is a sharp transition of the network behavior marked at a low level of connectivity *i.e* $K = 2$ for $N = 30$, $K = 3$ for $N = 60$, $K = 4$ for $N = 100$. So the threshold for K seems to scale almost linearly with N. Below the threshold value for K there is a high probability to find the network into an oscillatory behavior with, like in Kauffman's net, small clusters of oscillating neurons separated by large zone of resting neurons. Above this threshold the network nearly almost falls into a fixed point. This is an important difference with Kauffman's results obtained for Boolean net since strongly connected networks behave very simply as fixed point, to be

---

9. Kauffman S.A. (1989). Principles of Adaptation in Complex Systems. *Lectures in the Sciences of Complexity - SFI Studies in the Sciences of Complexity.* Addison-Wesley: Ed. D. Stein, 619-712.

contrasted with strongly connected Boolean nets showing more complicated dynamics. Amari[10] has shown why for large random networks with global connectivity, you can apply the law of large number and assimilate the HNN to a set of disconnected and isolated networks which thus become all convergent. However we think that such a threshold effect can be better explained by relying on the results obtained for linear networks.

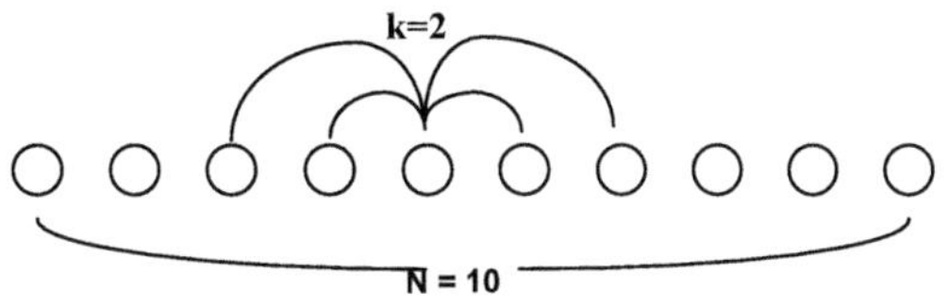

$$\begin{bmatrix} 0&1&1&0&0&0&0&0&1&1 \\ 1&0&1&1&0&0&0&0&0&1 \\ 1&1&0&1&1&0&0&0&0&0 \\ 0&1&1&0&1&1&0&0&0&0 \\ 0&0&1&1&0&1&1&0&0&0 \\ 0&0&0&1&1&0&1&1&0&0 \\ 0&0&0&0&1&1&0&1&1&0 \\ 0&0&0&0&0&1&1&0&1&1 \\ 1&0&0&0&0&0&1&1&0&1 \\ 1&1&0&0&0&0&0&1&1&0 \end{bmatrix}$$

Figure 3. A *NK* (10-2) Hopfield Network

20 years ago, Gardner and Ashby[11] followed by May[12], all three interested in the behavior and the stability of large ecosystems, have accomplished a seminal preliminary investigation simply by analyzing the stability conditions of a linear network:

$$\frac{da_i}{dt} = \sum_j m_{ij} a_j$$

In general the stability of all systems of differential equation can be studied by restricting this study to the behavior of the linearized system at an equilibrium point. This appears as a further motivation for studying the effect of the connectivity structure on the dynamics of the linear network.

10. Amari S. (1972). Characteristics of Random Nets of Analog Neuron-like Elements. *IEEE Transactions on Systems, Man and Cybernetics*, Vol. SMC **2**, 5, 643-657.
11. Gardner M.R. and Ashby W.R. (1970). Connectance of Large Dynamic (Cybernetic) Systems: Critical Values for Stability. *Nature*, Vol. **228**, 784.
12. May R.M. (1972). Will a Large Complex System be Stable? *Nature*, Vol. **238**, 413- 414.

The three authors reached the same non obvious conclusion that large networks with randomly fixed connection matrix are stable up to a certain degree of connectivity. Beyond this degree, scaling linearly with the number of units, there is a sharp transition and divergent dynamics characterize the network behavior. In brief, local networks *i.e.* when the degree of connectivity (for a given interconnectivity strength) is below a well-defined threshold are stable while global ones are unstable. In his paper, May makes the following claim "Applied in an ecological context, this ensemble of very general mathematical models of multi-species communities, in which the population of each species would by itself be stable (m_{ii}= -1 in his modeling), displays the property that too rich a web connectance or too large an average interaction strength leads to instability. The larger the number of species, the more pronounced the effect."

The relation between fixed point behavior in HNN and divergent behavior in linear system comes from the fact that a positive eigenvalue (easier to obtain with global network) would be responsible for a fixed point in Hopfield net: If $\lambda a_i = \sum_j m_{ij} a_j$ with λ the eigenvalue, then λa_i^2 = $\sum_j m_{ij} a_j a_i$ and since a fixed point implies $\sum_j m_{ij} a_j a_i \geq 0$, the fact that $\lambda \geq 0$ has higher probability for global network implies also a greater probability of fixed point for global connectivity.

3. FRUSTRATION AND CLUSTERING IN IMMUNE IDIOTYPIC NETWORKS

3.1 Frustration

The possibility that frustration turns an homeostatic idiotypic network into an oscillatory one was already observed with circumspection by Hiernaux[13]. He believed that this remarkable sensitivity of the dynamics of the network to its connectivity ought to make questionable the idiotypic network structure of the immune system, another negative perception of the frustration effect. In this section, the next qualitative transition will be investigated: from oscillations to chaos. The system of coupled differential equations showing this interesting frustration induced phenomenon was originally meant to study the dynamics of one particular immune idiotypic

---

13. Hiernaux J. (1977). Some Remarks on the Stability of Idiotypic Network. *Immunochemistry*, Vol. **14**, Pergamon Press, 733-739.

network first proposed by Varela *et al.*[14], Stewart and Varela[15] and largely studied and described in the literature[16].

The interest for the dynamical behavior of the immune network arose from the observation that the concentration of natural antibodies displays fluctuation patterns that are believed to be related to the connectivity of the immune cells instead of the result of encounters with external antigens. Moreover these antibodies in normal and auto-immune individuals have been shown to fluctuate in a different way, and hypothetically this might suggest to relate these two regimes with different structures of connectivity and to explain the disease by a structural alteration. Up to date, the most interesting dynamical pattern exhibited by the network simulations is an oscillatory regime (some of the fluctuation patterns observed in biological experimental data have in fact a strong oscillatory tendency[17]) in which the units always separate in two groups oscillating in counterphase. As a consequence, this type of network turns out to be susceptible to a frustration phenomenon that we indeed observed in the presence of odd loops.

The immune idiotypic network model under study contains N units ($i = 1...N$) (such a unit is often called a clone in the immunology literature). In contrast with other more familiar biological structures like neural networks, a unit is representative of two different immune cells: the antibody f_i and its associate producer the B lymphocyte b_i. On account of the very high specificity of B lymphocytes which only produce antibodies sharing this same specificity, a unique index i serves as reference mark to one type of antibody and its B lymphocyte associate producer. For each clone i, the system of equations accounts for B lymphocyte proliferation and

---

14. Varela F.J. and Coutinho A. (1991). Second Generation Immune Network. *Immunology Today*, Vol. **12**, 5, 159-166.
15. Stewart J. and Varela F. (1990). Dynamics of a Class of Immune Networks. II. Oscillatory Activity of Cellular and Humoral Components. *Journal of Theoretical Biology*, **144**, 103-115.
16. Bersini H. and Calenbuhr V. (1995). Frustration Induced Chaos in a System of Coupled ODE's, *Chaos, Solitons and Fractals,* Vol. **5**, 8, 1533-1549; Calenbuhr V., Bersini H., Stewart J. and Varela F.J. (1995). Natural Tolerance in a Simple Immune Network. *J. Theoretical Biology,* **177**, 199-213; Detours V., Calenbuhr V. and Bersini H. (1995). Clustering Phenomena in Idiotypic Network, *IRIDIA Internal Technical Report;* Stewart J. and Varela F. (1990). Dynamics of a class of immune networks. II. Oscillatory activity of cellular and humoral components. *Journal of Theoretical Biology*, **144**, 103-115; Varela F.J., Coutinho A., Dupire B. and Vaz N.N. (1988). Cognitive Networks: Immune, Neural and Otherwise - In A.S. Perelson (ed.), *Theoretical Immunology,* Part Two, SFI Studies in the Sciences of Complexity, vol. 3, Reading, MA: Addison-Wesley, 377-401; Varela F.J. and Coutinho A. (1991). Second Generation Immune Network. *Immunology Today,* Vol. **12**, 5, 159-166.
17. Varela F.J. and Coutinho A. (1991). Second Generation Immune Network. *Immunology Today,* Vol. **12**, 5, 159-166.

maturation, antibody production, the formation and subsequent elimination of antibody-antibody complexes, the natural "death" of antibodies and *B* lymphocytes and finally a supply of *B* lymphocytes (named the "source") coming from the bone marrow. Antibody-antibody and antibody-*B* lymphocyte interactions are determined by a so-called affinity matrix *m*, which is symmetric and reflects the network structure of the model. An entry m_{ij} is called the affinity between clone *i* and *j* and, for the present study, only takes value 1 if affinity exists between clone *i* and *j* and 0 if not. The evolution in time of the concentration of two immune actors f_i and b_i is described by the two differential equations:

$$\frac{df_i}{dt} = -k_1\sigma_i f_i - k_2 f_i + k_3 \mathrm{mat}(\sigma_i) b_i$$

$$\frac{db_i}{dt} = -k_4 b_i + k_5 \mathrm{prol}(\sigma_i) b_i + k_6 \qquad i = 1 \ldots n$$

k_1 to k_6 are six time constants. The extent to which two clones interact in the network is thus determined by *m* and the concentration of the antibodies. The integral impact of the whole network on a specific clone *i* is measured by a value σ_i which is called the field: $\sigma_i = \sum_{j=1}^{j=n} m_{ij} f_j$

mat and prol are two log-normal functions which determine how *B* lymphocytes maturate and proliferate upon activation by the field:

$$\mathrm{mat}(\sigma_i) = \exp\left\{-\frac{\ln(\sigma_i/\mu_m)}{s_m}\right\}^2 \qquad \mathrm{prol}(\sigma_i) = \exp\left\{-\frac{\ln(\sigma_i/\mu_p)}{s_p}\right\}^2$$

The parameter values for the simulations described in this paper are: $k_1 = 0.0016[\mathrm{conc}^{-1}\mathrm{d}^{-1}]$; $k_2 = 0.02[\mathrm{d}^{-1}]$; $k_3 = 2.0[\mathrm{d}^{-1}]$; $k_4 = 0.1[\mathrm{d}^{-1}]$; $k_5 = 0.2[\mathrm{d}^{-1}]$; $k_6 = 0.1[\mathrm{d}^{-1}]$; $m_m = 80[\mathrm{conc}]$; $s_m = 0.5$; $m_p = 120[\mathrm{conc}]$; $s_p = 0.5$.

The biological motivations behind such a modeling is outside the scope of this paper (see Varela and Countinho (1991) for these motivations). Also the value for each parameter was determined in agreement with biological data which need not be discussed here. Basically the parameters were tuned so as to obtain an oscillatory behavior for the simplest possible network containing 2 complementary clones with the affinity matrix given by: $\begin{bmatrix} 0 & 1 \\ 1 & 0 \end{bmatrix}$

In order to produce frustration in the structure of connectivity, attention will be paid only to affinity matrix which reflects the complementarity of the clones (two complementary clones which can be seen as two opposite spins), thus in the absence of self-affinity (m_{ii}=0) and with affinity restricted to complementary clones ($m_{ij} = m_{ji}$ takes its value in [0,1]). Such a structure will indeed lead to the open and closed chains which are precisely our objects of interest.

In Figure 4, simulations are shown for the network sizes 2 and 3 with indicated for each case the corresponding affinity matrix. Only the temporal evolution of the antibody concentrations is shown. The figure clearly shows the periodic nature of the regime obtained for both the two-clone and the non frustrated three-clone situation. Interestingly enough the 3-clone open chain situation is very close to the 2-clone situation when substituting one of the clones in the 2-clone case by a couple of them in the 3-clone case. It is remarkable to see for the 3-clone situation and for whatever number of clones in general how much the 2-clone dynamics prevails and the attractive effect it exerts on all other configurations. In the 3-clone open chain situation described by six differential equations the network behaves nearly in the same way as in the 2-clone case with clones 1 and 3 oscillating in perfect synchronization so as to form a double-clone equivalent to one of the two clones in the 2-clone case (except for the amplitude which is not surprisingly half the value of clone 2 equal to each of the clones of the 2-clone case). Since taken individually clone 1 and 3 are in a situation indistinguishable from the 2-clone situation i.e. they present affinity with one and only one clone (*i.e.* clone 2), they tend to behave just like in the 2-clone situation with as direct consequence their mutual coupling and the appearance of the double-clone.

When closing the chain ($1 \leftrightarrow 2 \leftrightarrow 3 \leftrightarrow 1$) and so doing obtaining the very same frustrated triangle already encountered in the section dedicated to Hopfield networks, the periodic regime switches to an aperiodic one. Since now, taken individually, each clone presents the same local connectivity (they are all connected to two neighbors) and then appears indistinguishable from the others (as will be confirmed below), none of them can assume the privilege to oscillate alone (then differently) in counterphase with the double-clone. Accordingly, the double-clone is continuously and erratically changing the nature of its members. This perfect clonal equivalence is obviously a very basic reason and original feature of the complicated regime which typifies the closed chain case. As expected this type of aperiodicity disappears for four clones and in general for an even number of clones. In the 4-clone either closed or open chain, the two double clones oscillate in counterphase, in contrast with the 5-clone closed chain where the aperiodic regime re-appears. We have extended the observation of how chains of interconnected clones (only the neighboring elements of the diagonal of the connectivity matrix are non-zero) behave up to 19 clones. As we expected, first return maps and the calculated power spectra indicate the presence of chaos for any odd loop while even chains are responsible for oscillatory behavior.

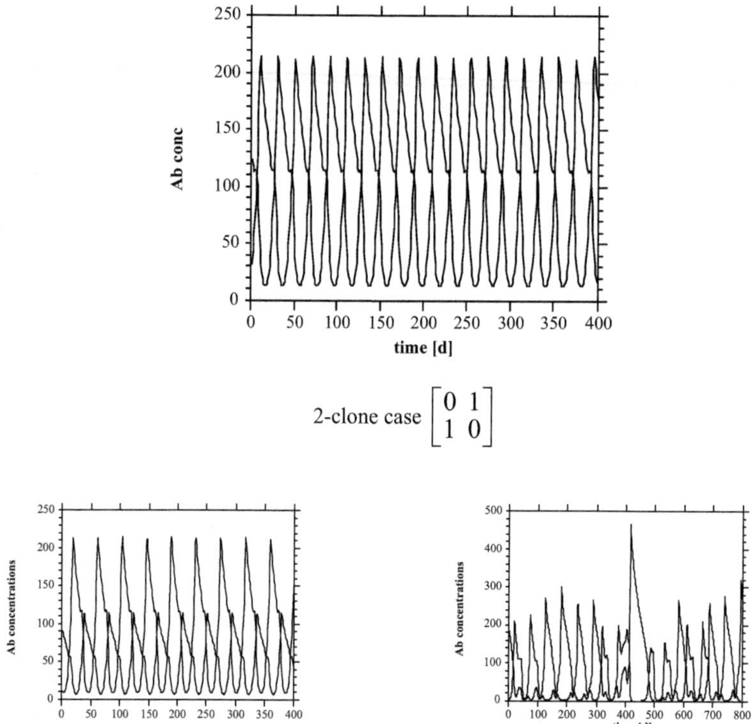

$$\text{2-clone case} \begin{bmatrix} 0 & 1 \\ 1 & 0 \end{bmatrix}$$

$$\text{3-clone open chain} \begin{bmatrix} 0 & 1 & 0 \\ 1 & 0 & 1 \\ 0 & 1 & 0 \end{bmatrix} \qquad \text{3-clone closed chain} \begin{bmatrix} 0 & 1 & 1 \\ 1 & 0 & 1 \\ 1 & 1 & 0 \end{bmatrix}$$

Figure 4. Time series of the 2 and 3-clone cases, respectively.
In all three cases, two clones are displayed

An important question to be addressed in this paper is the nature of the 3-clone closed chain regime. The computer experiments of the dynamics show strong evidence of an aperiodic behavior. A simple reasoning may help to eliminate some well-known possible regimes intermediary between chaos and periodicity like a toroidal attractor. Indeed this type of attractor is generally due to the merging of distinct periodicities, each associated to different variable actions. However, the main characteristic of the 3-clone closed chain is the perfect equivalence among the three clones and their respective concentration. As a consequence it is unlikely that this irregular behavior be attributed to a toroidal attractor characterized by an heterogeneity in the dynamics of its associated variables. In previous papers[18], we have performed and described technical analysis: power spectrum, first return Poincaré map, Lyapunov exponents, symbolic dynamics with all results converging to testify the presence of an original chaos in the time series.

Although this dynamics presents the typical signs of chaos it is hard to fit it into the well known chaotic regimes. Intuitively, the network perfect clonal equivalence *i.e.* the homogeneity in the variables dynamics makes the classical stretch-and-fold interpretation characterizing the largest family of chaotic dynamics more delicate to apply here. Rather the frustrated chaos behavior is a succession of attempts to decouple the system in two groups of oscillators, an impossible achievement making the dynamics rambling over very brief and successful configurations. Since each cyclic behavior shows fractal basin boundaries, the wavering among these cyclic attractor is beyond all predictability. You can't predict when a point in the phase space will be attracted by one of the three cycles since the attractive regions have frontiers themselves impossible to draw with finite precision. In brief, we attend an on-line and continuous manifestation of the final state sensitivity discussed in Ott *et al.*[19] and imputable to the fractal basin boundaries of the cyclic attractors.

3.2 Clustering

Always aiming at a better characterization of how the three-clone dynamical regimes scale up when increasing the network size, together with Detours and Calenbuhr[20] we have recently launched a systematic study which again was largely inspired from Kauffman's *NK* analysis for Boolean networks.

18. Bersini H. and Calenbuhr V. (1995). Frustration Induced Chaos in a System of Coupled ODE's, *Chaos, Solitons and Fractals,* Vol. **5**, 8, 1533-1549.
19. Ott E., Sauer T. and Yorke J.A. (eds) (1994). Coping with Chaos. Wiley Series in *Nonlinear Science,* John Wiley and Son, Inc.
20. Detours V., Calenbuhr V. and Bersini H. (1995). Clustering Phenomena in Idiotypic Network, *IRIDIA Internal Technical Report.*

The *NK* analysis reduces to the chain analysis described above for $K = 1$ and the network is fully connected for $K = N$. A complete study of the results is under progress but so far some general tendencies can be drawn. Chaos is found for most of the *N-K* values and a significant outcome appears to be the fragmentation of the network into clusters of 2, 3 or sometimes 4 activated clones separated by resting clones. The number of separating resting clones is related to the degree of connectivity. Clusters of 2 and 4 clones are oscillating and clusters of 3 clones shows the frustration chaos presented above. The clones within the clusters fluctuate with an average of $f_i = 40$ whereas the resting clones separating the clusters fluctuate around mean concentration three orders of magnitude less. So like in the HNN case, clustering is a natural self-organized tendency of large INN networks.

Notice however that in order to obtain such a fragmentation, the network needs not be diluted from the very beginning, quite the contrary is true, the dilution comes to be a consequence, and no longer a necessary condition, of the fragmentation. During its time evolution, the idiotypic network spontaneously tunes its connectivity to low value. It is known that the embrionnary idiotypic network shows larger connectivity that the adult one. This is indeed found also in simulations together with a reinforcement of the selectivity of the new clones to be recruited in the network.

The works of Gardner, Ashby and May discussed above seem to suggest that a biological ecosystem in order to be stable must be organized into a set of separated sub-networks where species in one sub-network are insulated from interactions with species in another sub-network. This self-selection for local type of connectivity was also observed in a coupled-map-lattice computer simulation of ecosystems[21] where again the degree of interconnectivity appeared as an emergent property regulated by the network itself on the road to its equilibrium states. Compartmentalization of species communities into independent clusters were experimentally validated by Stuart Pimm[22] and seems to be characteristic of animal communities. In their natural quest for stability the ecological network tunes autonomously their connectivity to low threshold value.

---

21. Solé R.V., Bascompte J. and Valls J. (1992). Nonequilibrium Dynamics in Lattice Ecosystems: Chaotic Stability and Dissipative Structures. *CHAOS,* **2**, 3, 387-395.
22. Pimm Stuart (1991). *The Balance of Nature.* University of Chicago Press; Keley K. (1994). *Out of Control. The Rise of Neo-biological Civilization.* Addison Wesley.

4. CONCLUSIONS: FREE SPECULATIONS ON THE GOODNESS OF FRUSTRATION AND CLUSTERING IN BIOLOGICAL NETWORKS

The main motivation of this paper was a qualitative overview of two biological networks modeling in an attempt to spot similar form of emergent dynamical sensitivity to structural aspects. These biological networks are Hopfield Neural Network and Immune Idiotypic Network. The two structural influences we observed are the great sensitivity the networks dynamics present to frustrated connectivity and the tendency for regularly connected networks to fragment into small clusters. Frustrated connectivity is, in few cases, responsible for enlarging the diversity of equilibrium regimes but, more generally, for provoking instability in network: fixed points turn into oscillation while oscillation turns into chaos. This instability is due to the "wavering" of the network unable to settle into one of the equally possible equilibrium regime. As a benefic outcome of frustration, the network is able to recurrently propose a large repertoire of potential behaviors which can be triggered in response to external interaction. The fact that frustration has been detected in the great majority of physical and biological networks studied so far: spin glass[23], genetic[24], neural[25], oscillatory[26] and immune[27], whose mathematical description can be quite different, is pleading for an understanding of this frustration as just emergent from the structure of connectivity. Such a generic phenomenon should not be restrictively construed as an insignificant artefact of our mathematical and computer modeling but rather as a real biological effect,

---

23. Toulouse G. (1977). *Commun. Phys.,* **2**, 115.
24. Thomas R. (1991). Regulatory Networks Seen as Asynchronous Automata: A Logical Description. *J. Theor. Biol.,* **153**, 1-23.
25. Amit D.J. (1989). Modelling Brain Function: *The World of Attractor Neural Networks.* Cambridge: Cambridge University Press; Sherrington D. (1990). Complexity Due to Disorder and Frustration, *Lectures in the Sciences of Complexity - SFI Studies in the Sciences of Complexity,* Lect. Vol. II, Addison-Wesley: Ed. Erica Jen, 415-455; Marcus C.M., Waugh F.R. and Westervelt R.M. (1991). Nonlinear Dynamics and Stability of Analog Neural Networks. *Physica D,* **51**, 234-247; Atiya A. and Baldi P. (1989). Oscillations and Synchronization in Neural Networks: An Exploration of the Labeling Hypothesis. *International Journal of Neural Systems,* Vol. **1**, 2, 103-124; Sherrington D. (1990). Complexity Due to Disorder and Frustration. *Lectures in the Sciences of Complexity – SFI Studies in the Sciences of Complexity,* Lect. Vol. II, Addison-Wesley, Ed. Erica Jen, 415-455.
26. Daido H. (1992). Quasi Entrainment and Slow Relaxation in a Population of Oscillators with Random and Frustrated Interactions. *Physical Review Letters,* Vol. **68**, 7, 1073-1076.
27. Bersini H. and Calenbuhr V. (1995). Frustration Induced Chaos in a System of Coupled ODE's, *Chaos, Solitons and Fractals,* Vol. **5**, 8, 1533-1549.

playing yes or not a benefic role (if not a natural way to un-frustrate the network must exist), that further investigation will have for goal to better characterize.

This ubiquity is also true for the clustering effect that we observed when both networks were regularly connected. First we have shown a sharp behavioral transition in increasing the HNN network connectivity and that a dilute form of connectivity is often responsible for more interesting regimes. As a matter of fact, as far as our knowledge of natural networks go, dilute type of connectivity, simpler and more economical, appears to be the rule in nature. Actually, clustering by fragmentation is only possible in dilute Boolean or Hopfield networks. We need to distinguish further between two forms of clustering: "clustering by fragmentation" and "clustering by synchrony". We have shown that clustering by fragmentation is likely to occur only in regularly connected network. Clustering by synchrony could substitute it in networks randomly interconnected and thus more realistic. Clustering is important as a way of assigning a label or a meaning to any form of external interaction. For instance, clustering by synchrony seems to be of great interest in neural networks to support cognitive mechanisms such as labeling and variable binding. On the other hand, in immunology some authors are convinced that immune memory and locality are unseparable aspects[28]. The use of frustration to easily generate diversity together with clustering for labeling any interaction are two emergent phenomena which should deserve both increasing attention and more formal analysis in the future.

Acknowledgments: Thanks to V. Calenbuhr and V. Detours for their essential contribution in shaping the ideas that are presented in this paper.

REFERENCES

Amari S. (1972). Characteristics of Random Nets of Analog Neuron-like Elements. *IEEE Transactions on Systems, Man and Cybernetics.* Vol. SMC. **2**, 5, 643-657.

Amit D.J. (1989). *Modeling Brain Function: The World of Attractor Neural Networks.* Cambridge: Cambridge University Press.

Atiya A. and Baldi P. (1989). Oscillations and Synchronization in Neural Networks: An exploration of the Labeling Hypothesis. *International Journal of Neural Systems,* Vol. **1**, 2, 103-124.

---

28. Neumann A.U. and Weisbuch G. (1992). Dynamics and Topology of Idiotypic Networks. *Bulletin of Mathematical Biology,* Vol. **54**, 5, 699-726.

Bersini H. and Detours V. (1994). Asynchrony Induces Stability in Cellular Automata Based Models. In R. Brooks and P. Maes (eds). *Artificial Life IV.* MIT Press, 382-387.

Bersini H. and Calenbuhr V. (1995). Frustration Induced Chaos in a System of Coupled ODE's. *Chaos, Solitons and Fractals,* Vol. **5**, 8, 1533-1549.

Calenbuhr V., Bersini H., Stewart J. and Varela F.J. (1995). Natural Tolerance in a Simple Immune Network. *J. Theoretical Biology*, **177**, 199-213.

Chaté H. and Manneville P. (1992). Collective Behaviors in Coupled Map Lattices with Local and Non Local Connections. *CHAOS,* **2**, 3, 307-313.

Daido H. (1992). Quasi-entrainment and Slow Relaxation in a Population of Oscillators with Random and Frustrated Interactions. *Physical Review Letter.* Vol. **68**, 7, 1073-1076.

De Boer R.J. and Perelson A. (1991). Size and Connectivity as Emergent Properties of a Developing Immune Network. *J. Theoretical Biology,* **149**, 381-424.

De Boer R.J., Perelson A.S. and Kevrekidis I.G. (1993). Immune Network Behavior, I. From Stationary States to Limit Cycle Oscillations. *Bulletin of Mathematical Biology,* Vol. **55**, 4, 745-780.

Detours V., Calenbuhr V. and Bersini H. (1995). Clustering Phenomena in Idiotypic Network. *IRIDIA Internal Technical Report.*

Gardner M.R. and Ashby W.R. (1970). Connectance of Large Dynamic (Cybernetic) Systems: Critical Values for Stability. *Nature,* Vol. **228**, 784.

Hiernaux J. (1977). Some Remarks on the Stability of Idiotypic Network. *Immunochemistry,* Vol. **14**, 733-739.

Hirsch M.W. (1987). Convergence in Neural Networks. *Proc. 1987 Int. Conf. Neural Networks,* San Diego, CA.

Hopfield J.J. (1982). Neural Networks and Physical Systems with Emergent Collective Computational Abilities. *Proc. Nat. Acad. Sci. USA,* Vol. **79**, 2554-2558.

Kaneko K. (1989). Pattern Dynamics in Spatiotemporal Chaos. *Physica D,* **34**, 1-41.

Kaneko K. (1992). Overview of Coupled Map Lattices. *CHAOS,* **2**, 3, 279-282.

Kauffman S.A. (1989). Principles of Adaptation in Complex Systems. *Lectures in the Sciences of Complexity - SFI Studies in the Sciences of Complexity,* Addison-Wesley, Ed. D. Stein, 619-712.

Keley K. (1994). *Out of Control. The Rise of Neo-biological Civilization.* Addison Wesley.

Lewis J.E. and Glass L. (1992). Non-linear Dynamics and Symbolic Dynamics of Neural Networks. *Neural Computation,* Vol. **4**, 5, 621-642.
Lourenço C. and Babloyantz A. (1994). Control of Chaos in Networks with Delay: A Model of Synchronization of Cortical Tissue. *Neural Computation,* **6**, 1141-1154.
Lumer E.D. and Nicolis G. (1994). Synchronous Versus Asynchronous Dynamics in Spatially Distributed Systems. *Physica D,* **71**, 4, 440-452.
Marcus C.M., Waugh F.R. and Westervelt R.M. (1991). Non-linear Dynamics and Stability of Analog Neural Networks. *Physica D,* **51**, 234-247.
May R.M. (1972). Will a Large Complex System be Stable? *Nature,* Vol. **238**, 413-414.
Neumann A.U. and Weisbuch G. (1992). Dynamics and Topology of Idiotypic Networks. *Bulletin of Mathematical Biology,* Vol. **54**, 5, 699-726.
Omata S. and Yamaguchi Y. (1988). Entrainment Among Coupled Limit Cycle Oscillators with Frustration. *Physica D,* **31**, 397-408.
Ott E., Sauer T. and Yorke J.A. (eds) (1994). Coping with Chaos. Wiley Series in *Nonlinear Science,* John Wiley and Son, Inc.
Perelson A.S. (1990). Theoretical Immunology. *Lectures in Complex Systems, SFI Studies in the Sciences of Complexity,* Lect. Vol. II, Addison-Wesley, Ed. Erica Jen.
Pimm Stuart (1991). *The Balance of Nature.* University of Chicago Press.
Sherrington D. (1990). Complexity Due to Disorder and Frustration, *Lectures in the Sciences of Complexity - SFI Studies in the Sciences of Complexity,* Lect. Vol. II, Addison-Wesley, Ed. Erica Jen, 415-455.
Solé R.V., Bascompte J. and Valls J. (1992). Nonequilibrium Dynamics in Lattice Ecosystems: Chaotic Stability and Dissipative Structures. *CHAOS,* **2**, 3, 387-395.
Stewart J. and Varela F. (1990). Dynamics of a Class of Immune Networks. II. Oscillatory Activity of Cellular and Humoral Components. *Journal of Theoretical Biology,* **144**, 103-115.
Strogatz S.H., Mirollo R.E. and Matthews P.C. (1992). Synchronization of Pulse-Coupled Biological Oscillators. *SIAM Journal on Applied Mathematics,* Vol. **50**, 6, 1645-1662.
Thomas R. (1991). Regulatory Networks Seen as Asynchronous Automata: A Logical Description. *J. Theor. Biol.,* **153**, 1-23.
Toulouse G. (1977). *Commun. Phys.,* **2**, 115.
Varela F.J., Coutinho A., Dupire B. and Vaz N.N. (1988). Cognitive Networks: Immune, Neural and Otherwise. In A.S. Perelson (ed.),

Theoretical Immunology, Part Two, SFI Studies in the Sciences of Complexity, Vol. 3, Reading, MA: Addison-Wesley, 377-401.
Varela F.J. and Coutinho A. (1991). Second Generation Immune Network. *Immunology Today,* Vol. **12**, 5, 159-166.
Winfree A.T. (1987). *The Timing of Biological Clocks,* Scientific American Library.

I. SCIENTIFIC APPROACH

B. Self-Organization and Biology: Thematic Standpoints

RENÉ THOMAS

ANALYSIS AND SYNTHESIS OF REGULATORY NETWORKS IN TERMS OF FEEDBACK CIRCUITS

SUMMARY

Studies on the biological role of feedback circuits were initially centered on systems with sigmoid or stepwise interactions. It turned out recently that reasoning in terms of feedback circuits (rather than of individual interactions) can be used in a more general context, and, in particular, help understanding the behavior of weakly non-linear systems "à la Rössler" (with a single non-linear term) known to generate multiple periodicity or deterministic chaos.

The obvious way to formalize biological and other regulatory systems consists of using sets of differential equations. Since most regulatory interactions are non-linear, these differential systems usually cannot be treated analytically. In many cases, the shape of these non-linearities is sigmoid, *i.e.*, the effect of a regulator is negligible below a "threshold" value and it rapidly levels off beyond this threshold value. For this reason, it is tempting to caricature these interactions as step functions. This is the justification of the efforts to develop logical methods, hoping for qualitative, yet analytical tools. It is our experience that differential and logical methods nicely complement each other, and often gain to be used in conjunction.

The purpose of this paper is to show that although the logical approach has been developed to treat systems with step- or steep sigmoid interactions, the type of reasoning used can be fruitfully applied to weakly non-linear systems (the Rössler type of differential systems) which can generate complex behavior, including deterministic chaos.

This paper includes :

1) a brief account of recent developments in the *logical description* of regulatory systems;
2) a section on the properties and roles of *feedback circuits*;

B. Feltz, M. Crommelinck and P. Goujon, (eds.), Self-organization and Emergence in Life Science, 63-73.

3) a section on the recent concept of *circuit-characteristic states*;
4) a discussion on how differential systems can be treated in terms of feedback circuits;
5) an application to the Rössler-type differential systems.

It is realized that the brief description of items 1 to 4 is not self-sufficient, but these matters have been amply discussed elsewhere (bibliography below).

1. DEVELOPMENTS IN THE LOGICAL DESCRIPTION OF REGULATORY NETWORKS

In logical descriptions, variables and functions are treated as if they could take only a limited number of values; in simple cases, two only (1 and 0).

The state of a system can be symbolized by a logical vector, whose elements describe the level of relevant variables (Kauffman 1969).

1.1 Asynchronous *vs* Synchronous Description

Classically, time is inserted in the logical description by giving the state vector at time "t + 1" as a function of the state vector at time t. This so-called "synchronous" description is not appropriate for biological systems, for two reasons (i) each state has one, and only one possible follower and this prevents any possibility of differentiation from one logical state to two or more possible followers; (ii) in this description, if, for example, two genes are switched on together, one has to assume that their products will reach their threshold concentration in exact synchrony. This leads to severe artefacts.

This is why we developed an *asynchronous* description, in which each process can have its own timing (Thomas 1973, 1979, 1991; Thomas and D'Ari 1990). The set of logical equations behaves as an operator which, when applied to any state (x y z), provides the *image* (XYZ) of this state. For instance, if the image of state 0 1 0 (x absent, y present, z absent) is 1 0 0 (x present, y absent, z absent), it means that the interactions tend to drive the system from 0 1 0 toward 1 0 0; however, there is no reason why the commutations of x from 0 to 1 and of y from 1 to 0 should take place together. If we represent this situation by 0 1 0 (x is 0 but it has a command to switch to 1; y is 1 but it has a command to switch to 0), we will have:

+ -
0 1 0 → 1 1 0
OR
0 1 0 → 0 0 0

depending on the time delays involved in the commutations. In the synchronous description, one would have 0 1 0 ⇒ 1 0 0, which is in fact a marginal possibility.

When a state and its image are equal, one deals with a *stable logical state.*

Most authors continue to use an asynchronous approach, because it is easier to handle. This is reminiscent of the man who looks for his key below a street lamp because *there*, there is light (although he knows that the key is in fact elsewhere). Originally, the asynchronous description was suspected to be impracticable and to generate *any*thing; in fact it turned out to be practicable and to generate predictions that fit with those of the differential description.

1.2 When Should we Use More Than the Classical Two (0, 1) Logical Values? (see Van Ham 1979)

Usually, our criteria are quite simple. If a variable acts in two ways (for example, product y prevents the synthesis of x and activates its own synthesis) there is no reason why the thresholds for these two actions should be the same. Thus, we ascribe variable y two thresholds and consequently three logical values (0, 1, 2). More generally, if a variable acts at n levels, it has n thresholds and thus $n + 1$ logical values (0, 1, 2, ..., n).

1.3 Logical parameters (Snoussi 1989)

With each variable, one associates logical parameters, which can take the same range of logical values as the variable itself. An extremely simple but relevant example is a three-element negative feedback circuit which can be described as follows in what we now call "naïve" logics:

$$X = \overline{z}$$
$$Y = \overline{x}$$
$$Z = \overline{y}$$

Applying these logical equations from any initial state will result in a periodic behavior of all three variables. However, one knows from the differential description that a stable periodic behavior will take place only for proper kinetic parameters; outside this range, the system will proceed to a stable steady state. If instead we write

$$X = K_1 . \bar{z}$$
$$Y = K_2 . \bar{x}$$
$$Z = K_3 . \bar{y}$$

(in this simple case each K can take one of the values, 0 or 1), the apparent contradiction between the differential and logical description disappears; when all the K's = 1 we have the periodic behavior, but otherwise the system is blocked in a stable state, different according to the individual values of the K's.

The resulting logical description is much more subtle and general; our original logical description turned out to be a particular case of a more general description.

1.4 Inclusion of the thresholds as logical values. (Thomas and D'Ari 1990)

Classically, the logical description writes $x = 0$ if the real value of x is $< s$ (below the threshold, subliminal) and $x = 1$ if the real value is $> s$; however, the marginal situation in which the real value $= s$ is ignored. A number of steady states of the differential description are not "seen" in classical logical descriptions, for the simple reason that in these states one or more variable is located on a threshold value. For this reason, I introduced threshold values as logical levels; thus, instead of a scale 0, 1, 2, ... our present logical scale is 0, $s^{(1)}$, 1, $s^{(2)}$, 2, ... We thus have "regular" logical states which occupy a box in the space of the variables and "singular" states which are located on one or more thresholds, on a plane, an edge or a vertex between the boxes. Thanks to this improvement, *all* the steady states of the differential description can be identified in logical terms. This generalization required an extension of the concept of steady state to logical systems.

2. FEEDBACK CIRCUITS

When element x influences the rate of production of element y, which influences the rate of production of z, which in turn influences the rate of production of x, we say that x, y, z form a feedback circuit. There are two types of feedback circuits; either *each* element of the circuit exerts (*via* the other elements if any) a positive influence on its own further production, or *each* element exerts a negative influence on its own further production. Accordingly, one denotes these circuits positive or negative. Whether one deals with a *positive* or a *negative* circuit, depends on the parity of the number of negative elements; a circuit with an even number of negative interactions is a positive circuit, with an odd number of negative interactions, a negative circuit.

For example $x \underset{-}{\overset{-}{\rightleftarrows}} y$ is a *positive* circuit.

The properties of positive and negative feedback circuits are deeply different; negative circuits are responsible for *homeostasis,* with or without oscillations, while positive circuits are responsible for *multistationarity,* a general phenomenon whose biological modality is differentiation (Thomas 1981 and Plahte *et al.* 1995).

3. THE CONCEPT OF CIRCUIT-CHARACTERISTIC STATE

Consider the feedback circuit $x^{-(2)}\, y^{+(1)}\, z^{+(1)}$ (x exerts a negative action on y if its level exceeds its second threshold $s^{(2)}$, y exerts a positive action on z if its level exceeds its first threshold $s^{(1)}$ and similarly z exerts a positive action on x if its level exceeds its first threshold $s^{(1)}$). The logical state located at $s^{(2)}\, s^{(1)}\, s^{(1)}$ (*i.e.,* $x = s^{(2)}$, $y = s^{(1)}$, $z = s^{(1)}$), thus, at the level of the thresholds involved in the circuit, plays a special role in the operation of the circuit. For this reason, it has been called *circuit-characteristic state.*

It has been realized (Thomas 1991), and subsequently demonstrated (Snoussi and Thomas 1993), that among the (often very many) singular states of a system, only those that are circuit-characteristic can be steady (see section 1.4). Inversely, when one considers a circuit-characteristic state, there are parameter values for which it is steady (at least in the subspace of the variables involved in the circuit). In practice, when a circuit is functional — *i.e.,* it actually generates homeostasis (if negative) or multistationarity (if positive) — its characteristic state is steady (in the subspace...), and *vice versa.*

This introduces a surprisingly simple relation between feedback circuits and singular steady states; instead of having to scan through all the singular logical states of a system and check for each of them whether (or for which parameter values) it is steady, one just has to identify the circuits (or, more generally, unions of disjoint circuits), consider for each of them the (unique) characteristic state and see whether (or in which range of parameter values) it is steady. This process can be lead "by hand" without problem for up to 3 variables, but it had of course to be computerized for more variables (Thieffry *et al.* 1993).

4. DIFFERENTIAL SYSTEMS SEEN IN TERMS OF FEEDBACK CIRCUITS

There may have been some ambiguity in the past concerning the precise definition of "interactions" and of "feedback circuits" (should one consider only the regulatory interactions, etc.). Any ambiguity disappears if one simply considers the jacobian matrix (the equivalent of the derivative for an n-dimensional system) of the system and state that variable j exerts a positive (*vs* negative) action on variable i if element a_{ij} of this matrix has a positive (*vs* negative) value. Feedback circuits (or more generally unions of disjoint circuits) are identified as sets of non-zero elements of the matrix whose indices i and j are permutations of each other; in particular, each non-zero diagonal term (a_{ii}) denotes an one-element circuit. A circuit is positive or negative according to the parity of the number of negative elements it comprises.

The typical[1] behavior of simple feedback circuits can be described as follows:

- an one-element circuit (direct autoregulation; direct autocatalysis if one deals with a positive circuit) generates (in the subspace of the variable considered) a steady state which is attractive or repulsive depending on whether the circuit is negative or positive;
- a two-element *positive* circuit forces the variables involved in the circuit to choose between two attractors (it generates a saddle point, located on a separatrix which divides the space of the variables considered into two attraction basins);
- a two-element *negative* circuit generates a periodic approach to a steady state (a stable focus), but in the presence of an autocatalytic term the focus can be destabilized, thus resulting in a periodic departure from its vicinity;

---

1. "Typical" means that the behavior described is found for a wide range of parameter values, such that the interactions that constitute the circuit considered are strong enough and not hindered by other interactions.

– a three-element circuit can generate a saddle-focus which, for a negative circuit, is attractive along one direction and periodically repulsive along a normal surface, and, for a positive circuit, is attractive on a (separatrix) surface and repulsive along a normal direction.

5. APPLICATION TO THE RÖSSLER-TYPE SYSTEMS

An admirable (because astonishingly simple) system of differential equations giving rise to deterministic chaos was discovered by Rössler (1976). His system consists of a set of three ordinary differential equations with only one non-linearity. In a form slightly modified by Gaspard and Nicolis (1983) in order to have a (convenient) steady state with coordinates (0, 0, 0), it writes:

$$\begin{aligned}\dot{x} &= -y - z\\ \dot{y} &= x + ay\\ \dot{z} &= bx + xz - cz\end{aligned} \qquad (1)$$

(*a, b, c* are positive coefficients). The jacobian matrix is:

$$\begin{pmatrix} 0 & -1 & -1 \\ 1 & a & 0 \\ b+z & 0 & x-c \end{pmatrix}$$

A relevant aspect of the system is the existence of two[2] unstable steady states (saddle-foci); one is attractive following the z axis and periodically repulsive in x y, the second is periodically attractive in x z and repulsive along the y axis (Gaspard and Nicolis 1983).

In spite of the remarkable structural simplicity of the system, the exact role of each term in the equations is by no means obvious. With the hope of better understanding this aspect, one may reason in terms of feedback circuits and try to re-build a system which would display the two types of steady states just mentioned.

Using what we know (see the end of section 4) about the properties of feedback circuits, the first steady state can be built as follows. In order to be

2. As forcasted by Nicolis, a single appropriate steady state is sufficient. Using the same type of reasoning as above, I found indeed a variant of this system which generates deterministic chaos with a single steady state (Thomas 1999). See also Goldbeter (1995) for a more intricate system with a single steady state.

attractive along the z axis, we need a negative diagonal term in z, in order to be periodic in $x\ y$ we need a negative circuit in $x\ y$, and in order for this focus to be repulsive, we need a diagonal autocatalytic term in x or y. One of a few equivalent possibilities is described by the qualitative jacobian matrix:

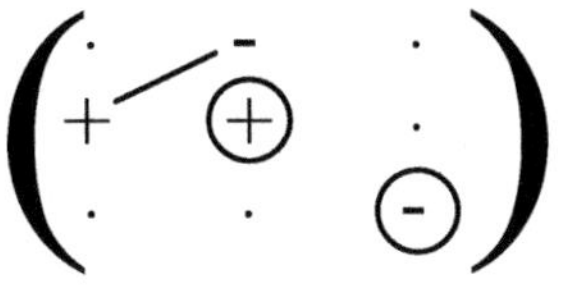

One can easily check[3] that even a *linear* system of this structure can generate the first type of saddle-focus.

Similarly, in order to generate the second steady state, one needs a negative circuit in $x\ z$ in order to have a periodic attractivity in plane $x\ z$, and a positive circuit of y on itself in order to be repulsive in y;
for example:

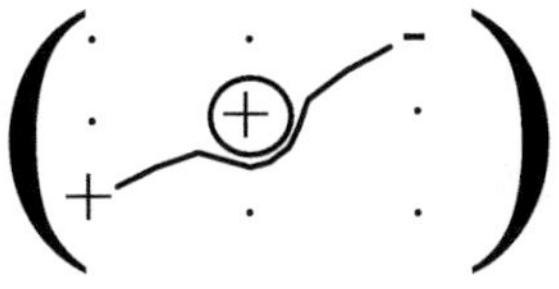

Combining the two matrices, we get:

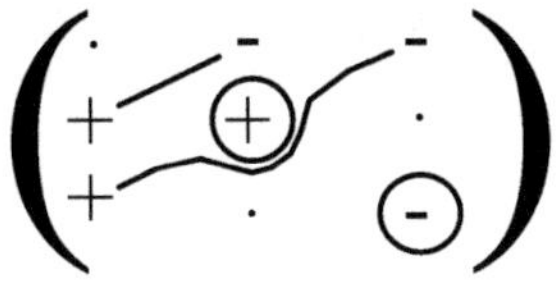

Note that term a_{22} is common to the two matrices and indeed serves two purposes: destabilize the first focus and generate repulsivity along the y axis

3. Take, for example, the system
$x = -2y$
$y = 2x + 0.5y$
$z = -10z$
The roots of the characteristic equation are –10 and +0.25 +/– 1.98 i.
Starting, say, from (0.1, 0.1, 2), the trajectory has the expected shape.

in the second focus. As a matter of fact, this term represents the only positive circuit in the system, and it is thus responsible for the existence of two distinct steady states.

Comparing this matrix with the jacobian matrix of the Rössler system one can remark that the term $x\,z$ of the Rössler system is not present here (if it were, there would be a term $+z$ in element a_{31} and a term $+x$ in element a_{33}). This lead me to ask whether in Rössler's equations there is really a *structural* need for an $x\,z$ term, or whether it is there simply because there must be at least one non-linear term in the differential equations in order to have a complex behavior.

In order to investigate this point, I checked what happens when one deletes the $x\,z$ term in Rössler's equation and renders another term non-linear.

For example:

$$\begin{aligned} \dot{x} &= -y - z \\ \dot{y} &= x + ay \\ \dot{z} &= bx^2 - cz \end{aligned} \tag{2}$$

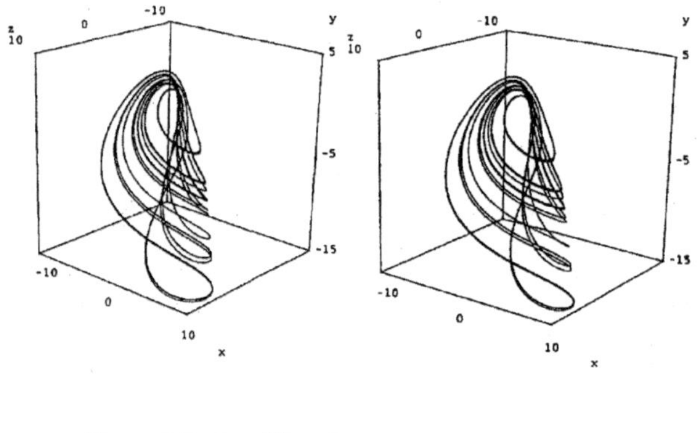

Figure 1. System (2): $\dot{x} = -y - z$
$\dot{y} = x + ay$
$\dot{z} = bx^2 - cz$

$a = 0.385$, $b = 0.3$, $c = 2$. Initial state was (.1, .1, .1) but integration was run (step: 0.01) from time 0 to time 100 without recording the trajectory (in order to eliminate transitories), then from time 100 to 200. The chaotic character was checked by determining the Lyapunov exponents.

For a wide range of values of the three parameters, this system displays a deterministic chaotic behavior (Figure 1). As a matter of fact, any of a number of nonlinear functions can be used: x^2, x^3, $x/(1+x)$, $\sin x$, $\operatorname{tg} x$, tanh x,... Note that some of them yield three unstable steady states and two symmetrical chaotic attractors.

Also, the non-linearity can be located elsewhere (for example in position a_{13}) although not anywhere, for reasons which are understood. As expected from the properties of the feedback circuits, one can also permute the signs of the terms forming the negative loops (thus inverting the way of rotation) without loosing the characteristic behavior of the system.

The interest of the above is that we now can build systems of an extremely simple structure, which generate deterministic chaos or multiple periodicity, and in which the exact role of each term is understood.

Acknowledgments: I wish to thank Professor G. Nicolis for stimulating discussions, Rob De Boer for kindly providing his GRIND Program and Pascale Gyserman who provided me with her program giving Lyapunov exponents and kindly helped me using it.

REFERENCES

Gaspard P. and Nicolis G. (1983). What Can we Learn from Homoclinic Orbits in Chaotic Dynamics? *J. Stat. Phys.*, **31**, 499-517.

Goldbeter A. (1995). *Biochemical Oscillations and Cellular Rythms.* Cambridge University Press.

Kauffman S.A. (1969). Metabolic Stability and Epigenesis in Randomly Constructed Genetic Netw. *J. Theoret. Biol.*, **22**, 437-467.

Plahte E., Mestl T. and Omholt S.W. (1995). Feedback Loops, Stability and Multistationarity in Dynamical Systems. *J. Biol. Syst.*, **3**, 1-9.

Rössler O.E. (1976). An equation for continuous chaos. *Phys. Letters,* **57A**, 397-398.

Snoussi E.H. (1989). Qualitative Dynamics of Piece-linear Differential equations: a discrete mapping approach. *Dyn. Stability Syst.* **4**, 189-207.

Snoussi E.H. and Thomas R. (1993). Logical Identification of all Steady States; The Concept of Feedback Loop Characteristic States. *Bull. Math. Biol.*, **55,** 973-991.

Thieffry D., Colet M. and Thomas R. (1993). Formalisation of Regulatory Networks: a Logical Method and its Automatization. *Math. Modelling and Sc. Computing*, **2**, 144-151.

Thomas R. (1973). Boolean Formalisation of Genetic Control Circuits. *J. Theoret. Biol.*, **42**, 565-583.

Thomas R. (ed.) (1979). Kinetic Logics. Lecture Notes in *Biomathematics,* **29**, 507 pp.

Thomas R. (1981). On the relation between the logical structure of systems and their ability to generate multiple steady states or sustained oscillations, Springer Series in Synergetics 9, 180-193.

Thomas R. (1983). Logical vs Continuous Description of Systems Comprising Feedback Loops: The Relation between Time Delays and Parameters. *Studies in Physical and Theoretical Chemistry*, **28**, 307-321.

Thomas R. (1991). Regulatory Networks Seen as Asynchronous Automata: a Logical Description. *J. Theor. Biol.*, **153**, 1-23.

Thomas R. (1994). The Role of Feedback Circuits: Positive Feedback Circuits are a Necessary Condition for Positive Real Eigenvalues of the Jacobian Matrix. Ber. Bunzenges. *Phys. Chem.*, **98**, 1148-1151.

Thomas R. (1999). The Rossler Equations Revisited in Terms of Feedback Circuits. *J. Biol. Syst.*, **7**, 2, 225-237.

Thomas R. and D'Ari R. (1990). *Biological* Feedback. CRC Press, Boca Raton, 316 pp.

Van Ham P. (1979). Lecture notes in *Biomathematics*, **29**, 326-343.

P. LEFÈVRE, C. TU, M. MISSAL AND M. CROMMELINCK

PROPERTIES EMERGING FROM SENSORIMOTOR INTERFACES. INTERACTION BETWEEN EXPERIMENTATION AND MODELING : IN NEUROSCIENCES

1. INTRODUCTION

In the last two decades much progress has been made in the field of neurosciences. This is due, on the one hand, to the spectacular development of very performing investigation techniques of the nervous functions (functional imagery, cellular recording in awake animals...), and on the other hand, to the explanatory power of models stemming from various fields (molecular biology, control theory, system analysis, neural networks, signal processing and modeling of cognitive processes...). Some did not hesitate to compare these last advances with those made in physics, in the beginning of this century, or in molecular biology in the mid-century. However, in spite of these significant progress, there is at the moment no unified theory to account for the way nervous activities give rise to integrated perception and action, underlined by intentionality giving sense to the human behavior. In the latest theoretical propositions — some of them dating back to the mid-century — some important concepts have been put forward, among which the concept of "population coding". The concept implies the implementation of relatively important sets of cells in behaviors or in mental activities. Inside these sets, particular modes of co-operation between the processing units, the neurons, ensure the coding of specific information by the central nervous system. In a certain number of cases, these cellular sets form functional "maps" — sensory and motor maps —, whose activity controls the concrete modalities of interaction between the individual and the environment. These maps are dynamic: many experimental approaches proved indeed their property of structural and functional plasticity. Thus, not

B. Feltz, M. Crommelinck and P. Goujon, (eds.), Self-organization and Emergence in Life Science, 75-95.

only the "topography" of these maps may be modified through training (for instance, extension of the area corresponding to the trained modality or segment), but also their functioning rules (for instance, emergence of synchronic cellular activities).

Today it seems clearly established, according to this principle of "population coding" and the empirical data related to it, that nervous information is contained not only in the activation level of individual neurons but also in the values of the "weights" of the synaptic connections (excitatory and inhibitory) characterizing these neural networks. In a way, nervous functions appear as properties emerging from these interactions within the functional maps; it is then the integration of these local properties of the "microlevel" that organizes and structures the "macrolevel" of the behavior and the mental states, with loops feeding back from the "macro" to the "microlevel".

Besides, cerebral maps are structures liable to embody internal models inside of which representations are stored in a more or less long term. These representations concern not only the characteristics of the outer world (as, for instance, perceptive categories), but also of the interaction procedures between the organism and the environment (as motor schemes), as well as of the sensory consequences expected during the processing of these motor programs (as anticipation schemes of the references). The concept of "functional maps" may thus be associated with another important theoretical concept, the "internal model". Some theoreticians have developed, for instance, the idea that perception could be considered as the result of an interaction between a sensory input and the "simulation of an internal model" (see among others Berthoz 1997). This internal model corresponds not so much to a static image of reality but rather to a dynamic field of interrelations between the organism and its environment, including cycles of "perception-action" (see among others Neisser 1976). The eye's exploration of the visual field is, for instance, clearly guided by schemes which act as plans anticipating the perception-action or by expectations resulting from a base of acquired or innate knowledge (Yarbus 1967).

In the present paper we would like to explain, on the basis of a relatively elementary experimental paradigm, how certain properties emerging from the neural maps processing solve elegantly complex problems concerning the sensorimotor integration, an important function of the central nervous system. More specifically, these problems are related to the information transfer from the sensory modules to the motor modules. The experimental paradigm at stake here consists of the study of nervous control mechanisms of the ocular saccades, a particular type of eye motor functions. The problem which, within these control mechanisms, seems liable to be solved in an original way by the properties emerging from the neural maps and by the implementation of internal models is the spatiotemporal processing occurring during the information transfers between the visual maps and the oculomotor maps.

2. THE MOVEMENTS OF EYE ORIENTATION AS A STUDY PARADIGM OF SENSORIMOTOR INTEGRATION

The mechanisms of eye orientation represent an interesting experimental model to study the modalities of sensorimotor representations within the central nervous system as well as the issue of the spatiotemporal transformations in the sensorimotor integration. Indeed, not only the inputs (sensory information, mainly visual) and the outputs (eye movements) of the system are easily mastered experimentally, but also the cellular assemblies controlling these "visuo-oculomotor" integrations are likely to be recorded (direct recording in awake animals, thanks to the techniques of cellular electrophysiology, and indirect recording in human beings, thanks to the techniques of functional mental imagery). From a very schematic angle, we distinguish two types of eye orientation movements in the superior mammals. On the one hand, ocular saccades permit to move very rapidly the gaze axis from a stationary target to another: this type of movement is used when exploring complex forms or when reading. On the other hand, the ocular pursuit (which is sometimes called "smooth pursuit") permits to maintain the visual axis on a moving target. Saccades are thus controlled by a position error produced by signals present in the outer environment or simply mentally represented, whereas the smooth pursuit is generally controlled by an error in the velocity domain. These two types of orientation movements appear essentially in species having a frontal vision as well as a fovea, i.e. a small area, like a depression, of the central retina characterized by a very good acuity, compared with that of the retinal periphery. So as regards human beings, it has been estimated that as soon as the target image is displaced by 1 degree from the fovea center, the visual acuity decreases by a factor 3. In other words, we see only the ten-thousandth of our visual field with a maximum acuity. One understands then why, as regards foveate species, complex mechanisms of eye orientation have developed: they either permit to acquire rapidly, with a saccade, the image of a stationary target inside the "foveal tunnel", ensuring a high resolution analysis of the image, or allow to stabilize the image on the fovea, with a movement of smooth ocular pursuit, when the target is moving in the visual field.

We shall limit ourselves to the study of some aspects of the saccadic control.

3. OCULAR SACCADES

During the saccades, there cannot be any accurate vision: indeed, the image of the visual world slips over the retina, bringing about a hazy image. Actually, the fuzziness due to the eye rotation is not perceived, and this is due to several factors. First of all, the saccades are extremely rapid: maximal angular speeds of about 600 to 800 deg/s may be reached, ensuring a short movement time, in the range of 20 to 150 ms according to the amplitude. Moreover mechanisms of active inhibition ("saccadic suppression") of the visual afferences during the saccade were highlighted.

Although ocular saccades may be triggered off, as a reflex or intentionally, from external or internal signals (mental representations of the target), no voluntary control on the speed and the duration of these movements is possible: these variables mainly depend on the amplitude. The position and the movement of each eye in its orbit are controlled by 6 extraocular muscles laid in antagonist pairs. These muscles are particularly powerful in proportion to their size. Since most of the eye movements are combined and the position of each eye must be adjusted precisely, there is an extremely strict co-ordination of the activity of the 12 muscles. These ocular movements are often accompanied by the combined rotation of the head. We shall evoke further how the saccades are controlled by the central nervous system.

4. THE COLLICULAR CONTROL OF THE EYE SACCADES: A MODEL OF SENSORIMOTOR INTERFACE

Ocular saccades are controlled by a complex network of subcortical structures and cortical areas. We will focus, within the network, on one structure of the brainstem and more precisely of the mesencephalon, the superior colliculus (SC). This structure appears as a remarkable module participating not only in the multisensory integration and in the processes of sensorimotor co-ordination, but also in the mechanisms of attention control. Let us first recall some elements relating to the wiring and the structural organization of the SC. The axons of the ganglion cells of the retina, making up the optic nerve, are subdivided into two main pathways: one pathway reaches the primary visual cortex (the area V1 of the occipital cortex) after a relay at the level of the thalamus (the lateral geniculate body), the other heads for the SC. It should be noticed that the SC receives an important number of fibers coming from numerous cortical areas (occipital, parietal, frontal cortices...) and subcortical structures (basal ganglia, reticular formation, cerebellum, thalamus...). Besides, the SC consists of seven layers (cells and fibers alternatively) that may be classified in two main categories: the superficial layers and the deep-intermediate layers. Let us now summarize the physiology of the collicular neurons belonging to those layers (see among others the reviews of Guitton 1991; Crommelinck and Guitton 1994).

The superficial layers receive sensory afferences from visual origin exclusively which cover the entire contralateral visual hemifield and, importantly, are "retinotopically" organized. Thus, the visual neurons of the superficial layers are arranged in such a way that they form a map of the retina (and thus of the visual field) organized topographically in two dimensions. The fovea is represented on the rostral part of the SC, and the retinal periphery, on the caudal part (the collicular rostro-caudal axis represents the azimut); the upper retinal quadrant is represented medially and the lower quadrant, laterally (the collicular medio-lateral axis represents the elevation). As it is the case for other visual areas, the collicular retinal map is not homogeneous: a given surface of the central visual field takes up a more important portion of the map than the equivalent surface of the peripheral visual field ("magnification factor" of the foveal and perifoveal areas). Note that the coding of the input variables of the structure is a place coding: it is precisely the *place* where the cellular activity takes place on the map which makes up the relevant sensory information for the system.

The deep-intermediate layers are closely linked to the premotor and motor structures responsible for the orientation movements of the eye, the head and even the entire body. We shall now examine closely the neurophysiology of these deep-intermediate layers. Different kinds of sensory and/or oculomotor activity were recorded in the monkey and the cat. Projections coming from the auditory and somatic areas are superposed on the visual representation, all of them being topographically organized and are in spatial coincidence with each other. As far as the oculomotor activity is concerned, we will only describe two forms of activity, produced by two cellular groups playing an important part in the mechanisms of saccadic control.

The neurons whose phasic burst, consisting in a set of high frequency action potentials, is closely related to the saccade (SRBNs for *saccade-related burst neurons,* see Sparks 1978) are characterized by a motor field : for a given neuron, the most important burst precedes — latency about 20 ms — saccades with a specific amplitude and direction. For non optimum saccadic vectors, the burst is weaker. If the amplitude and/or direction of the saccade deviate sufficiently from the optimum saccade, the neuron remains silent. So a given cell is active for a certain range of saccadic vectors, with a maximum activity for the preferential saccade. Thus, for a given saccadic vector, there is a synchronic activity in a set of collicular neurons ("population coding"), with an activity peak centred on a precise point of the layer. For other saccadic vectors, these activity maxima will be located in other places. So the SRBNs are distributed within the SC in such a way that they form a topographically organized motor map. The relevant information (related to the characteristics of the saccadic vectors) is, here too, spatially coded ; the SRBNs neurons form a mototopic map.

The tecto-reticular neurons (TRNs) and the tecto-reticulo-spinal neurons (TRSNs) represent the principal output pathway of the SC controlling the saccadic movement (we group them in the same category: TR(S)Ns). These neurons (at least for the cat) are linked to areas of the reticular formation of the brainstem where the premotor neurons, responsible for the velocity and position eye signals, are situated, as well as to areas intervening in the control of the head movements. These TR(S)Ns were studied on the cat from a functional point of view (Grantyn and Berthoz 1985) and two types of them were highlighted (Guitton and Munoz 1991). The TR(S)Ns situated in the rostral part of the SC next to the representation of the fovea, are tonically activated as long as the animal fixates at a real or imaginary target; these neurons are silent during the saccade. They are identified as fixation neurons (or fTR(S)Ns). Besides, the oTR(S)Ns (o = orientation) show a visual response followed by a sustained activity which leads to a phasic burst preceding a gaze saccade, the head being mobile or immobile. These neurons form within the deep layers a mototopic map according to topographical principles equivalent to those described for the SRBNs neurons.

During an ocular saccade towards a visual target, the motor activity in the deep layers is preceded by a visual activity in the superficial layers. Given the correspondence of the visual (superficial layers) and motor maps (deep layers), the activity on the visual map is produced just above the activity on the motor map. This spatial matching property of the superposed maps in the collicular network gave some weight to the "foveation model" in which the visual information makes its way straight up from the sensory layers to the motor command layers.

5. RELATIONS BETWEEN THE ACTIVITY IN THE DEEP LAYERS AND THE OCULOMOTOR CIRCUITS: THE ISSUE OF THE SPATIOTEMPORAL PROCESSING

We will start with a question which is still largely discussed today: how is the nervous message, elaborated in the SC, correctly reorganized by the modules which, downstream, work out the motor program for the ocular saccade? These output modules form a very complex network made up of numerous cell populations (among others, the oculomotor nucleus forming the final common pathway, the premotor nucleus elaborating the eye velocity and position signals, or certain signals ensuring an inhibitive control). As we underlined it in the previous paragraph, the collicular signals are functionnally characterized by spatial properties ("place coding"): their position on the map represents an important part of the conveyed information.

The premotor and motor neurons of the eye code the output information in a temporal dimension essentially ("time coding"): they specify precise activation or inhibition durations, or frequency modulations (for instance, accurate control of the inhibition duration of inhibitory neurons corresponding to a signal allowing the production of a saccade of a given amplitude, or accurate control of the duration of an excitation phasic burst corresponding to the eye velocity signal, etc.). The issue of the spatiotemporal processing may be worded this way: "how is the space coded information in the SC converted into a time coded information in the eye premotor and motor modules?" In other words and in more concrete terms, "how can a nervous activity, coming from a particular point of the SC — space coding — receiving an input from an area of the visual field situated 20° to the right on the horizontal for instance, create a motor message having the exact duration (in the range of 55 ms) — time coding — required to move the eye 20° to the right?" (see Robinson 1975). Through which mechanism and in which interface is such a transformation ensured?

6. THE ROLE OF THE FEEDBACK LOOP IN SPATIOTEMPORAL PROCESSING

The notion of feedback is widely used in models of gaze orienting movements (Laurutis and Robinson 1986; Droulez and Berthoz 1991; Tweed and Vilis 1990; Lefèvre and Galiana 1992; Van Opstal and Kappen 1993; Optican 1995). However, though saccades are clearly controlled by a feedback signal, it is not clear how and where this control is performed. This issue is even more controversial for combined eye-head orienting movements.

Recent experimental data show evidence for dynamic gaze error coding inside the intermediate layers of the SC in the cat (Munoz *et al.* 1991) and in the monkey (Waitzman *et al.* 1990; Munoz and Wurtz 1995a,b). Based on these data, several recent models have proposed that gaze velocity feedback was applied in the Superior Colliculus (SC), so that dynamic gaze error is continuously updated within the SC itself (Droulez and Berthoz 1991; Lefèvre and Galiana 1992; Van Opstal and Kappen 1993; Optican 1995).

Even though there has been recent experimental evidence, both in the cat and the monkey (Munoz *et al.* 1991; Munoz and Wurtz 1995a,b), for a spreading wave of activity in the output cells of the SC, the issue of whether this spread is due to the feedback or not is still open. Two different approaches for the form of update in SC activity are proposed in existing models.

The first hypothesis is based on recordings of Saccade Related Burst Neurons (SRBNs) in the monkey (Waitzman *et al.* 1991).The location of activity on the caudal SC map codes the initial size of the gaze shift, while its intensity decays with the size of the remaining motor error or movement velocity. In this case, the same population is active during the saccade and the position of the locus of activity does not vary on the SC map. Hence, the location of SC activity can be considered open-loop during the movement, but the intensity of activity is coded in closed-loop fashion, so that it is correlated with the dynamics of the current gaze trajectory (Van Opstal and Kappen 1993).

We will base our model on the second hypothesis; in a first step, an intuitive formulation of the model will be presented, the mathematical description will come after.

This second hypothesis is based on data from Tecto Reticulo Spinal Neurons (TRSNs) in the cat (Munoz *et al.* 1991) and from Build up neurons (BUNs) in the monkey (Munoz and Wurtz 1995a,b). In the "spreading wave" hypothesis, not only does the intensity of activity at the initial site decay with motor error, but it is also proposed that the site of the peak of activity moves accordingly on the map in a continuous fashion towards the rostral (foveation) zone. Thus, in the wave context, both the site and intensity of SC activity on the map can be considered to be controlled in closed-loop fashion by the movement parameters (Droulez and Berthoz 1991; Lefèvre and Galiana 1992; Optican 1995).

7. INTUITIVE DESCRIPTION OF THE MODEL

Figure 1a represents two motor maps that are assumed to coexist in the deep layers of the SC. The first one is made of the SRBNs and the second one of TR(S)Ns. oTR(S)Ns activate brainstem excitatory burst neurons (EBNs) and reticulo spinal neurons (RSNs) that are premotor relays to ocular motoneurons (MNs). fTR(S)Ns are located in the rostral SC (fixation zone) and activate brainstem omnipause neurons (OPNs), that in turn inhibit EBNs. In this way, the SC can either trigger a saccade (excitation of EBNs) or stop it (excitation of OPNs), depending on the location of the peak of activity on the TR(s)N map. Several anatomical and electrophysiological arguments (see Guitton 1991 for a review) suggest that SRBNs inhibit OPNs and inhibit rostral fTR(s)Ns.

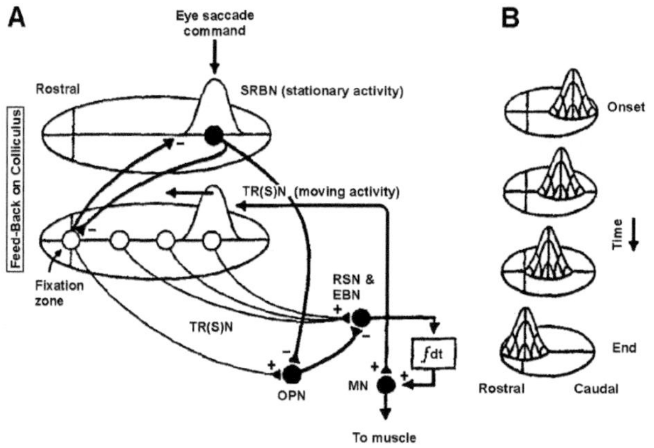

Figure 1. Control mechanisms of the gaze saccade. A.: feedback loop on the superior colliculus controlling premotor and motor circuits of the saccade. B.: schematic representation of the displacement of the neural activity on the collicular map as a function of time. Time is the vertical axis, from the top to the bottom.

The "hill" on each map represents the discharge of collicular cells. It is known that the frequency and duration of cell discharge (for instance SRBNs) is bigger for saccades having a specific amplitude and direction. Discharge frequency is thus a function of the saccadic vector: it is maximal for a specific vector (population coding). On each map, the hill represents SC activity at different instants of a saccade. A saccade is the consequence of the excitation of brainstem structures by SC output cells.

The hypothesis of a feedback loop controlling saccades is illustrated in figure 1a. On the basis of experimental results, it is assumed that a saccade is triggered when a population of SRBNs and TR(s)Ns is activated on the SC motor map. SRBNs inhibit OPNs, that release their inhibition on EBNs. The activity on TR(s)N map moves then across SC map toward the rostral (fixation zone) of the SC. This displacement is due to a feedback signal of eye velocity coming from the brainstem. As long as oTR(s)Ns are activated, they excite EBNs and the saccade keeps going. But as soon as the activity on the TR(s)N map reaches the rostral zone, fTR(s)Ns are reactivated and excite OPNs that stop the saccade. This mechanism can be compared to the simulation of an internal model of the gaze movement in real time.

This model will now be presented in a more rigorous way in the next paragraphs.

8. MATHEMATICAL DESCRIPTION OF THE MODEL

8.1 The Collicular Network

The Superior Colliculus (SC) is represented by a 2D neural network (Figure 3), which is connected to downstream brainstem structures (Figure 2). The SC is inside the dynamic feedback loop controlling gaze saccades. The feedback is based on gaze velocity, and the model controls head-free gaze saccades. Figure 2 shows the collicular part of the model (2D sheet of interconnected cells), with a zoom on one part of the model on the right. For simplicity, only one quadrant of the visual field is represented (25 x 25 cells),where the dashed lines correspond to the vertical and horizontal meridians. The black circle is the fixation zone (FNs), the projection of the fovea (cell N(9,9)).

In this model, the SC coordinates are cartesian and we did not address explicitely the question of the polar-like SC coordinates, like Optican (1995) did in his recent model. There exists a topographic correspondance between both coordinate systems. However, the so-called magnification factor is present as an emerging property of the model when the SC is placed inside the feedback loop.

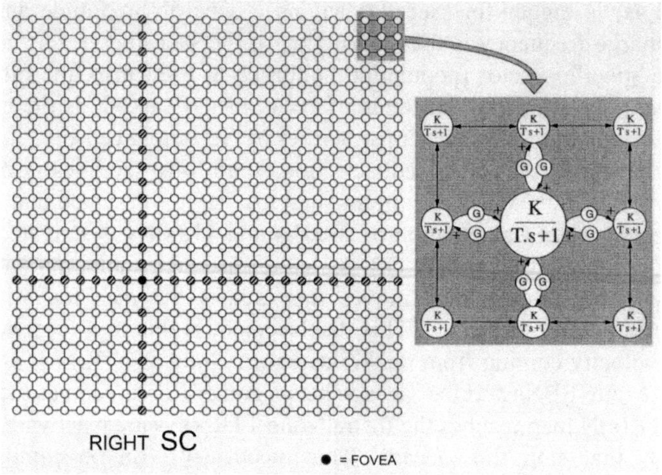

Figure 2. The collicular model. The right SC motor map, made up of an array of interconnected cells. For simplicity, only one quadrant of the visual field is represented. The black dot represents the fixation zone of the SC. On the right part, there is a zoom of one part of the network, illustrating reciprocal excitatory connections between each cell and its neighbours (T = 2 ms, K = 1, G = 0.25)

On the right part of Figure 2, there is a zoom of one part of the network. Each cell is modeled by a low pass filter (T = 2ms) and connected to its 4 immediate neighbors by excitatory connections. This connectivity is homogeneous throughout the map. When a saccade occurs, a graded gaze velocity feedback is applied on the border area of the array (from row and column 16 to the borders (25).

8.2 The Brainstem Structures

Figure 3 shows the complete gaze control model, with the collicular motor layer of Figure 2 on the upper part. The lower part represents horizontal and vertical saccade generators in the brainstem. These are the 2D extension of the model proposed by Lefèvre and Galiana (1992). They are characterised by three important properties: these premotor circuits generate both eye and head movements, they are based on internal models of eye and head plants, and they are controlled by a gaze velocity feedback loop. The SC is inside the loop and receives inhibitory gaze velocity feedback from premotor circuits. On the other hand, the weighted average of SC motor activities provides horizontal and vertical motor errors to saccade generators. These two motor errors are not independent, but the smallest saccade component is stretched, to generate straight oblique saccades (common source model).

8.3 Methods

In simulations, initial conditions on the collicular network were imposed according to equation 1:

$$IC(i,j) = 0.61 * \exp^{-\frac{(MaxH-i)^2+(MaxV-j)^2}{30}} \tag{1}$$

where $IC(i,j)$ is the initial condition on cell (i,j) and (MaxH, MaxV) are the indices of the cell carrying the initial maximum of activity.

Horizontal and vertical errors projecting to the brainstem are evaluated by equations 2 and 3.

$$ErrV = \sum_{ij}^{21} WV(i,j)*[u(i,j)+1] \text{ for } [u(i,j)+1] > 0 \tag{2}$$

$$ErrH = \sum_{ij}^{21} WH(i,j)*[u(i,j)+1] \text{ for } [u(i,j)+1] > 0 \tag{3}$$

Where $WV(i,j)$ and $WH(i,j)$ are weights of projections to the vertical and horizontal parts of the brainstem respectively (see equation 4 and 5) and $u(i,j)$ is the activity of cell (i,j). ($wtfac = 0.016667$).

$$WV(i,j) = (j-5) * wtfac \tag{4}$$

$$WH(i,j) = (j-5) * wtfac \tag{5}$$

Horizontal and vertical errors are then normalized (common source model) following equations 6 and 7.

$$ErrV' = \frac{(ErrV-10)*|ErrV|}{\sqrt{ErrV^2+ErrH^2}} \tag{6}$$

$$ErrH' = \frac{(ErrH - 10) * |Err\,H|}{\sqrt{ErrV^2 + ErrH^2}} \tag{7}$$

ErrV' and *ErrH'* are the two error signals feeding the brainstem (Figure 3).

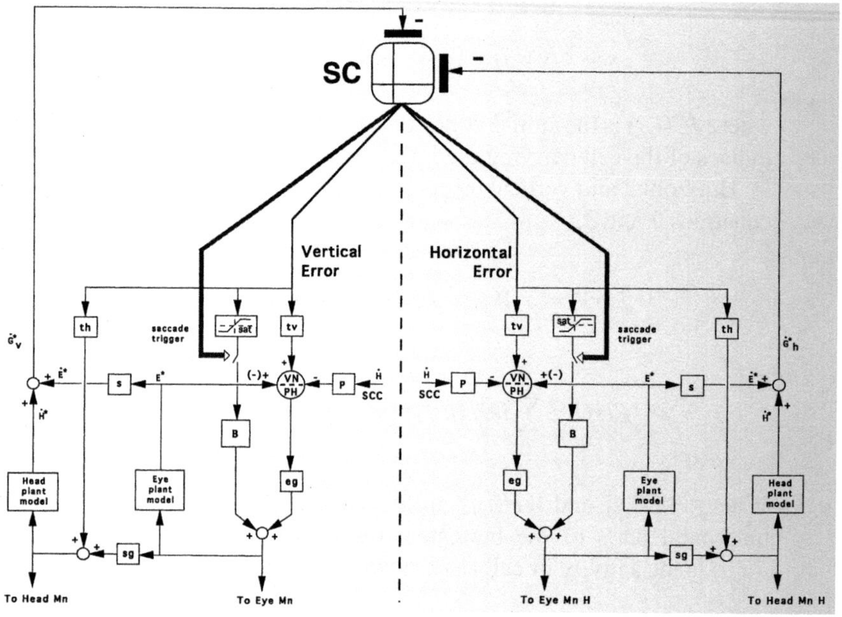

Figure 3. Complete gaze control model, with the SC motor layer on the upper part. The lower part represents the 2D extension of the Lefèvre and Galiana (1992) model, the brainstem circuits controlling eye and head movements.

During the saccades, each collicular cell receives a feedback signal which is the sum of two terms, proportional to horizontal and vertical gaze velocities (equation 8). This feedback is concentrated on the border zones of the network (for i and $j \geq 16$) and follows equations 9 and 10. ($fbfac = 0.0000003$).

$$fb(i,j) = fbv(i,j) + fbh(i,j) \tag{8}$$

$$fbv(i,j) = \begin{cases} -(j-5)^2 * \dot{G}v * fbfac & j \geq 16 \ \& \ i \geq 6 \\ 0 & else \end{cases} \tag{9}$$

$$fbh(i,j) = \begin{cases} -(i-5)^2 * \dot{G}h * fbfac & i \geq 16 \ \& \ j \geq 6 \\ 0 & else \end{cases} \tag{10}$$

For the brainstem circuits, model parameters are the same as in Lefèvre and Galiana (1992) for Horizontal (H) and Vertical (V) systems, except for sg and SAT gains, which were adapted to fit cat data (sg = 6, SAT = 25).

Simulations were done with Matlab and Simulink softwares on a Sun Sparc 20 workstation. In the simulations, integration steps of 0.5 ms were used.

8.4 Simulation Results
Simulation of Natural Oblique Gaze Shifts

A large amplitude oblique eye-head gaze shift is illustrated in Figures 4 and 5.

Figure 4 shows SC motor activities at four different times during an oblique gaze shift. The left-top part of Figure 4 corresponds to the initial SC activity at saccade onset. Red is low activity and white is high activity. Here, the movement coded has a larger horizontal component. White lines are horizontal and vertical meridians. The three other parts of Figure 4 show SC activity later during the movement. As the saccade progresses, the gaze velocity feedback shapes the spreading wave, causing displacement of both the peak and the center of gravity of SC activities. This reactivates the fixation zone of the SC and brainstem Omni-Pause Neurons (OPNs), stopping the saccade. This occurs when fixation activity reaches one third of the peak of collicular activity.

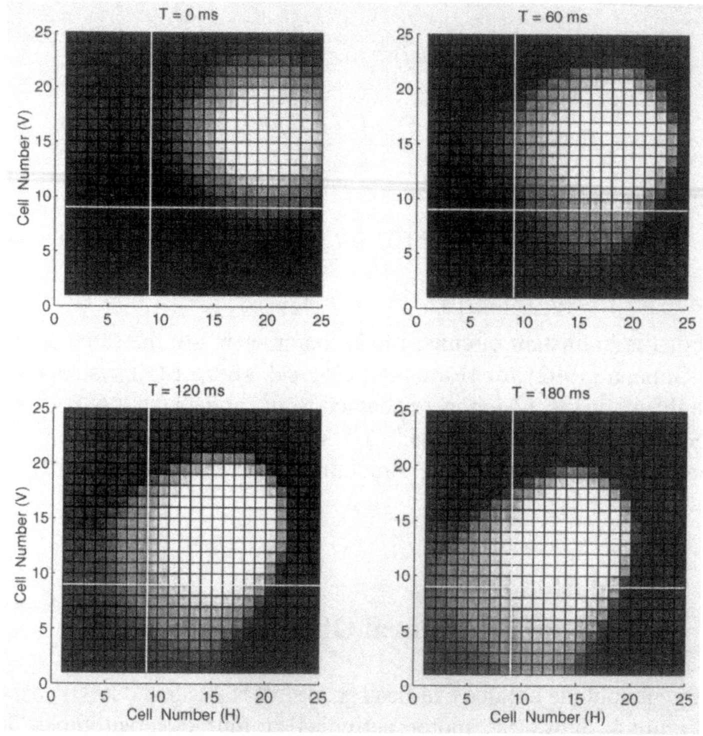

Figure 4. Simulation of a large oblique gaze shift in the head free condition. The figure represents SC motor activities at 4 different times during the gaze saccade. From left to right and top to bottom: at saccade onset, 60 ms, 120 ms and 180 ms after saccade onset. Red is low activity and white is high activity. White lines are horizontal and vertical meridians.

Figure 5 shows eye and head movements generated by these SC activities. In Figure 5a (5b) are horizontal (vertical) eye, head and gaze positions (sum of eye and head positions). Figure 5c represents gaze trajectory in X-Y coordinates. Gaze position is sampled every 10 ms. The smaller vertical component is stretched and the model generates straight oblique saccades.

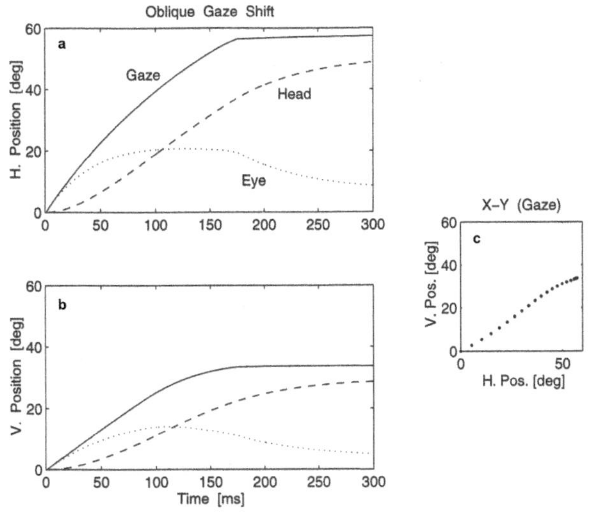

Figure 5. Simulation of a large oblique gaze shift in the head free condition, for the same simulation as in figure 4. [a] Horizontal gaze (solid), eye (dotted) and head positions (dashed line). [b] Vertical components (same line symbols). [c] Gaze trajectory in X-Y coordinates. Initial peak of SC activities on cell (15,19).

8.5 Simulation of Interrupted Gaze Shifts and Slow Correcting Eye Movements

In addition to its basic properties in the control of head free gaze saccades, the model shows interesting emerging properties. In 2-20% of movements with cats trained to fixate visual targets, eye saccades are followed by peri-saccadic SCMs (and sometimes a corrective saccade). These SCMs were active corrective movements (and not passive eye drifts due to a pulse-step mismatch). Indeed, their direction was always toward visual target and their dynamics was variable (time constant between 100 and 400 [ms]). Their amplitude was correlated with residual error and with their velocity.

Figure 6 shows the model simulation of a 2D head-free gaze shift in the cat, with a first saccade, an intermediate SCM, a second corrective saccade and a terminal drift. In this simulation, the saccade was artificially interrupted by the reactivation of OPN cells 70 ms after saccade onset, during 50 ms. The saccades are synchronised on the horizontal (Figure 6a) and vertical components (Figure 6b). In the model, SCMs are controlled by residual gaze error. Both saccades and SCMs contribute to gaze error reduction, in the same direction (X-Y plot, Figure 6c). The intermediate SCM is faster than the terminal SCM, as experimentally reported (Missal *et al.* 1993).

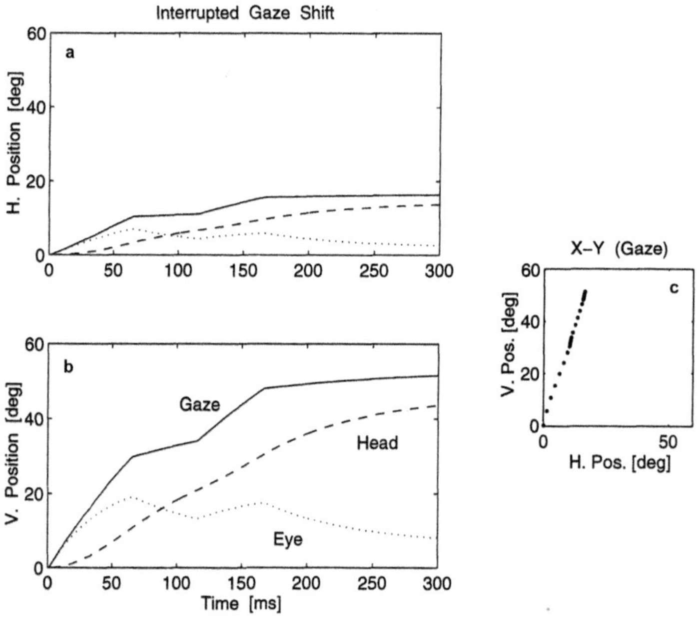

Figure 6. Simulation of an interrupted gaze shift. The gaze shift was artificially interrupted at time T = 70 ms for 50 ms. The model generates both saccades and SCMs to reduce gaze error. Initial peak of SC activities on cell (19,13). For [a], [b] and [c], same as in Figure 5.

Moreover, this model can simulate saccades evoked by electrical stimulation of the SC and predicts the alternance of saccades and slow eye movements that were confirmed experimentally (Missal *et al.* 1996).

9. CONCLUSIONS

Recent experimental data show evidence for dynamic motor error coding in the Superior Colliculus (SC) (Munoz *et al.* 1991; Munoz and Wurtz 1995a,b). We propose here a new 2D model of gaze orientation that has the following properties:

1. In addition to the properties of classical head fixed models, the model generates realistic gaze trajectories in the head free condition in two dimensions.
2. Moreover, this model has several emerging properties. It can generate and explain peri-saccadic Slow Correcting Movements (SCMs) and interrupted saccades (Missal *et al.* 1993). The SC is active both during saccades and SCMs. The same SC output signal reduces residual gaze error, either with fast movements (saccades), or with slow movements (SCMs). Also, several other predictions of the model, related to movements elicited by electrical stimulations of the SC were recently confirmed experimentally.
3. Emerging properties of neuronal networks that were described in this paper are important. They perform very precise tasks in the complex processing of spatially coded visual inputs into temporally coded neuronal outputs.

Acknowledgments: This work was supported by the E.U.'s ESPRIT contract MUCOM B.R. 6615. P. Lefèvre was supported by FNRS (Belgium). This paper presents research results of the Belgian Programme on Interuniversity Poles of Attraction, initiated by the Belgian State, Prime Minister's Office for Science, Technology and Culture. The scientific responsibility rests with its authors.

REFERENCES

Arai K, Keller E.L. and Edelman J.A. (1994). Two-dimensional Neural Network Model of the Primate Saccadic System. *Neural Networks*, **7**, 1115-1135.

Berthoz A. (1997). *Le sens du mouvement*. Paris: Éditions Odile Jacob.

Crommelinck M. and Guitton D. (1994). Oculomotricité. In M. Richelle, J. Requin and M. Robert (eds), *Traité de Psychologie expérimentale*, Paris: Presses Universitaires de France, Vol. 1, 657-728.

Droulez J. and Berthoz A. (1991). A Neural Network Model of Sensoritopic Maps with Predictive Short-term Memory Properties. *Proceedings of National Academy of Sciences* (USA), **88**, 9653-9657.

Guitton D. (1991). Control of Saccadic Eye and Gaze Movements by the Superior Colliculus and Basal Ganglia. In R.H.S. Carpenter (ed.), *Vision and Visual Dysfunction. Eye Movements.* London: MacMillan, Vol. 8, 244-276.

Guitton D. and Munoz D. (1991). Control of Orienting Gaze Shifts by Tecto-reticulo-spinal System in the Head-free Cat. I: Identification, Localization and Effects of Behavior on Sensory Responses. *Journal of Neurophysiology*, **66**, 1605-1623.

Grantyn A. and Berthoz A. (1985). Burst Activity of Identified Tecto-reticulo-spinal Neurons in the Alert Cat. *Experimental Brain Research*, **57**, 417-421.

Laurutis V.P. and Robinson D.A. (1986). The Vestibulo-ocular Reflex During Human Saccadic Eye Movements. *Journal of Physiology*, **373**, 209-233.

Lefèvre P. and Galiana H.L. (1992). Dynamic Feedback to the Superior Colliculus in a Neural Network Model of the Gaze Control system. *Neural Networks*, **5**, 871-890.

Missal M., Crommelinck M., Roucoux A. and Decostre M.F. (1993). Slow Correcting Eye Movements of Head Fixed, Trained Cats Toward Stationary Targets. *Experimental Brain Research*, **96**, 65-76.

Missal M., Lefèvre Ph., Delinte A., Crommelinck M. and Roucoux A. (1996). Smooth Eye Movements Evoked by Electrical Stimulation of the Cat's Superior Colliculus. *Experimental Brain Research*, **107**, 382-390.

Munoz D.P., Pélisson D. and Guitton D. (1991). Movement of Neural Activity on the Superior Colliculus Motor Map During Gaze Shifts. *Science,* **251**, 1358-1360.

Munoz D. and Wurtz R.H. (1995a). Saccade Related Activity in Monkey Superior Colliculus. I. Characteristics of Burst and Buildup Cells. *Journal of Neurophysiology*, **73**, 2313-2333.

Munoz D. and Wurtz R.H. (1995b). Saccade Related Activity in Monkey Superior Colliculus. II. Spread of Activity During Saccades. Characteristics of Burst and Buildup Cells. *Journal of Neurophysiology*, **73**, 2334-2348.

Neisser U. (1976). *Cognition and reality*. San Francisco: W. Freeman.

Optican L.M. (1995). A Field Theory of Saccade Generation: Temporal to Spatial Transform in the Superior Colliculus. *Vision Research*, **35**, 3313-3320.

Robinson D.A. (1975). Tectal Oculomotor Connections. *Neurosciences Res. Prog. Bull.*, **13**, n° 2, 238-244.

Sparks D.L. (1978). Functional Properties of Neurons in the Monkey Superior Colliculus: Coupling of Neuronal Activity and Saccade Onset. *Brain Research,* **156**, 1-16.

Tweed D.B. and Vilis T. (1990). The Superior Colliculus and Spatio-temporal Transformation in the Saccadic System. *Neural Networks*, **3**, 75-86.

van Gisbergen J., van Opstal J., Berthoz A. and Lefèvre Ph. (1993). Models of Gaze Orienting System: a Brief Survey ». In A. Berthoz (ed.), *Multisensory control of movement.* Oxford: Oxford University Press, 213-225.

van Opstal A.J. and Kappen H. (1993). A Two-dimensional Ensemble Coding Model for Spatial-temporal Transformation of Saccades in Monkey Superior Colliculus. *Network*, **4**, 19-38.

Waitzman D.M., Ma T.P., Optican L.M. and Wurtz R.H. (1991). Superior Colliculus Neurons Mediate the Dynamic Characteristics of Saccades. *Journal of Neurophysiology,* **66**, 1716-1737.

Yarbus A (1967). *Eye movements and vision.* New York, Plenum Press. Bibliography.

FRANCISCO J. VARELA †

NEURONAL SYNCHRONY AND COGNITIVE FUNCTIONS

ABSTRACT

This paper present a novel reading of ideas on temporal binding as key for cognitive operations by means of fast neuronal synchrony. I advocate a view of binding between widely distributed cell assemblies, transiently locked in a neural hypergraph which serves as a reference point to incorporate (or interpret) other less coherent concurrent, neural events. The paper concludes with some implications for the constitution of a unified cognitive-mental space.

1. THE CONTEXT: CELL ASSEMBLIES AND COGNITION

I wish to present here a new view about cognitive-mental functions based on a large-scale integrating brain mechanism that has been slowly emerging with increasing plausibility. A long standing tradition in neuroscience, dating back to the days of cybernetics, looks at the brain basis of cognitive acts (perception-action, memory, motivation and the like) in terms of *cell assemblies* or, synonimously, of *neuronal ensembles.*

> *Definition: A Cell Assembly (CA) is a distributed subset of neurons with strong reciprocal connections.*

Thus a CA will comprises distributed neuronal populations (very likely neo-cortical pyramidal neurons, but not limited to them) requiring active connections. Because of their assumed strong interconnections a cell assembly can be activated or ignited from any of its smaller subsets,

B. Feltz, M. Crommelinck and P. Goujon, (eds.), Self-Organization and Emergence in Life Sciences: 95-108.

sensori-motor, or internal. Notice also that the term reciprocal is crucial here: it is one of the main results of modern neuroscience that brain regions are indeed interconnected in reciprocal fashion (this is what I like to call the Law of Reciprocity). Thus, whatever the neural basis for interesting cognitive tasks turns out to be, it necessarily engages vast and geographically separated regions of the brain. Furthermore, these distinct regions cannot be seen as organized in some sequential arrangement as if a cognitive act could emerge from a gradual convergence from various sensory modalities, into association or multimodal regions, and further into higher frontal areas for active decision and planning of behavioral acts. This traditional sequentialistic idea derives from the time of the dominance of the computer metaphor with its associated idea of information flow going in an up-stream direction. Here, in contrast, I emphasized a strong dominance of reciprocal network properties where sequentiality is replaced by reciprocal determination and relaxation time.

The genesis and determination of CAs can be seen as having three distinct causal and temporal levels of emergence.

1. First, a very basic *onto-genetic* level which sets the anatomical architecture of a given brain into circuits and subcircuits.
2. It has been widely suspected that beyond the basic genetic wiring, neurons develop a variable degree of effective interconnectivity by strengthening or weakening their synaptic contacts. This is a second, strictly *developmental-learning* level and time-scale: sets of neurons that are frequently co-active strengthen their synaptic efficacies. Known generically as Hebb's rule the notion has suffered many theoretical formulations and additions in the recent connectionist movement. More importantly, a substantial amount of evidence shows that Hebb's rule in some form is the case during learning and early life (*e.g.* Ahissar *et al.,* 1992; Bonhoeffer *et al.,* 1989).
3. A third and final level of determination for CA is our concern here. This is the faster time scale at the *perception-action* level of fractions of a second when a CA is ignited and it either reaches a distributed coherence or is swamped by the competing ignitions of overlapping CAs. As Braitenberg puts it, the CA must "hold" after its activation (1978). In the language of the theoretician the CA must have a relaxation time. This holding time is bounded by two simultaneous constraints: (1) it must be larger than the time for spike transmission between neurons either directly or through a small number of synapses; (2) it must be smaller than the time it takes for a cognitive act to be completed, which is of the order of fraction of a second (*e.g.* Varela *et al.,* 1981; Dennett, 1992).

In other words: the relevant neuronal processes are not only distributed in space, but they are also distributed in an expanse of time that cannot be compressed beyond a certain fraction of a second.

2. THE HYPOTHESIS: SYNCHRONY AS NEURONAL GLUE

In view of the above, I wish to propose two interlinked (but logically independent) working Hypothesis.

Hypothesis I: A singular, specific cell assembly underlies the emergence and operation of every cognitive act.

In other words, the emergence of a cognitive act is a matter of coordination of many different regions allowing for different capacities: perception, memory, motivation, and so on. They must be bound together in specific grouping appropriate to the specifics of the current situation the animal is engaged in, and are thus necessarily transient, to constitute meaningful contents in meaningful contexts for perception and actions. Further, Hypothesis I predicts that all the physiological correlates associated with CA (*i.e.* multi-unit activity, local field potentials, MEG/EEG scalp recordings, frequency coherences, etc.) should be repeatedly detected for a repeated cognitive act, say, in an odd-ball discrimination task conducted in the laboratory, in an otherwise intact awake human or animal.

Notice that the Hypothesis I is *strong* in the sense that it predicts that only *one* dominant or mayor CA will be present during a cognitive act. We will come back to this below, but it highlights a basic problem opened by Hypothesis I: How is a specific cell assembly selected in successive moments? Although this will be main topic in the rest of this article, I wish to formulate it as the second part of my working Hypothesis. The basic intuition to answer the problem just raised is that a specific CA emerges through a kind of temporal "glue". More specifically, the neural coherency-generating process can be understood as follows:

Hypothesis II: A specific CA is selected through the fast, transient phase locking of activated neurons belonging to sub-threshold competing CAs.

Since in recent literature the notion of neuronal synchrony and binding has received a wide attention, I do not need to provide many empirical details (see Singer, 1993; Varela, 1995).

3. THE MECHANISM: PHASE-LOCKING IN RECIPROCAL CIRCUITS

It is well known that oscillations and rhythms are quite natural to neurons and neural circuits, and they have been explored widely (*e.g.* Glass and Mackey, 1988; Levan Quyen, Schuster and Varela, 1996). Given that there are finite transmission times in the nervous systems oscillations and cycles are to be expected just on the basis of reciprocal connectivity, as already popularized by Lorente de No in his well-known "reververating" circuits. This entails that one should expect that patterned activity of neurons will display spatio-temporal regularities. A further quite different universal mechanism for generating rhythms of interest to us here is the introduction of inhibition within a population of reciprocally connected excitatory elements, as clearly analyzed by Wilson and Cowan (1973).

A different matter which is my central concern here, is the precise manner in which such coherence can be established. According to Hypothesis II the key idea is that ensembles arise because neural activity forms transient aggregates of *phase locked* signals coming from multiple regions. Synchrony (via phase-locking) must *per force* occur at a rate sufficiently high so that there is enough time for the ensemble to "hold" together within the constraints of transmission times and cognitive frames of a fractions of a second. However if at a given moment several competing CAs are ignited, different spatio-temporal patterns will become manifest and hence the dynamics of synchrony may be reflected in several frequency bands. The neuronal synchronization hypothesis postulates that it is the precise coincidence of the firing of the cells that brings about unity in mental-cognitive experience. If oscillatory activity promotes this conjunction mechanism, it has to be relatively fast to allow at least few cycles before a perceptual processes is completed (*e.g.,* head orientation followed by face recognition).

Now, how fast is fast? Consider the following reasoning: There are numerous connections between cortical regions, and a recent study puts their conduction velocities at over 10 m/sec (Aboitiz *et al.,* 1992). This means that, roughly, one cycle of spike exchanges between two hemispheres would be about 40 ms. If we assume that a CA needs at least one round trip of spike to synchronize, this puts the minimum relevant associated frequency at over 25 Hz, that is, in the so-called gamma band (say 35-60 Hz). In other words, if Hypothesis II holds, then large numbers of neurons should give indications of increased activation in local field potentials, EEG/MEG, or single cell in this range, although not necessary at the exclusion of slower rhythms.

This simple reasoning illustrates one of the many avenues one can use to conclude that looking further into these non-classical, fast rhythms may be of cognitive interest. Oscillatory activity in the gamma range was, in fact, already described by Adrian in 1942 in the in the olfactory bulb of the hedgehog, work that was followed by the research line of W. Freeman (1975) using macro-potential in awake animals. Similarly, work with humans, using EEG, MEG and ERPs led Sheer and Galambos early on to similar ideas. Observations from neuropsychology also prompted Damasio and others to select phase locking as crucial (Damasio, 1990; Bressler *et al.,* 1993; Jolliot *et al.,* 1994; Varela *et al.,* 1995). Most recently, work with single units in the visual systems in animals (for review see Singer, 1993; for our own work Neuenschwander *et al.,* 1993; 1996) have made the idea quite popular. I will delve more in detail in this empirical evidence below, but for the moment let us stay at the general level of the Hypothesis itself.

In these studies the main idea is that fast oscillations in the gamma-beta range serve as *carriers* for a phase synchronization of neuronal activity, thus allowing for a process of selection by resonance into a transient coherent ensemble that underlies the unity of cognitive act in a fraction of a second. The substantial experimental support for the hypothesis makes it clear that we are dealing with a *bona fide* candidate for the synthesis of a cognitive space. At the same time I haste to add that the empirical support is far from being limpid, and that the credibility and interpretation of the available observation is not unanimous.

This focus on gamma band, though restrictive, is not meant to imply that fast rhythms are the sole correlates of cognitive processes. The literature provides numerous examples of theta and alpha rhythms in cortex, hippocampus, thalamus and brain stem which are induced by sensory stimulation or motor behavior (see Basar 1992 for a review). It has been show that alpha-like oscillations are present in visual evoked potentials in humans (Mangun 1992; Basar *et al.,* 1992), and alpha-rhythms can desynchronize during complex behavioural tasks, like reading, or planning of finger movements (Pfurtscheller and Klimesch 1992; Pfurtscheller and Neuper 1992). Rhythmic slow activity may operate in the spread of activity over the hippocampus and even facilitate or promote synaptic modifications, ultimately stabilizing memory traces in the limbic cortex (Lopes da Silva 1992). However, slow rhythms generally involve large neural masses, locked in a global state of hyper-synchrony (as in delta sleep or barbiturate-induced spindles). It is hard to conceive how such a slow rhythmic activity could provide the necessary dynamics for attention, perception and purposive motor behavior, which are continually evolving, non-stationary processes that self-organize into cognitive aggregates in a fraction of a second.

4. THE CORE HYPOTHESIS

I would like to come back to my initial, more general point: what could this large-scale binding do for us? For the sake of stating my ground as clearly as possible, let me now *rephrase* the main idea presented above in Hypothesis I+II this time phrased as the emergence of mental-cognitive states in general.

> *Core Hypothesis: Mental-cognitive states are interpretations of current neural activity, carried out in reference to a transient coherency-generating process generated by that nervous system.*

To clarify, let direct the reader to the following comments:

I am referring to "primary" consciousness only:

I am restricting my discussion here to the kind of mental-cognitive events shared by non-verbal creatures. In all of us, the ongoing constitution of a mental space makes possible a selection and internal evaluation of multiple, concurrent neural events. For example, a visual recognition is surely lived differently depending on conditions related to the overall state of arousal and motivation, and depending on associative memories unique to that individual.

What do I mean by "interpretations":

In this sense it is clear that the neural events accompanying the recognition are not taken at face value but shaped and modified in the context of the rest of the neural events related to, say, limbic and memory activation. This is what I mean by an "interpretation": the generation of a mental-cognitive state corresponds to the constitution of an assembly which incorporates or discards into its coherent components other concurrent neural activity generated exogenously or endogenously[1]. In other words, the synchronous glue provides the reference point from which the inevitable multiplicity of concurrent potential assemblies is evaluated until one is transiently stabilized and expressed behaviorally. This is a form of neural hermeneutics since the neural activity is "seen" or "evaluated" from the point of view of the cell assembly that is most dominant at the time. Dynamically this entire process takes the form of a bifurcation from a noisy background to conform a transiently stable, distributed structure bound by synchrony.

---

1. I have been influenced by Chiel (1993) for this unusual approach to neural activity.

Ongoing neural activity assimilated in the dominant assembly:
It should be also clear that the neural events that participate in this process of synthetic interpretation are derived indistinctly from sensory coupling and from the intrinsic activity of the nervous system itself, *i.e.* levels of activation, memory associations and the like. It is also clear that whatever the mental state thus arrived it will *ipso facto* have neural consequences at the level of behavior and perception. For instance, if a visual recognition is interpreted in the context of an evasive emotional set and in conjunction with painful memory association, it can lead to a purposeful plan for avoidance behavior complete with motor trajectories and attention shifts to certain sensory fields. This illustrates one key dimension of the view of mental states I am offering here: there is a level-crossing reciprocity in that a mental state as such (*i.e.* as a global interdependent pattern) can effectively *act* on neural events (that is, it can have downward causation as the phrase goes). For this to be more than a simple dualistic rehash it is essential that the dominant interpretation be itself an emergent neural event. Whence the odd-looking part of my definition that requires a neural events to be the basis of interpretation of another class (of non synchronous, less coherent) neural events.

Mental events are a distributed hypergraph:
It is also clear that what I am proposing is related to a process which is, by definition, distributed since it involves a variety of dispersed neural activity. Thus a basic cognitive-mental space is topological object, and not topographical one, it is a question of a hypergraph of synchronous relationships rather than one of localization. The process underlying this cerebral hermeneutics itself is, by hypothesis, an ongoing phenomenon, providing a continual emergence. Notice that this process demands that it operates by a distributed coupling of groups of oscillators, it will exhibit a characteristic relaxation time. Thus, we expects that cognitive (and experiential) time will manifest in the manner of discontinuous aggregates over a horizon on ongoing, continuos activity.

Synchronous assemblies are universal:
The key in all of this is, then, that we can identify a neural process which can be a credible support for the transitory coherence from whose vantage point a neural interpretation can happen. The alleged process must be universal enough to be supported and present in a the nervous system of animals at least for all higher vertebrates, and its presence or absence can help identify where sentiency is present in the sense presented here. The evidence discussed above make it plausible that we actually have a good candidate for a neural mechanism. The specificity of a synchronous

hypergraph present for every mental-cognitive state can in principle be studied by the new techniques combining MEG-EEG (see Tiitinen *et al.,* 1993) and fMRI-PET studies. Only a systematic study of this global functional aggregate, followed at the millisecond level during mental experience, will give a definitive answer to the extent to which the Core Hypothesis is valid.

How is this related to our own mental experience?:
By their very nature, mental states make reference to our own experience and thus require a phenomenological account, which we can carry out as sentient humans. That we are both cognitive creatures and self-conscious is both an advantage and a difficulty. Advantage because we can rely on human phenomenology of mental states as valid data. Disadvantage because we have to be careful to address the appropriate primary dimensions of mental life common to all animals, and not those dimensions which are properly human. An adequate phenomenology of mental states in this sense needs to be done by some explicit phenomenological pragmatics, and not just the "It seems to me" method. This has been notoriously lacking in cognitive science, and it is not surprising since it entails a radical turn to examine the texture of our field of experience (see Varela *et al.,* 1991; Varela 1996) for more on phenomenology and neuroscience). I will not attempt to enter into this essential topic here, but let me at least provide some pointers.

Some basic dimensions of mental experience that need to be brought in for this discussion are the following:

1. Mental events occur in a unitary space: there is no fragmentation in the manner in which, for instance, different modalities appear to experience or a disjointness between sensations and memories and body tone.
2. Mental states are transitory in the most obvious sense that no one state lasts for a sustained duration beyond a limit. Conversely it does not seem possible to experience a mental state without a span of duration which is non-vanishing. Thus, mental states are finite, and have an incompressible and inextendible duration.
3. Mental states are always body-bound, embedded in a particular field of sensation. In fact most of the time a mental state has a dominant sensory modality which colors its texture.
4. Mental states can be causally triggered by endogenous events. It is also the case that a mental state can be seen as having a distinct perceptual or behavioral consequence. (If this seems strange, think of the classic example of the "voluntary" inversion of the two faces of an ambiguous visual figure). Thus, the kind of neural events underlying a mental state

must be distinct and distinguishable from other kinds of neural events so that this two-relation relationship holds.

These basic phenomenological dimensions of a mental states must enter as an arbiter in the validation of any approach to mental-cognitive processes (Varela *et al.*, 1991; Varela, 1996). In other words, we need to satisfy what we know about neuroscience and come up with a mechanism that is a convincing counterpart to these four dimensions of a mental experience. We need to advance a cognitive science where there is a true circulation between lived experience and the biological mechanisms in a seamless and mutually illuminating manner, as we have discussed elsewhere (Varela *et al.*, 1991), and it has recently been claimed by others from their own perspective (see *e.g.* Flannagan (1992) and his notion of a "unified theory"). Mental states as viewed through the Core Hypothesis provide an explicit avenue to conduct research in cognitive science as if both brain physiology and mental experience mattered.

REFERENCES

Aboitiz F., Scheibel A.B., Fisher R.S. and Zeidel E. (1992). Fiber Composition of the Human Corpus Callosum. *Brain Res.*, **598**, 143-153.

Ahissar E. and Vaadia E. (1990). Oscillatory Activity of Single Units in a Somatosensory Cortex of Awake Monkey and Their Possible Role in Texture Analysis. *Proc. Natl. Acad. Sci. (USA)*, **87**, 8935-8939.

Ahissar E., Vaadia E., Ahissar M., Bergman H., Aireli A. and Abeles M. (1992). Dependence of Cortical Plasticity on Correlated Activity of Single Neurons and on Behavior Context. *Science*, **257**, 1412-1415.

Basar E. (1992). Brain Natural Frequencies are Causal Factors for Resonances and Induced Rhythms. In E. Basar and T.H. Bullock (eds), *Induced Rhythms in the Brain.* Berlin: Birkhäuser, 425-467.

Basar E., Basar-Eroglu C., Parnefjord R., Rahn E. and Schürmann M. (1992). Evoked Potentials: Ensembles of Brain Induced Rhythmicities in the Alpha, Theta and Gamma Ranges. In E. Basar and T.H. Bullock (eds), *Induced Rhythms in the Brain.* Berlin: Birkhäuser, 425-467.

Bonhoffer T., Staiger V., Aertsen A. (1989). Synaptic Plasticity in Rat Hippocampal Slice Culture. *Proc. Natl. Acad. Sci. (USA)*, **86**, 8113-8117.

Bouyer J.J., Montaron M.F., Vahnée J.M., Albert M.P. and Rougeul A. (1987). Anatomical Localization of Cortical Beta Rhythms in Cat. *Neuroscience*, **22**, 863-869.

Bouyer J.J., Montaron M.F. and Rougeul A. (1981). Fast Fronto-Parietal Rhythms During Combined Focused Attentive Behaviour and Immobility in Cat: Cortical and Thalamic Localizations. *Electroenceph. Clin. Neurophysiol.*, **51**, 244-252.

Braitenberg V. (1978). Cell Assemblies in the Cerebral Cortex. In Heim R. and Plam G. (eds), *Theoretical Approaches to Complex Systems*, Lecture Notes in Biomathematics n° 21. Berlin: Springer Verlag, 171-188.

Bressler S., Coppola R. and Nakamura R. (1993). Episodic Multiregional Cortical Coherence at Multiple Frequencies During Visual Task Performance. *Nature*, **366**, 253-155.

Bringuier V., Frégnac Y., Debanne D., Shultz D. and Baranyi A. (1992). Synaptic Origin of Rhythmic Visually Evoked Activity in Kitten Area 17 Neurons. *NeuroReport*, **3**, 1065-1068.

Canu M.-H. and Rougeul A. (1992). Nucleus Reticularis Thalami Participates in Sleep Spindles, not in β Rhythms Concomitant with Attention in Cat. Paris: *C. R. Acad. Sci.*, **315**, 513-520.

Carter G.C. (1987). Coherence and Time Delay Estimation. *Proc. IEEE,* **75**, 236-255.

Chatila M., Milleret C., Buser P. and Rougeul A. (1992). A 10 Hz 'alpha-like' Rhythm in the Visual Cortex of the Waking Cat. *Electroenceph. Clin. Neurophysiol.*, **83**, 217-222.

Chiel H.J. (1993). Cognitive Neuroethology: an Approach to Understanding Biological Neural Networks. In L.S. Sterling (ed.), *Intelligent Systems.* New York: Plenum Press, 143-167.

Damasio A. (1990). Synchronous Activation in Multiple Cortical Regions: a Mechanism for Recall. *Semin. Neurosciences,* **2**, 287-296.

Delagrange P., Tadjer D., Bouyer J. J., Rougeul A. and Conrath M. (1989). Effect of DSP4, a Neurotoxic Agent, on Attentive Behavior and Related Electrocortical Activity in the Cat. *Behav. Brain Res.,* **33**, 33-44.

Dennett D. (1992). *Consciousness Explained.* New York: Little Brown.

Dumenko W.N. (1961). Veränderungen der eketrischen Rindenaktivität bei Hunden bei der Bildung eines Stereotyps motorischer bedingter Reflexe. *Pavlov Zeitsch. Hšhere NerventŠtigkeit*, **11**, 184-191.

Eckhorn R., Bauer R., Jordan W., Brosch M., Kruse W., Munk M. and Reitboeck H.J. (1988). Coherent Oscillations: a Mechanism of Feature Linking in the Visual Cortex? Multiple Electrode and Correlation Analysis in the Cat. *Biol. Cybern.*, **60**, 121-130.

Eckhorn R., Frien A., Bauer R., Woelbern T. and Kehr H. (1993). High Frequency (60-90 Hz) Oscillations in Primate Visual Cortex of Awake Monkey. *NeuroReport*, **4**, 243-246.

Engel A.K., König P., Gray C.M. and Singer W. (1990). Stimulus-Dependent Neuronal Oscillations in Cat Visual Cortex: Inter-Columnar

Interaction as Determined by Cross-Correlation Analysis. *Eur. J. Neurosci.,* **2**, 588-606.

Flannagan O. (1992). *Consciousness Reconsidered.* Cambridge: MIT Press.

Freeman W.J. (1975). *Mass Action in the Nervous System.* New York: Academic Press.

Freeman W.J. (1992). Predictions on Neocortical Dynamics Derived from Studies in Paleocortex. In E. Basar and T.H. Bullock (eds), *Induced Rhythms in the Brain.* Berlin: Birkhäuser, 183-199.

Freeman W.J. and Dijk B.W. (1987). Spatial Patterns of Visual Cortical Fast EEG during Conditioned Reflex in a Rhesus Monkey. *Brain Res.,* **422**, 267-276.

Galambos R. (1992). A Comparison of Certain Gamma Band (40-Hz) Brain Rhythms in Cat and Man. In E. Basar and T.H. Bullock (eds), *Induced Rhythms in the Brain.* Berlin: Birkhäuser, 201-216.

Galambos R., Makeig S. and Talmochoff P.J. (1981). A 40-Hz Auditory Potential Recorded from the Human Scalp. *Proc. Natl. Acad. Sci. (USA)* **78**, 2643-2647.

Gawne T.J., Eskandar E.N., Richmond B.J. and Optican L.M. (1991). Oscillations in the Responses of Neurons in the Inferior Temporal Cortex are not Driven by Stationary Visual Stimuli. *Soc. Neurosci. Abstr.,* **17**, 180.18.

Ghose G.M. and Freeman R.D. (1992). Oscillatory Discharge in the Visual System: Does it Have a Functional Role?. *J. Neurophysiol.,* **68**, 1558-1574.

Glass L. and Mackey M. (1988). *From Clocks to Rhythms.* Princeton: Princeton Univ. Press.

Gray C.M. and Singer W. (1989). Stimulus-Specific Neuronal Oscillations in Orientation Columns of Cat Visual Cortex. *Proc. Acad. Sci. (USA),* **86**, 1698-1702.

Gray C.M., Engel A.K., König P. and Singer W. (1990). Stimulus-Dependent Neuronal Oscillations in Cat Visual Cortex: Receptive Field Properties and Feature Dependence. *Eur. J. Neurosci.,* **2**, 607-619.

Gray C.M., Engel A.K., König P. and Singer W. (1992). Synchronization of Oscillatory Neuronal Responses in Cat Visual Cortex: Temporal Properties. *Eur. J. Neurosci,.* **8**, 337-347.

Hebb O. (1949). *The Organization of Behavior: A Neuropsychological Theory.* New York: J. Wiley.

Joliot M., Ribary U. and Llinas R. (1994). Human oscillatory brain activity near 40 Hz Coexists with Cognitive Temporal Binding. *Proc. Natl. Acad. Sci. (USA),* **91**, 11748-11751.

Jagadeesh B., Gray C.M. and Ferster D. (1992). Visually Evoked Oscillations of Membrane Potential in Cells of Cat Visual Cortex. *Science,* **257**, 552-554.

Kreiter A.K. and Singer W. (1992). Oscillatory Neuronal Responses in the Visual Cortex of the Awake Monkey. *Eur. J. Neurosci.,* **4**, 369-375.

Laurent G. and Davidowitz H. (1994). Encoding of Olfactory Information with Oscillating Neural Assemblies. *Science,* **265**, 1872-1875.

Llinas R. and Ribary U. (1993). Coherent 40-Oscillation Characterizes Dream State in Humans. *Proc. Natl. Acad. Sci. (USA),* **90**, 2078-2081.

Lopes da Silva F. (1992). The Rhythmic Slow Activity (Theta) of the Limbic Cortex: An Oscillation in Search of a Function. In E. Basar and T.H. Bullock (eds), *Induced Rhythms in the Brain.* Berlin: Birkhäuser, Berlin, 83-102.

Mangun G.R. (1992). Human Visual Evoked Potentials: Induced Rhythms or Separable Components? In E. Basar and T.H. Bullock (eds), *Induced Rhythms in the Brain.* Berlin: Birkhäuser, 217-231.

Montaron M.F., Bouyer J.J. and Rougeul-Buser A. (1979). Relations entre l'attention et le rythme mu chez le chat et le singe. *Rev. EEG Neurophysiol.,* **9**, 333-339.

Montaron M.F., Bouyer J.J. and Rougeul-Buser A. (1982). Ventral Mesencephalic Tegmentum (VMT) Controls Electrocortical Beta Rhythms and Associated Attentive Behavior in the Cat. *Behav. Brain Res.,* **6**, 129-145.

Nakamura K., Mikami A. and Kubota K. (1991). Unique Oscillatory Activity Related to Visual Processing in the Temporal Pole of Monkeys. *Neurosci. Res.,* **12**, 293-299.

Nakamura K., Mikami A. and Kubota K. (1992). Oscillatory Neuronal Activity Related to Visual Short-Term Memory in Monkey Temporal Pole. *NeuroReport,* **3**, 117-120.

Neuenschwander S. and Varela F. (1993). Visually Triggered Neuronal Oscillations in the Pigeon: an Autocorrelation Study of Tectal Activity. *Eur. J. Neuroscien.,* **5**, 870-881.

Neuenschwander S., Engel A., König P., Singer W. and Varela F. (1996). Synchronous Activity in the Optic Tectum of Awake Pigeons. *Vis. Neuroscien.,* **13**, 575-584.

Nicolelis M.A., Baccala L., Lin R.C. and Chapin J. (1995). Sensorimotor Encoding by Synchronous Neural Ensemble Activity at Multiple Levels of the Somatosensory System. *Science,* **268**, 1353-1358.

Pantev C., Makeig S., Hoke M., Galambos R., Hampson S. and Gallen C. (1991). Human Auditory Evoked Gamma-Band Magnetic Fields. *Proc. Natl. Acad. Sci. (USA),* **88**, 8996-9000.

Perkel D.H., Gerstein G.L. and Moore G.P. (1967). Neuronal Spike Trains and Stochastic Point Process I. The Single Spike Train. *Biophys. J.,* **7**, 391-418.

Perez-Borja C., Tyce F.A., MacDonald C. and Uihlein A. (1961). Depth Electrographic Studies of a Local Fast Response to Sensory Stimulation in the Human *Electroenceph. Clin. Neurophysiol.,* **13**, 695-702.

Pfurtscheller G. and Klimesch W. (1992). Event-Related Synchronization and Desynchronization of Alpha and Beta Waves in a Cognitive Task. In E. Basar and T.H. Bullock (eds), *Induced Rhythms in the Brain.* Berlin: Birkhäuser, 117-128.

Pfurtscheller G. and Neuper C. (1992). Simultaneous EEG 10 Hz desynchronization and 40 Hz synchronization during finger movements. *NeuroReport,* **3**, 1057-1060.

Pöppel E. (1971). *Oscillations as Possible Basis for Time Perception.* First Conference of the Society for the Study of Time. Oberwolfach.

Raether A., Gray C.M. and Singer W. (1989). Intercolumnar Interactions of Oscillatory Neuronal Responses in the Visual Cortex of Alert Cats. *Eur. J. Neurosci.,* Suppl. **2**, 72.5.

Rall W. and Shepherd G.M. (1968). Theoretical Reconstruction of Field Potentials and Dendrodendritic Synaptic Interactions in the Olfactory Bulb. *J. Neurophysiol.,* **31**, 884-915.

Ribary U., Ioannides A.A., Singh K.D., Hasson R., Bolton J.P.R., Lado F., Mogilner A. and Llinas R. (1991). Magnetic Field Tomography of Coherent Thalamocortical 40-Hz Oscillations in Humans. *Proc. Natl. Acad. Sci. (USA),* **88**, 11037-11041.

Rougeul A., Bouyer J.J., Dedet L. and Debray O. (1979). Fast Somato-Parietal Rhytms during Combined Focal Attention and Immobility in Baboon and Squirrel Monkey. *Electroenceph. Clin. Neurophysiol.,* **46**, 310-319.

Schillen T.B., König P., Engel A.K. and Singer W. (1992). Development of Oscillatory Neuronal Activity in the Visual Cortex of the Cat. *Europ. J. Neurosci.,* Suppl. 5, 3043.

Searle J. (1992). *The Rediscovery of Mind.* Cambridge: MIT Press.

Sheer D.E. (1970). Electrophysiological Correlates of Memory Consolidation. In G. Ungar (ed.), *Molecular Mechanisms in Memory and Learning.* New York: Plenum, 177-211.

Sheer D.E. (1992). Sensory and Cognitive 40-Hz Event-Related Potentials: Behavioural Correlates, Brain Function, and Clinical Application. In E. Basar and T.H. Bullock (eds), *Induced Rhythms in the Brain.* Berlin: Birkhäuser, 339-374.

Sheer D.E. and Grandstaff N. (1970). Computer-Analysis of Electrical Activity in the Brain and its Relation to Behaviour. In H.T. Wycis (ed.), *Current Research in Neurosciences.* New York: Karger, 160-172.

Singer W. (1993). Synchronization of Cortical Activity and its Putative Role in Information Processing and Learning. *Ann. Rev. Physiol.,* **55**, 349-374.

Steriade M., Jones E.G. and Llinas R.R. (1990). *Thalamic Oscillations and Signaling.* New York: John Wiley.

Tallon C., Bertrand O., Bouchet P. and Pernier J. (1995). Gamma Range Activity Evoked by Coherent Visual Stimuli in Humans. *Europ. J. Neurosci.,* **7**, 1285-1291.

Tiitinen H., Sinkkonnen J., Reinikainen K., Alho K., Lavikainen J. and Näätänen R. (1993). Selective Attention Enhances the Auditory 40-Hz Transient Response in Humans. *Nature,* **364**, 59-60.

Tovée M.J. and Rolls E.T. (1992). Oscillatory Activity is not Evident in the Primate Temporal Visual Cortex with Static Stimuli. *NeuroReport,* **3**, 369-372.

Varela F., Toro A., John E.R. and Schwartz E. (1981). Perceptual Framing and Cortical Alpha Rhythms. *Neuropsychologia,* **19**, 675-686.

Varela F.J., Thompson E. and Rosch E. (1991). *The Embodied Mind: Cognitive Science and Human Experience.* Cambridge, Massachusetts: The MIT Press.

Varela F.J., Martinerie J., Müller J., Pezard L., Adam C. and Renault B. (1995). Frequency Coherence in Multi-Site Cranial Recordings in Humans during Cognitive Tasks. *Human Brain Mapp,* Suppl. 1, 207.

Wilson H.R.. and Cowan J.D. (1973). A Mathematical Theory of the Functional Dynamics of Cortical and Thalamic Nervous Tissue. *Kybernetik,* **13**, 55-80.

Young M.P., Tanaka K. and Yamane S. (1992). On Oscillating Neuronal Responses in the Visual Cortex of the Monkey. *J. Neurophysiol.,* **67**, 1464-1474.

PHILIPPE MEIRE

ABOUT BIOLOGY AND SUBJECTIVITY IN PSYCHIATRY

Among the various medical branches, psychiatry raises in the most acute form the question of the relations between mind and matter. The difficulty appears already in the very term "psychiatry": literally "medicine of the psyche" (Greek version) or "medicine of the soul" (Latin version). It is a very difficult topic. The expression "medicine of the soul" indicates that we shall have to use metaphors.

This vocabulary may seem to reflect an outdated dualism. We must underline the role played by the material brain in the operations of the spirit as in psychic disturbances. So, psychiatry would be included in the medicine of organs, more precisely in neurology.

But the failure of such tentative and the constant use of the term "psychiatry" are tokens of the specificity of its domain which covers altogether cerebral or social determinations and experiences link with the terms subject, soul, spirit. All this let appear the fundamental division from which psychiatry suffers when placed in the frame of classical or Cartesian science, a division which may become schizophrenia. Indeed, when considered as a part of the medical and scientific domain, psychiatry belongs to the sphere of objectivity (of the "res extensa" of Descartes). But, on the contrary, such words as "psyche" and "subject" accompanied by "freedom" and "responsibility" put clearly psychiatry on the side of subjectivity and thinking (the "res gogitans" of Descartes).

It is easy to criticize the Cartesian dualism but more difficult to go beyond it. We cannot discard a dual experience and the necessity to articulate two aspects. In psychiatry, the experience of their interactions is at the heart of its work. Quite often, the soul suffers from disturbances of the (physical or social) body, revealing the failing character of subjectivity and of our feeling of liberty. But, reciprocally, the body may suffer from disturbances of the soul.

B. Feltz, M. Crommelinck and P. Goujon,(eds.), Self-Organization and Emergence in Life Sciences: 111-122.

It is thus the psychiatrist himself who is in quest of understanding the psycho-physical articulation. He needs a model enabling him to render account of the links between a logico-empirical approach, of an operational and deterministic type, and a hermeneutic approach respecting the subjective and intentional character of the human being. There is a need to give meaning to his work and metaphors describing it as a kind of mechanics of the psyche leave him unsatisfied even if an operational dimension is undeniably present in his practice.

Surely, if anyone devotes himself to the study of a single aspect, dualism is no more a difficulty. But, on the contrary, for hermeneutical and phenomenological discourses, one must take both aspects into consideration. The recognition of the dual character of our experience is inevitable and necessary. However, it may favor an ontological dualism leading to oppose an objective description and the subjective experience of the human being, an opposition reflected in a series of classical antinomies: matter-spirit, brain-psychism, body-soul, res extensa-res cogitans, materialism-idealism.

There is no question to reject the two dimensions of our dualistic experience but this should not necessarily lead to consider these antinomies as reflecting an ontological fact and to establish a radical distinction between matter and psychism which would render any articulation logically "unthinkable". Our difficulties would not be solved, in particular those of psychiatrists obliged to face various disturbances caused by ever present interactions between our to aspects, manifested by the "symbolization of the body" and the "incorporation of the language".

In order to render "thinkable" these constant interactions between "body and soul", psychiatry requires to develop a conceptual frame enabling to understand the logic of pathological facts. Psychiatry should be tackled but through a preliminary anthropological approach. This would avoid sterile conflicts occurring presently too often.

Old debates have been spectacularly reactivated by progresses in psychopharmacology allowing to produce psychotropic drugs with weak secondary effects. There is a risk of manipulating psychic subjectivity in view of modifying our behavior or our mental performances, namely regarding memory or the stress caused by competitivity. Such new phenomena have replaced the questioning of the rational subject by the so-called "masters of suspicion", Nietzsche, Marx and Freud. Psychopharmacology dismiss in a radical way the Cartesian subject and compels us to put the question of the subject in the very heart of biology. And the development of new psychotropic drugs plays a significant role in the strong increase of interest for ethical questions in present psychiatry:

what is the idea of man upheld in present psychiatric theory. Human psychism is indeed the place where ethical questions are raised and answers will depend on our views on it. Psychiatry is necessarily linked with anthropology and many books are nowadays published dealing with the "Body-Mind Problem", with what is now sometimes called "neurophilosophy". Through debates on the so-called "cognitive sciences" and on artificial intelligence, the great classical and philosophical questions which are at the heart of psychiatric practice come again to the fore.

In spite of obvious difficulties, I am intended to offer some reflections on the dynamics of living beings and on my clinical experience of various pathological forms of consciousness.

After having shown the interest to fulfil the analytical approach of the living being by a theoretical reflection experimentally based, I shall similarly deal with propositions regarding cognitive sciences with respect to consciousness. Therefrom, the development of a reflective conscience will appear as emerging of a long process of autonomization, a continuous process not deprived of apparent paradoxes since it unites an increasing individualization and an opening on the world ever more radical. And, finally, this emergence appears also through a radical discontinuity.

1. TWO COMPLEMENTARY APPROACHES OF THE PHENOMENON OF LIFE

1.1 The Operational and Analytic Approach in Biology

Present biology is practically identified with the operational approach of the living being The study of the mechanics of living beings is more and more successful. Their organization is divided in sub-systems represented in a series of models and genetic transmission is explained by molecular biology. Interactions between chance and necessity render account of the dialectics between invariance and novelty.

Any specificity of living matter has been eliminated by today's biology as well as any "vital principle". Biological mechanisms are expressed in physico-chemical terms associated to the efficient metaphors of codes and programs. Szent-Gyorgy already said: "Life as such does not exist. Nobody has ever seen it." And François Jacob confirms: "The operational value of the concept <life> vanishes. In our laboratories, we only tempt to analyze living systems, their structure, how they work, their history." In the approach of living systems and of their cybernetical regulation, they are considered as "machines" whose mechanisms may be objectified. A living being becomes a particularly complex object but, through the operational approach, its own dynamics and its tendency towards auto-organization seems to have vanished.

1.2 A Bio-logic of the Phenomenon of Life

It appears important to consider another approach which is not in opposition but in complementarity with the operational and analytic one. It is a point of view aiming at integrating as well our experience as the benefits of the operational method. It should be broad enough to take into account the dimension of subjectivity already present, even in a primitive form, in any life and, further on, to sketch a reflection on the emergence of consciousness.

Starting from such an interrogation on the object of biology, André Pichot developed a "theory of biology" which is a logic of life leading him to renovate deeply our understanding and to throw light on the operational science of living systems. A quotation: "Quite paradoxically, biology (the science of life!) is today a science for which the concept of life does not mean anything. Such an expression as 'biochemistry' illustrates this elimination of the biological object, negating its originality by reducing it to the level of chemistry."

Nevertheless, there is a specific object for biology. Pichot notes:

1. Our first experience is that everything which is living is not inanimate. The living being and the inanimate are mutually exclusive.
2. Moreover, there is no life independent of matter. Living "objects" belong to the physico-chemical level but life is not an intrinsic quality of matter. It is, so to say, matter in "movement", altogether linked with but extrinsic to matter.
3. A living being is a material element defined by a barrier separating it from the inanimate. It is a totality. Any part detached from the being becomes inanimate (at least for it).

We may draw from these considerations a few propositions:

Proposition 1: A living being is defined by the capacity of its matter to constitute an entity distinct from its external surrounding with which various exchanges (of matter, energy, information) take place, carried out in a strictly defined way by the physico-chemical organization prevailing on both sides of the separating border. Pichot underlines that life is thus not a state inherent to a physico-chemical organization (as analytic biology postulates it) but a bipolar phenomenon, a dialectic relation implying two terms, the living being and its external surrounding. He clearly distinguishes between the living being (the "living matter", the actively auto defining physico-chemical entity) and life which is the dialectic region between the entity and its surrounding.

Starting from the first proposition, two complementary ones are drawn:

Proposition 2: The living being is its own finality, there is a circular determinism carrying into effect its "internal coherence". The living being is autonomous.

This proposition is rather largely accepted. Claude Bernard spoke already of "free and independent life" and recalled that a snake biting its tail was the symbol of life in Antiquity.

However, proposition 2 has to be immediately completed by a 3rd one:

Proposition 3: The existence of the living being requires not only internal coherence but also an external coherence, meaning the establishment of a circular determinism with the external surrounding.

Indeed a living being reacts globally to two stimuli. Through the circular determinism, which is a revert on itself, the living being is a kind of stimulus for itself to which it reacts. A necessary coherence is thus created between the living being and its external surrounding that Pichot calls the external coherence. He underlines also that, for traditional biology, external coherence is important for natural selection only, according to the Darwinian model of the evolution of species. In such a perspective, the living entity is not a "subject" but only the object of natural selection. While Pichot and others consider that living being manifest already a kind of "proto-subjectivity".

Proposition 4: Thanks to its internal coherence, the living being exists by itself. Thanks to its external coherence, it exists in relation with what is not itself for itself. Existing "by itself" and "for itself", it acquires the character of a "subject".

This existence by itself and for itself is surely but an "opaque" unconscious subjectivity, not resulting from the play of an external "anima" but from the very movement of matter just mentioned above. The character of "subject" does not rule this movement, it emerges from it even if it seems the finality of the movement.

Surely, all this is enigmatic but what is problematic should not be discarded by a biology pragmatically concentrated on the functioning of the living entity. Its operational efficiency cannot put into oblivion the undeniable "proto-subjectivity" of the living being. Kant, Hegel, and many recent authors did not hesitate to speak of "living subjects".

The following proposition goes in the same direction:

Proposition 5: The circular determinism (the "internal coherence") does never operate in a synchronous manner as the various reactions require a certain duration. A living being present always a "want of existence" that it tries to fill by its development but it never fully succeeds. A balance in this respect works as an in attainable "attractor" ("a theoretical attractor").

Most authors consider the living organism as a totality having a more or less clear finality (generally the invariance, the homeostasis). The process is rather mysterious. While, in present bio-logic, the circular determinism is never achieved, there is a finality which is a tendency towards reaching a balance. The finality lies in a process continuously seeking a balance which would be its "totalization".

As Dell suggests, one should speak about living systems of a "principle of coherence" rather than of a "homeostatic principle". But there are only local, partial and transitory balances. Since the origin of life, a fundamental unbalance prevails, there is a continuous process of transformation from which any local balance can but artificially cut off. One may speak of a single continuous process since the origin of life.

Among other consequences, a last proposition may be retained: "during its development, a living being becomes more and more autonomous, taking its environment more and more into account, that is broadening what is for it its external surrounding".

Pichot speaks of a simultaneous increase in autonomy and in dependence. Life has thus a history, the history of an evolution distinct from the inanimate and characterized by an increase of individuation and autonomization. But this history is achieved through a constant interaction in the surrounding, an active articulation representing a sketch of "protocognition" (Pichot). The individual emerges from a split between two poles, altogether linked and distinct. His actual experience should be that of such a split.

Paradoxically, the autonomy of the living being goes together with a tendency to take more and more into account its environment, up to the point that, at the psychic level, each human being has to do with the whole universe. But, "*stricto sensu*", there is here no project of Evolution but the gradual effects of a search for a balance under the twin constraints of internal and external coherence.

Such an interpretation sees the living being neither as an object of the world nor as a separate entity but a being actively distinct from and harmonized with the world, a "subject" of the world. It evolves with the world and, so to say, it is a memory of the world even if it does not know it. Being such a memory and open to ever more dimensions of reality, one might say with Ortega y Gasset: "I am a part of everything I have met."

2. TWO COMPLEMENTARY APPROACHES OF THE PSYCHIC LIFE

Classic science has favored the development of an analytic and operational approach in view of disclosing working mechanisms. The dimension of "becoming" is almost forgotten and the emergence of the conscience completely foreign. Nevertheless, the Cartesian approach had many valuable aspects and we should not throw away the baby with the water of the bath. However, in order to trace the progressive emergence of subjectivity within matter, our scientific knowledge should interplay with the phenomenological description of our experience regarding life and in particular the life of the conscience.

In this second part, I would like to suggest that the type of relation just described between the living being and its environment may be applied as well to the comprehension of psychic life. In the vast field of cognitive sciences, one may detect a tendency to insert cognitive processes in the dynamics of the living being in order to complete the models arising from operational approaches and to overcome certain conceptual difficulties.

2.1 Operational Approaches in Cognitive Sciences

Cognitive sciences have first developed models of performances trying to specify mechanisms in terms of representations and of computations. The descriptions were based on operations able to be experimentally tested and not on the personal experience of the subject.

In classical cognitivism, called "symbolic paradigm", elementary and essentially unconscious functions have been especially described by these models through the already classical metaphor of the computer working according to a sequential architecture. The representation was taken as an objective mirror of reality at the image of a computer reproducing the content of a magnetic tape. In such descriptions, datas are introduced in the brain, dealt with according to programs and stocked in memories.

This metaphor raises quite a number of questions:

- How are the datas elaborated and selected?
- Why can we note this tendency towards development and creativity in mental activity?
- Wherefrom do such precise and tractable programs come?
- How can we understand the feeling and our experience of life. Where from does the sense come?
- Wherefrom does the creativity of language come if it is but a translation?

According to Daniel Andler, "Homo cognitivus" appeared first as a man without a body, without subjectivity and without conscience. However, cognitive sciences with their tremendous development could not ignore for ever the problem of the conscience if a unified theory of cerebral and mental activities was to be elaborated. This became also necessary for a cognitive psychopathology aiming at a comprehension of important subjective phenomena such as hallucination, delirium or loss of self consciousness.

After that William James described the process of conscience at the end of the XIXth century, it was viewed as a series of functions bound with neuropsychological models. And in order to explain the continuity of our consciousness, one had recourse to various modes of memories. But many ambiguities and misunderstandings remained. The conscious experience is regularly the victim of reductionist schemes.

Surely, we cannot trust that a scientific approach inspired by natural sciences may render account of subjective conscious experience. In his famous paper "What is it like to be a bat?", T. Nagel holds that proposing a model for the cognizant functioning of a bat cannot give us any feeling on what may be the own experience of a bat. Subjective experience cannot be described in terms of operations. However, there are certain scientific formulations which negate subjective feeling and experience on the ground that they cannot be integrated in a theory; while others are looking for a model which, at least, might be compatible with such a primary experience or may even render account of its emergence.

Outstanding thinkers as Husserl or Merleau-Ponty knew well that any study of man going so far as to raise ultimate questions will inevitably meet deep phenomenological problems. Let us quote Merleau-Ponty: "Does one follow phenomenology or empirical psychology, there is always question of man ... and if empirical psychology pays sufficient attention to what it describes, it will always end by admitting that man is not a part of the world but the bearer of reflection."

2.2 The Auto-Organizing Dynamics of the Living Being

The reaction against the scarce room allowed to conscious experience in classical cognitivism came through ideas among neurobiologists impressed by evolutionist theories and studying specifically human pathologies in neuropsychiatry. The development of evolutionist theories in neurobiology marks the return of a reflection on consciousness in the frame of a temporal perspective which characterizes the auto-organizing dynamics of the living being, defining itself with respect to what is not itself but for itself.

The best known representative of this tendency is Gerald Edelman (who was awarded a Nobel prize) whose book "Bright Air, Brilliant Fire: On the Matter of Mind" is now famous. He is not a reductionist as Dennett. He describes the progressive constitution of a complex cerebral organization able to sustain subjectivity and the experience particular to each individual.

Edelman wondered at the scarce interest of a certain cognitivism for this brain. For him, the theory of evolution is an indispensable tool in order to understand the spirit. He refuses to assimilate the brain to a computer. In order to throw light on emergence of consciousness, one has to rely on specific properties of living beings that he calls "recognition systems", establishing a continuous and adaptative correspondence of the elements of a physical domain to novelties occurring in elements of another physical domain, more or less independent of the first, an adjustment operating without preliminary instructions.

Such characteristics belong to life. Thus physics cannot tackle with recognition systems which are essentially biological and historical systems. For Edelman, the brain functions as a selective system establishing correlations according to the (so-called) neuronal Darwinism. The theory of the selection of neuronal groups allows to understand the emergence of new morphological kinds in the brain, to be selected in function of the interaction with the surrounding. The constitution of interacting neuronal groups allows to compare memories of the self and of the non-self, a process leading, according to Edelman, to a primary consciousness. The incarnation of consciousness produces meaning for the individual as the possibility to anticipate and to correct errors.

The brain dynamics proposed by Edelman is not far away from the logic of the living being described in the first part of this paper. The brain, interacting with its environment, is continuously in search of a balance and this search is an auto-organizing one. With respect to other organs, the brain is at a superior level, in a position suggesting an analogy with what I proposed to call a "meta" type.

3. THE REFLEXIVE CONSCIENCE AND THE ANTHROPOLOGICAL DIFFERENCE

An interesting feature in Edelman's work is that the levels of primary and of superior consciousness (the last specific to humans) are clearly distinguished. Thanks to their memories and their consciousness, humans can situate themselves with respect to the past and to the future. However, the way he sees the passage to superior consciousness by means of the language remains altogether classical and enigmatic. It is like taking the effect for the cause. Surely, this is a boundless question that I can but lightly touch, taking notice that some stress the progressive evolution between superior primates and humans while others stress a radical cut, the anthropological difference implying a clear threshold.

Thanks to their cerebral development, there are animals possessing a primary conscience. They have a memory but they live in the present. They have no conscious reflexivity allowing to discriminate "qualia" (feelings of the subjective consciousness).

There are some indications regarding processes of superior consciousness which are provided by studying specifically human pathology, in particular those pathologies which dissociate explicit and implicit consciousness, including the recognition of the self or of its own values. A pioneer work on this "neurology of the subject" is to be credited to the Russian school of Alexander Luria. And recently, much work has been done on deficiencies of self-awareness.

In a recent book of the American neuropsychologist Antonio Damasio "Descartes's Error, Emotion, Reason and the Human Brain", it is acutely shown that our consciousness bases continuously itself on our global experience of the body at various levels of integration. In particular, there are pathologies of the image of the body, at a non-conscious level, distinct from those situated at a "meta" level. And this meta level seems to know more than our consciousness.

It is not possible to dwell more on such hypotheses. I would just indicate a track for approaching the emergence of reflexive consciousness in the dynamics of the living being. Evolution manifests a progressive complexification and, at a certain moment, a threshold is crossed leading, barring accidents, to introduce a human baby into the world of creativity, reflexivity and language. It is not impossible that a "meta" activity autonomizes itself, in the prefrontal region of the brain, in order to render an account of this reflexivity and this human creativity, at the origin of the jump into language.

This "meta" activity might function, with respect to the underlying levels, according to a logic of entangled hierarchies, as evoked by Hofstadter in his famous work "Godel, Escher, Bach": what is in a "meta" position cannot categorize itself but may interpret the underlying level. This should be a process of mediation, specifically human, necessary for understanding the origin of symbolic categories and of reflexive consciousness. And this reflexive consciousness should be an experience linked with backwards effects of unconscious subjectivity on cerebral activity. The experience of interiority then emerges, with an opening on the world....

REFERENCES

Andler D. (1989 et ss.). Cognitives (sciences). *Encylopaedia Universalis.*

Andler D. (dir.) (1992). *Introduction aux sciences cognitives.* Paris: Gallimard, coll. Folio.

Crommelinck M. (1990). La conscience en neuro-sciences et sciences cognitives: une conscience à la frontière des sciences de la nature et des sciences de l'esprit. *Revue québecoise de psychologie,* **11**, 1-2, 89-106.

Damasio A. (1994). *Descartes's Error. Emotion, Reason and the Human Brain.* Grosset/Putnam Books. (Trad. fr. (1995). *L'erreur de Descartes. La raison des emotions.* Paris: O. Jacob).

De Duve C. (1990). *Construire une cellule.* Bruxelles: De Boeck.

Dell P. (1982). Beyond Homeostasis: Toward a Concept of Coherence. *Family Process,* **21**, 21-41.

Dennett D. (1991). *Consciousness Explained.* Boston: Little, Brown and Cy.

Edelman G. (1992). *Bright Air, Brilliant Fire: On the Matter of Mind.* Basic Books. (Trad. fr. (1992). *Biologie de la conscience.* Paris: O. Jacob).

Engel P. (1994). *Introduction à la philosophie de l'esprit.* Paris: La Découverte.

Feltz B. et D. Lambert (eds) (1994). *Entre le corps et l'esprit.* Liège: Mardaga.

Haroche M.-P. (dir.) (1990). *L'Âme et le Corps. Philosophie et Psychiatrie.* Paris: Plon.

Hofstädter D. (1981). *Gödel, Escher, Bach: an Eternal Golden Braid.* New York: Basic Books.

Hofstädter D. et D. Dennet (1987). *Vue de l'Esprit. Fantaisies et Réflexions sur l'Être et l'Âme.* Paris: Interéditions.

Jacob F. (1970). *La logique du vivant.* Paris: Gallimard.

Lwoff A. (1969). *L'ordre biologique.* Paris: Laffont.

Meire Ph. (1995). Le cerveau-sujet et les boucles étranges de la démence. *Psychologie Médicale,* **27**, 3, 160-164.

Meire Ph. (1994). *Le sujet vivant. Entre autodifférence et ouverture, la dynamique du non-équilibre.* Louvain-la-Neuve: Thèse d'agrégation de la Faculté de Médecine.

Meire Ph. (1990). Demain, quelle éthique, quelle épistémologie en psychiatrie? *Psychoanalyse,* **6**, 187-200.

Merleau-Ponty M. (1967). Les sciences de l'homme et la phenomenology. Paris: Cours Sorbonne, Centre de Documentation universitaire, 16. In Merleau-Ponty M. *Résumés de cours (Collège de France, 1952-1960).* Paris: Gallimard, coll. Tel n° 71.

Monod J. (1970). *Le hazard et la nécessité.* Paris: Seuil.

Nagel T. (1974). What is it Like to be a Bat? *Philosophical Review,* **83**, 435-445.

Pichot P. (1988). Naissance et Vicissitudes du Concept de Maladie Mentale. *Acta psychiat. belg.,* **88**, 206-211.

Pichot A. (1980). *Éléments pour une théorie de la biologie.* Paris: Maloine.

Rorty R. (1979). *Philosophy and the Mirror of the Nature.* Princeton: Princeton Univ. Press.

Smith Churchland P. (1986). *Neurophilosophy. Toward a Unified Science of the Mind/Brain.* Cambridge (Mass.): Bradford Book.

Tinland F. (1977). *La difference anthropologique.* Paris: Aubier-Montaigne.

Varela F., Thompson E. et Rosch E. (1993). *L'inscription corporelle de l'esprit. Sciences cognitives et experience humaine.* Paris: Seuil.

HENRI ATLAN AND IRUN R. COHEN

SELF-ORGANIZATION AND MEANING IN IMMUNOLOGY

Except for monozygotic twins, each person is born with a unique assortment of genes. But one's genotype, like one's birth, is only a potentiality. Each of us realizes individuality in the practice of life through the exercise of two systems: the central nervous system, the seat of our psychological self, and the immune system, the adjudicator of our molecular self. These two systems help create individuality because they work to make each of us different from all other persons, including our monozygotic twin. Individuation results from the capacity of these two systems to organize themselves over time in response to the individual's unique environment. The self-organization of the central nervous system has been dealt with over the years by neurobiologists and cognitive scientists[1]. Our aim here is to consider the processes determining self-organization in the immune system. In doing this, we shall consider two formative principles: the creation of information and the creation of meaning.

1. LANGUAGE SORCERY

The concepts of "information" and "meaning" are not at home in immunology and we shall use them here in ways that might seem only metaphorical. In doing so, however we maintain a tradition in immunology, a science which is quite used to anthropomorphic and even-mentalistic metaphors in carrying out its scientific discourse. For example, "recognition" is the term applied by immunologists to the physical binding of an antigen to a receptor; "memory" is used to explain the differences in

1. Atlan H. (1972). *L'Organisation biologique et la théorie de l'information.* Paris: Hermann; Atlan H. (1974). On a Formal Theory of Organization. *J. Theoret. Biol.,* **45**, 295-304; Burnet F.M. (1959). *The Clonal Selection Theory of Acquired Immunity,* Cambridge: Cambridge University Press.

B. Feltz, M. Crommelinck and P. Goujon, (eds.), Self-Organization and Emergence in Life Sciences: 121-139.

the immune reactions noted between the first and second contacts of the immune system with a particular antigen; "response" refers to a measured reaction; "tolerance" denotes a lack of "response" to an antigen and implies "deletions" of specific lymphocytes; antigen "presentation" is beginning to acquire a molecular reality; "self", "foreign", "anergy", "mimicry", and other borrowed terms are used often, but often without precise molecular or mechanistic definitions. Immunology, of course, is only behaving like the other sciences that have to employ metaphors to get on with their business in the face of incomplete information. We need not fear metaphors as long as we are not beguiled into thinking that an ill-defined word suffices for understanding. A metaphor, like a theory, is most serviceable when it opens, rather than closes thinking.

2. INFORMATION

Claude E. Shannon developed a probabilistic theory of information based on the relative frequencies of the different letters of an alphabet used to write a set of messages[2]. Shannon provided a way to quantitate the information that any particular message could possibly bear. Although Shannon's practical concerns were related to the engineering of telephone communications, his principles can be applied to any discipline including biology. Information, as defined by Shannon, involves an arrangement of elements, the string of nucleotides in a molecule of mRNA for example, that is "just so" as distinct from any other possible arrangement. Indeed, the "just-so" structure of the message compared to all other possible arrangements of the nucleotides constitutes the content of information. Shannon's concern was that the information contained in the input be transmitted faithfully to the output. Protein synthesis, for example, can be seen as a channel of communication in which an mRNA molecule is the input message and the amino acid sequence of the protein is the output. If the process is free of errors, the amino acid sequence of the latter is a replica of the nucleotide sequence of the former through a deterministic rule of correspondence called the genetic code. Random perturbations in the communication process, what has been called "noise", can disrupt the information and produce errors in the amino acid sequence. Noise, in effect, can decrease the information content of the output, the protein, in reference to the information carried by the input, the mRNA. Shannon's classical formulation helped to establish principles

2. Shannon C.E. (1948). A Mathematical Theory of Communication. *Bell Syst. Tech. J.,* **30**, 50-64; Weaver W. and Shannon C.E. (1949). *The Mathematical Theory of Communication.* Urbana: University of Illinois Press.

governing the fidelity of the transmission of information. For example, making parts of the message redundant can preserve information over a noisy channel of communication. Our instinctual repetition of key words is for clarity, as well as for emphasis. The added cost of adding redundancy pays off in added fidelity. Shannon's formulation is particularly robust because it holds true irrespective of the nature of the information borne by the message and is even independent of any meaning the message might convey to the receiver. Shannon, a scientist at the Bell Telephone Laboratories, assumed that the meaning of the message was the concern of the sender and receiver of the message; the packaging and fidelity of the transmitted information were Shannon's (and the Bell Telephone Company's) concern. However, the very robustness of Shannon's theory turns out to be its major weakness for biology. Andre Lwoff[3] in 1962 pointed out that Shannon's formulation was of limited applicability to biology because function depends on how molecules work, and not only on how much information they bear. For example, the sequence of nucleotides in a molecule of mRNA is translated (another metaphor) into the amino acid sequence of a protein. However, a similar molecule of mRNA, now with a mutation, can bear the same quantity of Shannon-type information, but may not encode the same protein, or any protein at all. How the protein functions, how it works, endows the protein (and the mRNA) with biological meaning. Along with the function of information, Shannon's theory neglects the creation of information. Self-Organizing systems like the brain and the immune system create new information and do not merely transmit or preserve existing information.

3. CREATING NEW INFORMATION

Obviously, a diversity of mRNA molecules contains more information than does only one species of mRNA. So the creation of information amounts to an increase in diversity. Note, however, that increasing the diversity should not be at the expense of errors in the existing information, otherwise information would be lost. The only way to create new information out of the existing information is to disrupt the old "just-so" order in an unforeseen way, to introduce unforeseen changes. Noise, the cause of random change, is the only agent that can effectively open the way for new opportunities of organization. Thus, any increase in the diversity of information would seem, paradoxically, to obligate a loss of information. To resolve this paradox

3. Lwoff A. (1962). *Biological Order.* Cambridge, Mass.: MIT Press.

within the framework of Shannon's mathematical definitions, one of us (Atlan) developed a theory of the creation of information based on a principle of complexity from noise[4] (see Figure 1). This theory constitutes a formal theory of self-organization. According to it, two conditions must be fulfilled before noise could possibly generate re-organization and not mere disorganization of a system:

1. The system must have a hierarchical, multilevel organization so that a decrease in the information transmitted in a channel at one level can actually produce an increase in the content of information at a more global level that includes the noisy communication channel as one of its constituents. For example, a mutation in a particular gene, which disorganizes the gene's original string of nucleic acids, could produce a new gene encoding a new protein with survival value at the level of the organism as a whole.
2. The system must feature redundancy. Redundancy refers to the existence in the system of multiple copies of the same or similar information. Redundancy preserves the original information while furnishing expendable copies for random diversification. Hence for Atlan, redundancy is a prerequisite that allows the creation of the diversity without sacrificing the old information. Note, however, that successful diversification reduces redundancy; the extra copies of the old information disappear as they become new information. Therefore, a system's initial redundancy, which sas its potential for self-organization, must be large enough to allow the reduction of this redundancy. Alternatively, a mechanism for recharging the system's redundancy must operate; otherwise, the effects of noise would be only to destroy and not diversify information. For example, it is now clear that a process of gene reduplication, which creates redundant copies of a particular gene, is required to allow the safe generation of multigene families. Each redundant copy of the gene independently can mutate and diverge over evolutionary time to create the diversity of genetic information that produces complex organisms.

The differences between Shannon's original theory and its extension by Atlan to account for the creation of information can be clarified by considering the different roles they assign to noise and redundancy. For Shannon, who wants to preserve the message, noise is the destroyer and

---

4. Atlan H. (1972). *L'Organisation biologique et la théorie de l'information.* Paris: Hermann; Atlan H. (1974). On a Formal Theory of Organization. *J. Theoret. Biol.,* **45**, 295-304; Atlan H. (1983). Information Theory. In R. Trappl (ed.), *Cybernetics.* New York and Berlin: Hemisphere Publ. and Springer Verlag, 9-41.

redundancy of the message is a premium we must pay to assure fidelity despite noise. For Atlan's theory of self-organization aimed at the evolution of new messages, noise is the (blind) creator and redundancy is not the cost of fidelity, but an asset, the vehicle for change. Noise, for the former, is a bitter pill, for the latter it is the spice of life. Redundancy for communication engineers is a burden. It is a bonus for biologists.

Atlan's two conditions, redundancy and multilevel organization, are necessary for self-organization to take place. However, a mere increase in diversity in Shannon's sense of added information does not guarantee the functional character of any particular state of organization. Adding diversity alone may not involve any meaningful functionality, so diversity, though necessary, is not sufficient for self-organization.

4. THE RANDOM GENERATION OF IMMUNE DIVERSITY

The principle of complexity from noise has been confirmed at the molecular level in the immune system. The immune system can be said to record information about the structures of molecules called antigens. Antigens are defined operationally by the fact that an antigen receptor, which may be an antibody or a receptor on a lymphocyte, can bind to a part of the antigen called an epitope. The binding to the epitope is due to steric complementary between the epitope and the antigen receptor; the antigen receptor is a three-dimensional mirror image of the antigen epitope. It can be argued that the diverse array of antigen receptors expressed by the cells of the system constitutes the immune system's information about antigens, what is termed the immune repertoire. Therefore, the creation of the immune repertoire exemplifies the creation of specific information. We now know that the diversity of the antigen receptors is fashioned by processes of genetic recombination, mutation, and random insertion of nucleotides in the genes that encode the receptors. These random and near-random processes create the unique specificites of the antigen combining site of each lymphocyte clone[5].

---

5. Schatz D.G., Oettinger M.A., Schissel M.S. (1992). V(D)J Recombination: Molecular Biology and Regulation. *Annu. Rev. Immunol.*, **10**, 359-383.

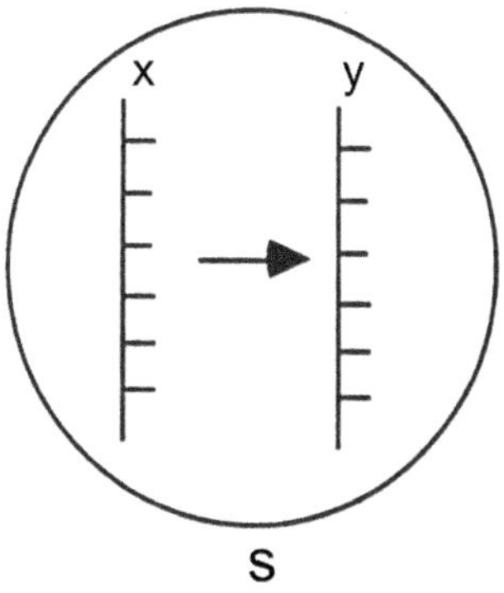

Figure 1. Shannon's function $H = - \Sigma p_i \log p_i$ is the inversed probability of occurrence of a given letter *i* of an alphabet in an ensemble of messages using that alphabet, averaged over all the letters of the alphabet. H expresses the average information content per letter, or per symbol, or per unit in an ensemble of ordered structures such as all possible sentences in a language, or all possible amino-acid sequences in a protein. H, also called message entropy, expresses an averaged a priori uncertainty about a given structure or message. In addition, H expresses the diversity of the class of messages or structures that could be built with this particular set of symbols or units. For example, when the N symbols of an alphabet occur with equal probability 1/N, then, H reduces to log N. In other words, the information borne by each symbol is equal to the log of the number of different symbols present in the alphabet; the more symbols you can use, the greater the amount of information in each symbol. That is why H has also been proposed as a measure of structural complexity.

In a communication channel where an input message X is transmitted into an output message Y, the transmitted information can be computed by the above formula using a term for the conditional probabilities $p(j/i)$, the probability of finding a letter j in output message Y at the location of letter *i* in input message X. Exact transmission with no error implies $p(j/i) = 1$ for $j = i$, and 0 if $j \neq i$. Errors of transmission due to noise produce conditional probabilities other than 0 or 1. That is why a conditional H function, $H(Y/X) = -\Sigma p_i p(j/i) \log p(j/i)$, called ambiguity, measures the average effect of noise. Shannon showed that the transmitted information from X to Y in such a case is equal to $T(X;Y) = H(Y) - H(Y/X)$. Ambiguity decreases the transmitted information.

The idea of complexity from noise is based on the following observation. If X and Y are seen as components of a system S, their joint contribution H(X,Y) to the information content of the whole system is $H(X,Y) = H(X) + H(Y/X)$. The ambiguity appears now with a + sign: no error means that Y is completely redundant. Being an exact duplicate of X, Y does not add any information about the structure of S. Maximum ambiguity, which is the absence of any correlation between X and Y, means maximum diversity or structural complexity for S as far as parts X and Y are concerned. This is true, obviously, only if S continues to exist and function, which it may do in a different fashion, with more diversity and less redundancy, despite the lack of communication between X and Y. Now, if the integrity of system S requires some form of communication between X and Y, then, their lack of communication could lead to the destruction of the system. Thus, the total amount of information in S would be optimized when X and Y have some connection, that is when the transmission of information is not zero, but not free of error.

Note that randomness characterizes not only the mechanism that generates the receptors of individual clones, but also the unedited collective of receptors that arise by chance, what might be called the primordial repertoire. Quite simply, the number of diverse antigen receptors that could be generated by any person's immune system is so large, perhaps 10^{10}-10^{20} different combining sites, that the chance realization of a large sample of the potential repertoire must produce an unmanageable and disorderly mob of functionally redundant clones. Redundant receptors are not strictly identical but they bind the same epitopes, sometimes with different affinities. To be serviceable, this primordial repertoire of clones has to be organized; from the primordial repertoire, an actual repertoire has to be generated that is limited in size and focused in a way that augments the frequencies of the more useful clones. In other words, the receptor repertoire can be made to work efficiently only by reducing the numbers of functionally redundant receptors and by establishing a hierarchy of dominant specificity. The mechanism responsible for this reduction in initial redundancy is the selection by specific antigens of the particular clones of lymphocytes bearing specifically complementary receptors. As the selected clones proliferate, their antigen receptors come to dominate the actual collective repertoire[6]. This process of clonal selection by antigens imposes a dynamic ordering of the receptor repertoire that reflects the actual antigenic experience of the individual with his particular antigen world. Thus, the frequency distribution of specific lymphocyte clones within the immune system is a measure of the antigen-specific information inherent in the system at any given time. This clonal distribution embodies self-organization, the creation of new information through selective experience with the antigenic world. Macfarlane Burnet, the chief proponent of the clonal selection theory of adaptive immunity, entertained the notion that some random process must underlay the generation of immune receptor diversity, although he never formalized this idea[7]. The discovery of the combinatorial basis for the clonal generation of receptors has confirmed Burnet's speculation, while it has provided a living example of complexity from noise as a principle for self-organization.

6. Mosmann T.R. and Coffman R.L. (1989). TH1 and TH2 cells: Different Patterns of Lymphokine Secretion Lead to Different Functional Properties. *Annu. Rev. Immunol.*, **7**, 145-173.

7. Burnet F.M. (1959). *The Clonal Selection Theory of Acquired Immunity*, Cambridge: Cambridge University Press.

5. THE CREATION OF MEANING

Information, as discussed above, is a property of the intrinsic organization of the message itself in the form of a frequency distribution of its elements. The meaning of the message, in contrast, is never intrinsic to the message; the meaning is the relationship of the message to some reference point outside of the information borne by the message. The referential status of meaning is concretely illustrated by considering the meaning of a word. The most precise meaning of the word depends not only on the reference of the word itself, but to a great degree on the sentence, the *context* of other words in which the particular word is used. The word's meaning also depends on the *history* of the word and of the situation, for example, do the communicators who transmit and receive the sentence speak the same language? The meaning of the word for the individual is also influenced by the *associations* that are triggered (what does the sentence call to mind?).

Is *meaning* — related as it is to *contexts, histories* and *associations* — a property limited to interactions between introspecting and verbally communicating humans, or can meaning be observed objectively as an attribute of non-human systems? At present, *meaning,* unlike information, is not quantifiable. The term "sophistication", a measure of meaningful complexity has been derived from computation and algorithmic complexity theory[8]. The classical Kolmogorov-Chaitin measure of complexity as a minimal description readable by a Universal Turing Machine does not relate the meaning of such a description. According to the Kolmogorov-Chaitin formulation, maximum complexity is achieved by randomness. In order to account for the meaning of a description, the theory must be accommodated to distinguish in a non-arbitrary fashion between the program and the data components present in a minimal description. Sophistication, then, is defined as the length of the minimal *program* component in a minimal description. This definition is akin to Bennett's notion of the "logical depth"[9] of an object based on the time needed by an evolutionary process to produce that object. These ideas are formal attempts to clarify ideas about logical properties of organized and Self-Organizing systems. However, they can hardly be applied to actual systems which are too complex to be

8. Koppel M. (1987). Structure. In R. Herken (ed.), *The Universal Turing Machine: A Half-Century Survey.* Oxford University Press, 435-452; Atlan H. and Koppel M. (1990). The Cellular Computer DNA: Program or Data. *Bull. Mathem. Biol.,* **52**, 335-348.

9. Bennett C. (1989). On the Logical 'Depth' of Sequences and their Reducibilities to Incompressible Sequences. In R. Herken (ed.), *The Universal Turing Machine: A Half-Century Survey.* Oxford University Press.

uniquely described by computable algorithms. Nevertheless, *meaning* is detectable, at least in principle through interpretation of its effects on a receiver.

The meaning of a message to an outside observer can be inferred by the effect of the message on the system that receives the message, irrespective of whether the receiver is human or not. The message's "objective" meaning can be seen as the reaction to the message, be it an observable change in behaviour or a covert change in state (of a human's mind, for example). Although we may not know mechanistically how the message is connected to the response, the correspondence of the response to the message is the objective meaning of the message.

Meaning is here defined without recourse to consciousness, intentionality or any of the other attributes of human mental activity. The generic sense of the *meaning* of a piece of information is the impact of that information on any chosen referent. *Meaning* is thus referential and contingent. A given immune response may mean one thing to the bacterium that bears the target antigen (death), another thing to the responding host (life), and yet another thing to certain cells within the infected tissue (healing). To the extent that a human has chosen a particular point of reference, there can be no rational *meaning* that is independent of a human observer. Nevertheless, human observation and human interpretation of the facts are not the same as human invention or human intervention to create facts. Meaning as we define it here is a functional relationship to a referent; what the information means to the referent is what the information does to the referent, no more and no less. Note that we define the information's *function* as what the information does, how it works. The teleological sense of *function* as *purpose* or *intention* is out.

Granted that a fundamental unit of information for the immune system is an antigen receptor, what reference points external to the antigen can serve the immune system to endow the antigen with meaning? An antigen is *perceived* by the immune system when an epitope of the antigen is bound by an antibody or a lymphocyte receptor; but what is the *meaning* of the antigen to the system? To reframe the question of *meaning,* how does the immune system "know" how to respond to the antigen? Is there a point of reference that could allow the immune system to vary its response to the antigen, to interpret as it were, the antigen's meaning?

6. THE CLONAL SELECTION OF MEANING: SELF-NOT-SELF DISCRIMINATION

Burnet, in his clonal selection theory, offered an external reference point to solve the problem of immune meaning, although he never formulated the concept of external reference or defined meaning as such. Nonetheless, for Burnet, the meaningful reference for any antigen was essentially one: was the antigen self or was it not-self[10]. The discrimination between self and not-self was seen by Burnet, and by most immunologists even today[11], as the primary function of the immune system. Molecules that originate from the body, the molecular self, are to be ignored by the system while molecules foreign to the body, the not-self, are to be rejected by an effective immune response. Thus the source of the meaning of any antigen depends on whether the antigen be self or not-self. But how is this self-not-self distinction to be made when an antigenic epitope is merely a fragment of molecular conformation unable, by itself, to declare its origin? Inherent in Burnet's theory is an answer that, in a wondrously thrifty way, links the creation of immune meaning (self-not-self discrimination) to the creation of immune information (the process of clonal selection). Burnet proposed that early in the course of their differentiation in the body, newborn lymphocytes are triggered to die whenever their receptors bind to an antigen. In contrast, the mature lymphocytes that meet their complementary antigens later in development do not die, but are stimulated to proliferate and produce their antibodies or other effector molecules. Now, the only antigens available to newborn lymphocytes are self antigens; therefore contact with self-antigens during ontogeny kills any lymphocyte clones that might possibly recognize the self. Hence, the reference point for self-not-self discrimination, the essence of immune meaning, is a product of the lymphocyte's history of being born into a context of self-antigens. Once past this filter, any antigen capable of being recognized by the mature lymphocyte must have been absent from the self-context into which the lymphocyte was born. Such an antigen, by definition, must be foreign and worthy of rejection. The filter of clonal suicide in ontogeny together with positive clonal selection in maturity both creates information and guarantees achievement of meaning, the discrimination between the self and the not-self. Burnet offers a view of the lymphocytes as an army that is taught during basic training what not to

---

10. Burnet F.M. (1959). *The Clonal Selection Theory of Acquired Immunity,* Cambridge: Cambridge University Press.
11. Klein J. (1982). *Immunology: the Science of Self-Nonself Discrimination.* New York: John Wiley.

shoot, but when at the front, to shoot all the rest. The correspondence between antigen stimulus and immune response is seen as a reflex. Meaning is built-in. The logic and parsimony of Burnet's theory has made it the paradigm that has influenced immunological research and immunological interpretation for three decades. The problem, however, is that the immune system is more complicated than anticipated by Burnet because it must solve more complex problems than dreamed of in Burnet's philosophy.

7. THE CHALLENGE OF NATURAL AUTOIMMUNITY

Self-not-self discrimination as the reference point for immunological meaning is rendered moot by the discovery of two facts: the immune repertoires of healthy individuals are filled with lymphocytes whose receptors can perceive self-epitopes and the infectious invaders, which must be rejected by the system, often express epitopes identical to or cross-reactive with self-epitopes of the host[12]. Perhaps the two facts are related; because of the evolutionary conservation of key genetic modules in multigene families, the immune system cannot afford to be blinded to epitopes that look like the self. Speculations aside, the fact is that natural autoimmunity does exist. Hence the self cannot be distinguished from the not-self solely on the basis of repertoire purging; the self is not antigenically unique and self-purging is not complete. Therefore, the problem of meaning cannot be solved by antigen receptors alone. Since an antigen might originate, for example, from an inert piece of food, from a virulent virus or bacterium, from a cancer cell, or from a healthy cell of the body, the immune system has to implement a response appropriate to the nature of the threat, or non-threat. We now know that there are many alternative types of response (including non-response) available to the immune system: cytotoxic T cells, helper T cells that can secrete different cytokines, a variety of antibody isotypes, anergy, "programmed" cell death, and more. In other words, the meaning of an antigen to the system is discernible in the type of immune response produced, not merely by whether or not the antigen is perceived by the receptor repertoire[13]. Because the meaning of the antigen is defined by the type of response that follows perception of the antigen, there is indeed a response repertoire and not only a receptor

---

12. Cohen I.R. (1992). The Cognitive Principal Challenges Clonal Selection. *Immunol. Today,* **13**, 441-444; Cohen I.R. (1992). The Cognitive Paradigm and the Immunological Homunculus. *Immunol. Today,* **13**, 490-494.
13. *Ibidem.*

repertoire. Contrary to the logic of clonal selection, the body's soldiers are capable of deploying an array of different weapons ranging from silence to knives to guided missiles against citizens of the body state as well as against the body's invaders.

There have been various attempts to qualify receptor behavior so that it might predicate the type of response and so prove meaningful. Melvin Cohn has developed the idea of the dependency of the effector response on reception by the lymphocyte of two signals rather than only one signal[14]. The first signal is the antigen epitope and the second signal is not an antigen, but a cytokine produced by a helper T cell that itself has been activated by an epitope. Cohn sees this two signal model as a way of achieving the goal of self-not self discrimination[15]. Cohn's model, however, suffers the same flaw as does Burnet's theory in ignoring the existence of natural autoimmunity. Niels Jerne proposed that the nature of the immune response could be regulated by a network of anti-idiotypes, clones with receptors capable of recognizing the receptors of other clones[16]. The Cohn and Jerne formulations complicate the behavior of the lymphocyte repertoire in that the lymphocytes interact with each other, as anti-idiotypes or as helpers, in addition to their interactions with the antigens. However, adding diversity and complexity to the organization of the repertoire may lead to more information, but never to meaning. Meaning must be created by relating the antigens seen by the receptor repertoire to something else, something outside of the repertoire.

8. THE COGNITIVE CREATION OF MEANING

If the lymphocyte soldiers are not merely instinctive shooters, but must exercise options about which target antigens should be shot and with what weapons, then some form of cognitive process is required to sort out the possible meanings of the antigen. The response is not a deterministic reflex. One of us (Cohen) has proposed a cognitive paradigm of the immune system[17]. The immune system can respond to a given antigen in various

14. Cohn M. (1994). The Wisdom of Hindsight. *Annu. Rev. Immunol.*, **12**, 1-62.
15. Jerne N.K. (1984). Idiotypic Networks and Other Preconceived Ideas. *Immunol. Rev.*, **79**, 5-24.
16. *Ibidem.*
17. Cohen I.R. (1992). The Cognitive Principal Challenges Clonal Selection. *Immunol. Today,* **13**, 441-444; Cohen I.R. (1992). The Cognitive Paradigm and the Immunological Homunculus. *Immunol. Today,* **13**, 490-494.

ways, it has "options". Thus, the particular response we observe is the outcome of an internal process of weighing and integrating information about the antigen. The flavor of (mechanical) cognition is evident in the probabilistic nature of the response. For example, 10 inbred, homozygotic mice of the same age and sex, raised in the same cage, and fed the same diet will each respond more or less differently to immunization with the same antigen; the response phenotype of the "identical" mice will show an apprecdeviation. The immune system exercises cognition by the interpolation of a level of information processing intrinsic to the system between the antigen stimulus and the immune response. A cognitive immune system organizes the information borne by the antigen stimulus within a given context and creates a format suitable for internal processing; the antigen and its context are transcribed internally into the "machine language" of the immune system.

The "machine language" of the central nervous system is composed of a network of electrical and chemical signals connecting neurons. What is the internal "machine language" of the immune system intercalated between the antigen and the response? As pointed out above, the receptor repertoire for antigens is somatically generated by random genetic recombinations and mutations of the receptor genes. In contrast, the molecules responsible for internal information processing are encoded in the individual's germ line. These molecules are many, but unlike the receptors for antigens, they are not diverse within the individual. These molecules include the enzymes and organelles for antigen uptake and processing, the major histocompatibility complex (MHC) molecules for antigen presentation, the cytokines that orchestrate inflammation and the suppression of inflammation, the cell-interaction and cell-adhesion molecules that organize cell-to-cell interactions and cell migrations. In short, the transcription of the antigens into processed peptides embedded in a context of germ-line ancillary signals is the "machine language" into which antigens are transcribed by the immune system.

An antigen's meaning is interpreted by the lymphocytes through the antigen's transformation into the string of ancillary germ-line molecular signals that accompany the antigen epitope[18]. The germ-line signals reflect the state of the tissues (inflamed or not) and critical features of the antigen (soluble or particulate; associated with a bacterial cell wall or virus or not; accompanied by other antigens or not; and so forth). The context of

18. Cohen I.R. (1992). The Cognitive Principal Challenges Clonal Selection. *Immunol. Today,* **13**, 441-444; Cohen I.R. (1992). The Cognitive Paradigm and the Immunological Homunculus. *Immunol. Today,* **13**, 490-494.

cytokines and cell-interaction molecules endows the antigen epitope with its meaning because these added signals influence strongly the nature of the immune response to the epitope. Immune meaning is thus created by the *association* of the antigen epitope with a *context* of molecular signals generated by the germ line. The germ line signals, of course, have evolved as a record of the evolutionary *history* of the species with infectious agents.

Meaning is thus created at the interface of the individual's private receptor repertoire with his species germ-line signals. The immune system "interprets" the meaning of antigens by connecting two subsystems, the individual and the species, that have each self-organized on vastly different scales of time and space. The central nervous system does the same; it appends the individual's experience with his environment to a set of "instinctual" behaviors encoded in the germ line of the species.

9. THE LANGUAGE METAPHOR

The grammatical structure of human language can be used as a metaphor to illustrate how meaning can be formed through the apposition of independently organized elements. A complete sentence, a universal coin of linguistic meaning, can be characterized as containing two elements: a noun phrase and a verb phrase[19]. The noun phrase is the designated *subject* of the action or description; the verb phrase is the predicate that is connected to the subject. A sentence bears its fullest meaning when properties or actions are predicated about a subject. Although individual words and even letters can have their own meanings, an unconnected subject or an unconnected predicate will tend to mean less than does the sentence generated by their connection. Just as a sentence creates meaning when it connects a subject to its predicates, an antigen gains meaning by its connection to a particular set of germ-line signals (MHC molecules, Cell Differencition markers, cytokines, cell-interaction molecules) that elicit the specific type of immune response (see Figure 2). Thus the antigen is like the noun, the subject of the immune sentence, while the germ-line signals are like the predicates, which function to allow the system to choose a particular type of response from among a relatively standardized set of responses. Meaning, the type of immune response, is the outcome of the concrete connection between the subject and the predicate signals.

19. Pinker S. (1994). *The Language Instinct.* New York: William Morrow and Company.

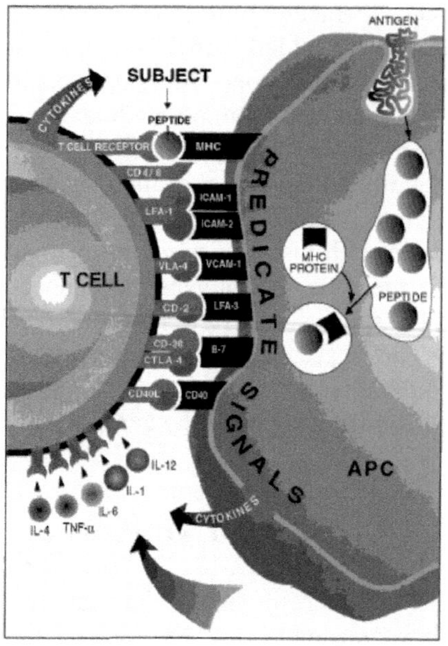

Figure 2. According to the language metaphor of the immune response, an informative immunological "sentence" uttered by an antigen presenting cell (APC) to a T-cell could be described as containing a subject, designated by the antigen's peptide epitope, and a predicate, created by the string of ancillary signals. These predicate signals can induce the T cell to produce various alternative biological effects in response to the subject antigen; become, in the parlance of immunology, a TH1 cell, become a TH2 cell, become a suppressor cell. The nature of the immune response will be markedly influenced by the proportion of each type of T cell active in the response. In this way, the APC, including the tissues, help predicate the outcome of the T-cell response to the invader. The T-cell response, in turn, predicates the behavior of the tissues. Immune communication is a dialogue.

The subject-predicate dichotomy is not only a convenient metaphor, the dichotomy is deeply structured within the immune system. The genetic basis for the perceptions of subjects and predicates is quite different. Subjects, the antigen epitopes, are recognized by a vast repertoire of T and B cell antigen receptors, a practically unlimited set of diverse molecules created and deployed anew in each individual by somatic recombination and mutations of mini-gene elements. In contrast, predicates, the ancillary tissue signals and their receptors, are encoded in a large, but limited number of genes inherited in the species germ line. In other words, the nouns of immune communication, an open class of signals, are the products of the individual's immune experience, whilst the predicates of immune communication are the more restricted products of the evolutionary experience of the species.

It is not yet known exactly how the string of predicate signals determines the type of immune response mounted by the immune system against an invader. A number of the predicate signals, particularly the cytokines, manifest pleiotropic effects on different cells and tissues; some cytokines seem to be redundant and act like other cytokine molecules, for example Interleukin-1 and Tumor Necrosis Factor; some cytokines inhibit the effects of other cytokines, for example Interleukin-4 and Interferon[20].

Although the number of these germ-line predicate signals is limited, it is large enough to generate large numbers of different patterns of activity by combinatorial associations. Such particular mixtures of predicate signals associated with changes in state of different populations of T and B cells, may be the substrate for a learning mechanism in networks of lymphocyte populations similar to those described by neural network computation[21] (see Figure 3).

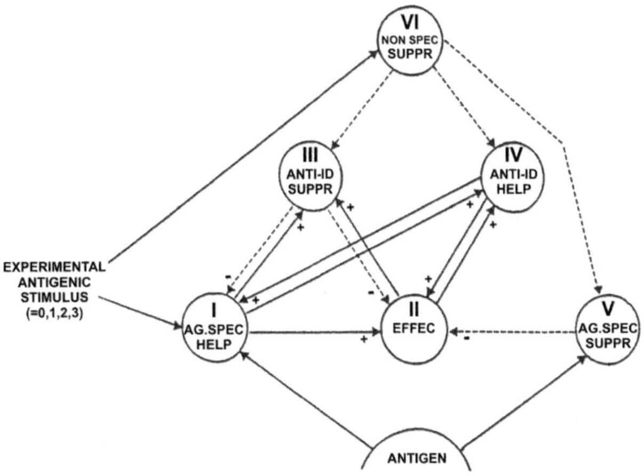

Figure 3. A network of six formal "neurons" representing six different populations of lymphocytes activated in an experimental autoimmune disease. These populations have been identified as: antigen specific helpers and antigen presenting cells (I); effector cells responsible for a pathogenic reaction with the self-antigen when activated (II); anti-idiotypic helper and suppressor cells (respectively III and IV); antigen-specific suppressors (V); and non-specific suppressors (VI) directly activated by the pathogenic state of inflammation. Experimentally manipulating the antigenic stimuli can drive the network into different stable states (attractors) corresponding to different clinical outcomes (healthy or diseased), which depend on the stable state of the effector cell population in resting, proliferating, or full

20. Mosmann T.R. and Coffman R.L. (1989). TH1 and TH2 cells: Different Patterns of Lymphokine Secretion Lead to Different Functional Properties. *Annu. Rev. Immunol.,* **7**, 145-173.

21. Romagnani S. (1994). Lymphokine Production by Human T Cells in Disease States. *Annu. Rev. Immunol.,* **12**, 227-257.

activity. In each stable atate of the network, the cells achieve a new pattern of activity and of cytokine production which constitute new connections in the network. This can be viewed as a possible mechanism for learning and for distributed memory: the new structure of the network is responsible for new stable states (such as resistance to disease) in response to repeated stimuli, while it is itself the result of the history of previous responses of the network to the antigen (for the details of the neural network computation, see ref. 24).

10. COGNITIVE SELF-NOT-SELF DISCRIMINATION

But what about the definition of the self? Given the existence of natural autoimmunity and the sharing of epitopes by host and parasite, how does the individual immune system discriminate functionally between the self and the not-self? How can the immune system adjudicate the molecular self?

The answer is simple in principle; in detail, it is still beyond a satisfying molecular description. In principle, we can say that the immune self is not a stable, "punctuate" chemical entity[22]. The self is rather the outcome of a dynamic process of continuing challenges and responses . The definition of the immune self is a process that takes place over time within the anatomic confines of one individual. Sometimes self antigens are attacked as a provisional measure to rid the body of invading organisms, aberrant tumor cells, or virus-infected cells. But the same autoimmunity, when properly turned on and then off, can cure disease. Autoimmunity, when not properly turned off, can cause autoimmune disease[23]. The turning on and turning off are cognitive events based on the integrations of strings of signals both at the level of the individual lymphocyte and at the level of populations of lymphocytes each capable of diverse behaviors[24]. The immune self does feature a collective of self antigens, the immunological homunculus[25], but this collective of self antigens is in dynamic flux and gains its meaning by associations of the self antigens with contexts of predicate signals. To use the language metaphor, the immune self is not an immutable subject defined by a fixed set of self-antigen nouns; the immune self is rather like a set of evolving immune sentences, self-antigens dynamically connected to

---

22. Atlan H. and Cohen I.R. (1992). Paradoxical Effects of Suppressor T Cells in Adjuvant Arthritis. In A.S. Perelson and G. Weisbuch (eds), *Theoretical and Experimental Insights into Immunology*. Berlin: Springer Verlag, 379-395.
23. Cohen I.R. (1992). The Cognitive Paradigm and the Immunological Homunculus. *Immunol. Today,* **13**, 490-494.
24. Cohen I.R. and Young D.B. (1991). Autoimmunity, Microbial Immunity and the Immunological Homunculus. *Immunol. Today,* **12**, 105-110.
25. Cohen I.R. (1992). The Cognitive Paradigm and the Immunological Homunculus. *Immunol. Today,* **13**, 490-494.

particular predicating signals[26]. In other words, the immune self is not the subject of a story, the immune self is the story, a story that writes itself meaningfully from cover to cover. Like the psycho-social I, it is self-organization of a self.

Aknowledgements: H. Atlan is director of the Human Biology Research Center and Ishaiah Horowitz scholar in residence at Hadassah University Hospital in Jerusalem. I.R. Cohen is the Mauerberger professor of immunology and the director of the Robert Koch-Minerva Center for research in autoimmune diseases at the Weizmann Institute of Sciences.

REFERENCES

Amit D.J. (1989). *Modeling Brain Function. The World of Attractor Neural Networks.* Cambridge: Cambridge University Press.

Atlan H. (1972). *L'Organisation biologique et la théorie de l'information.* Paris: Hermann.

Atlan H. (1983). Information Theory. In R. Trappl (ed.), *Cybernetics.* New York and Berlin: Hemisphere Publ. and Springer Verlag, 9-41.

Atlan H. (1992). Self-Organizing Networks: Weak, Strong and Intentional, the Role of their Underdetermination. *La Nuova Critica* I-II, Quaderno 19-20, Rome, 51-70.

Atlan H. and Cohen I.R. (1992). Paradoxical Effects of Suppressor T Cells in Adjuvant Arthritis. In A.S. Perelson and G. Weisbuch (eds), *Theoretical and Experimental Insights into Immunology.* Berlin: Springer Verlag, 379-395.

Atlan H. and Koppel M. (1990). The Cellular Computer DNA: Program or Data. *Bull. Mathem. Biol.,* **52**, 335-348.

Atlan H. (1974). On a Formal Theory of Organization. *J. Theoret. Biol.,* **45**, 295-304.

Bennett C. (1989). On the Logical 'Depth' of Sequences and their Reducibilities to Incompressible Sequences. In R. Herken (ed.), *The Universal Turing Machine: A Half-Century Survey.* Oxford University Press.

Burnet F.M. (1959). *The Clonal Selection Theory of Acquired Immunity,* Cambridge: Cambridge University Press.

Cohen I.R. (1992). The Cognitive Paradigm and the Immunological Homunculus. *Immunol. Today,* **13**, 490-494.

---

26. Cohen I.R. (1995). Treatment of Autoimmune Disease: to Activate or to Deactivate? *Chem. Immunol.,* **60**, 150-160.

Cohen I.R. (1992). The Cognitive Principal Challenges Clonal Selection. *Immunol. Today,* **13**, 441-444.

Cohen I.R. (1995). Language, Meaning and the Immune System. *Isr. J. Med. Sci.,* **31**, 36-37.

Cohen I.R. (1995). Treatment of Autoimmune Disease: to Activate or to Deactivate? *Chem. Immunol.,* **60**, 150-160.

Cohen I.R. and Young D.B. (1991). Autoimmunity, Microbial Immunity and the Immunological Homunculus. *Immunol. Today,* **12**, 105-110.

Cohn M. (1994). The Wisdom of Hindsight. *Annu. Rev. Immunol.,* **12**, 1-62.

Jerne N.K. (1984). Idiotypic Networks and Other Preconceived Ideas. *Immunol. Rev.,* **79**, 5-24.

Klein J. (1982). *Immunology: the Science of Self-Nonself Discrimination.* New York: John Wiley.

Koppel M. (1987). Structure. In R. Herken (ed.), *The Universal Turing Machine: A Half-Century Survey.* Oxford University Press, 435-452.

Lwoff A. (1962). *Biological Order.* Cambridge, Mass.: MIT Press.

Mosmann T.R. and Coffman R.L. (1989). TH1 and TH2 cells: Different Patterns of Lymphokine Secretion Lead to Different Functional Properties. *Annu. Rev. Immunol.,* **7**, 145-173.

Nossal G.J.V. (1993). Life, Death and the Immune System. *Scientific American,* **269**, 21-30.

Pinker S. (1994). *The Language Instinct.* New York: William Morrow and Company.

Romagnani S. (1994). Lymphokine Production by Human T Cells in Disease States. *Annu. Rev. Immunol.,* **12**, 227-257.

Rumelhart D.E. and McLelland J.L. (PDP Research Group) (1986-1987). *Parallel Distributed Processing,* I and II. Cambridge: MIT Press.

Schatz D.G., Oettinger M.A., Schissel M.S. (1992). V(D)J Recombination: Molecular Biology and Regulation. *Annu. Rev. Immunol.,* **10**, 359-383.

Shannon C.E. (1948). A Mathematical Theory of Communication. *Bell Syst. Tech. J.,* **30**, 50-64.

Tauber A.I. (1994). *The Immune Self: Theory or Metaphor?* Cambridge: Cambridge University Press.

Weaver W. and Shannon C.E. (1949). *The Mathematical Theory of Communication.* Urbana: University of Illinois Press. A shorter version of this article has been published: Atlan, H. and Cohen, I. (1998). Immune Information, Self-Organization and Meaning, in *International Immunology*, Vol. **10**, 6, 711-717.

II.HISTORIC APPROACH

A. Early Philosophical Conceptualizations

GERTRUDIS VAN DE VIJVER

KANT AND THE INTUITIONS OF SELF-ORGANIZATION

1. INTRODUCTION

For a number of years, historical studies of cybernetics (Beaune 1980; Dupuy 1985; Lévy 1985, 1985a, 1985b, 1985c; Livet 1985; Pask 1992; Van de Vijver 1991, 1992) and the morphodynamic approach in cognitive sciences (Petitot 1985, 1985a, 1985b, 1991, 1992) have sparked off renewed interest in the philosophy of Kant. In particular, Kant's approach of the issue of organization (the problem of purposiveness in nature) and the arguments adduced by him in support of his postulation of the impossibility of objectively apprehending natural purposes, are today being analysed and linked to current achievements in the fields of mathematics and philosophy. Studies in cybernetics *par là* provide an interpretation of the topical issue of self-organization, supplying a pertinent description in Kantian terms, through which major contemporary positions on the issue of purposiveness in nature are articulated.

In this article we propose to delineate the context of the issue of teleology in Kant, in order to acquire a better understanding of just what is at stake in certain current-day viewpoints, as well as in order to demonstrate the reasons for and the means of transcending the options as defined by Kant. Setting out from the Kantian context, among other things we shall be dealing with autopoietic interpretation and a number of current epistemological perspectives on the subject of teleology, in cybernetics and elsewhere.

2. KANT'S BASIC POSITION WITH REGARD TO THE ISSUE OF PURPOSIVENESS IN NATURE

In the second part of the *Critique of Judgment* (Critique of teleological judgment), Kant deals with purposiveness in nature and with the way in

B. Feltz, M. Crommelinck and P. Goujon,(eds.), Self-Organization and Emergence in Life Sciences: 143-161.

which we should understand the forms and the objects of nature[1]. Here, his chief concern is to understand the various aspects of the relationship between the organization of nature and our judgment.

As is commonly known, in the first part of this critique, Kant describes the issues of the beautiful and the sublime. Kant had said that beautiful forms seem to us as if they have been specifically designed for our judgment and in the introductory paragraph to the second part, he reminds us of this: they are forms that "are commensurate with our judgment because, as it were, their diversity and unity allow them to serve to invigorate and entertain our mental powers" (CJ, § 61, 235; Ak., Bd. V, 359). What is more, the representation of beautiful forms is something that resides within us, and hence it is readily conceivable, even *a priori,* "how such a presentation could be fit and suitable for attuning our cognitive powers in a way that is purposive within [us]" (*ibidem*).

But, says Kant, what of the objects of nature, that reciprocally make use of certain means with a view to purposes and of which the possibility cannot sufficiently be understood other than through the intervention of a kind of final causality? Indeed, the central issue is: "how purposes that are not ours, and that we also cannot attribute to nature (since we do not assume nature to be an intelligent being) yet are to constitute, or could constitute, a special kind of causality, or at least a quite distinct lawfulness of nature" (CJ, § 61, 235-236; Ak., Bd. V, 359).

As such, Kant deems organisms, the natural purposes, to be unknowable by means of the principles he described in the *Critique of Pure Reason:* natural purposes cannot be known through concepts and it is impossible to *explain* them proceeding from a final causality which would be described by *a priori* concepts. Nothing comes to ground *a priori* the existence of some form of final causality, any more than of a distinct lawfulness that would follow from it.

However, the fact that it is impossible to explain these forms of causality, in no way implies that they cannot be described in terms of final causes, in order to make them more intelligible: one has to proceed as *if* the development of natural purposes itself complies with certain purposes. Proceeding from the concept of natural purpose, it is possible to judge organized systems, but they cannot be known objectively. In other words,

1. We are using the English translation of W.S. Pluhar, *Critique of Judgment* (including the first introduction), Indianapolis, Cambridge, Hackett Publishing Company, 1987. From now on, we abbreviate the *Critique of Judgment* as CJ. References to the English translation are followed by the edition of the Academy, given as Ak., followed by the volume and the page number.

the concept of natural purpose may be employed by presuming a regulative principle, introduced there to judge phenomena, without however being able to appeal to a constitutive principle that allows us to infer these objects of nature, setting out from final causes. The concept of natural purpose pertains to reflective judgment, not to determinative judgment. The organisms escape all mechanical laws, they cannot be described adequately, nor be explained in such a way.

This is the basic position of Kant with regard to organisms, a position that is dictated by his way of broaching the issue of knowledge in the *Critique of Pure Reason,* which implies that organisms essentially elude know ability through concepts.

Kant is to apply this position, which testifies to a dualism between mechanism and teleology, in a twofold analogous operation: (i) the purposiveness in an organism cannot be understood, other than by analogy with an *a priori* concept, which is to say a plan or an intention, and (ii) the systematic unity of empirical knowledge is organic in nature.

We shall be dealing with these two aspects in five steps. Indeed, it is imperative that we understand in a more detailed manner (i) that which Kant labels a *natural purpose,* (ii) how he introduces a *principle,* which he considers as the definition of a natural purpose, and which brings with it the epistemological consequences that we have sketched above, (iii) *why,* with regard to the teleological judgment, he introduces the distinction between explaining and judging, that is, what is the status of the difference between reflective judgment and determinative judgment, between the regulative use of concepts and their constitutive usage, or still, what is his basic argument in considering organisms as highly particular systems, (iv) what role does the *Idea of the unity of empirical knowledge* play, and (v) which ontological implications may be inferred from this conception of unity, in other words, what does the *Idea of the unity of nature* represent and what is the function of this Idea. Here, we must take into account the issue of ontology and as such we shall be considering Kant's elaborations in his *Opus Postumum.*

3. NATURAL PURPOSES

To Kant, organisms must be seen as "Naturzwecke", as forms that are characterized by an intrinsic purposiveness. That these are natural purposes, may be inferred from the fact that these organisms present themselves as systems that hold within them the principle of their organization. From the fact that they present themselves to us as unified entities, as autonomous totalities, we must recognize that they are only conceivable as purposes.

In the Analytic of teleological judgment, more particularly in § 64 and § 65, which are respectively entitled "On the Character Peculiar to Things Considered as Natural Purposes" and "Things Considered as Natural Purposes Are Organized Beings", Kant provides us with the following specifications:

1. "(...) a thing exists as a natural purpose if it is both cause and effect of itself (although in two different senses)" (CJ, § 64, 249; Ak., Bd. V, 370, italics supplied).

A natural purpose must therefore refer to itself as cause and as effect. Two requirements are added further on.

2a. "*First,* the possibility of its parts (as concerns both their existence and their form) must depend on their relation to the whole" (CJ, § 65, 252; Ak., Bd. V, 373).

It is impossible to conceive of an organism while supposing that no single purpose is at work in it. It is because the thing itself is a purpose that we must think it on the basis of a concept or an Idea which *a priori* determines all that must be comprised in it. However, if this were the sole way to think the natural purpose, the latter would naturally be comparable to a work of art, since the causality that intervenes in the work of art (in the production of and the liaison between parts) is determined by the Idea of a whole. So, something additional is required to have a natural purpose, that is, the fact that a natural purpose holds within itself, in its intrinsic possibility, a relation with purposes. Hence the second requirement:

2b. "This second requirement is that the parts of the thing combine into the unity of a whole because they are reciprocally cause and effect of their form" (*ibidem*).

It is in this way that Kant perceives the distinction between a work of art and a natural purpose. This second condition indeed implies that the Idea of the whole does not determine the form and the relations between parts as a cause, but simply as a principle of knowledge for whomsoever judges natural purposes. The Idea of the whole is not to be thought along the lines of efficient causality. "In such a product of nature, just as each part exists only *as a result* of all the rest, so we also think of each part as existing *for the sake* of others and of the whole, *i.e.* as an instrument (organ). But that is not enough (for the part could also be an instrument of art, in which case we would be presenting its possibility as depending on a purpose as such). Rather, we must think of each part as an organ that *produces* the other parts (so that each reciprocally produces the other). Something like this cannot be an instrument of art, but can be an instrument only of nature, which supplies all material for instruments (even for those of art). Only if a product meets that condition, and only because of this, will it be both an *organized* and *Self-Organizing* being, which therefore can be called a *natural purpose*" (CJ, § 65, 253; Ak., Bd. V, 373-374, italics supplied).

3.1 Autopoietic Parenthesis: the Paradox of Self-Organization

It is interesting to briefly compare what Kant says here with Maturana's and Varela's definition of autopoiesis which, as we shall see, with respect to the Kantian view is not so original. Let us set out from Varela's definition in his book on the principles of biological autonomy: "An autopoietic system is organized (defined as a unity) as a network of processes of production (transformation and destruction) of components that produces the components that: (1) through their interactions and transformations continuously regenerate and realize the network of processes (relations) that produced them; and (2) constitute it (the machine) as a concrete unity in the space in which they exist by specifying the topological domain of its realization as such a network" (Varela 1979, 13).

In either case, that of natural purposes or that of autopoiesis, we are faced with a *paradoxical* situation. The paradox consists in the fact that in apprehending the natural purposes that present themselves to us in nature, we can only think them in terms of a reciprocal determination between the parts and the whole. Only in this way may we conceive their *possibility*. The parts seem to develop according to a certain plan, according to a conception of what the whole should be like, but at the same time, we are unable to affirm the presence or the existence of this determination by the totality at the moment when a dynamic comes about between the parts with a view to a purpose. In affirming the existence of a final cause for the things of nature, we run the risk of seeing efficient causality and final causality along the same lines. Indeed, we would in that case have to accept that final causes are required to explain what takes place during the development of an organism, so inverting cause and effect in such a way that future events could determine that which happened in an anterior time. To Kant, this is wholly unacceptable, amounting to the introduction of a determinative judgment in place of a reflective judgment, thereby making a regulative principle into a constitutive principle.

So, for Kant, to understand the enigma of intrinsic purposiveness, is to understand the mysterious reciprocal determination between the parts and the whole. Today, in a similar vein, we are looking at understanding self-organization in the strong sense, that is, understanding how the purpose to be achieved by a system, the purpose which defines its form or its structure, is an emergent property in the evolution of the system (Atlan 1991, 77 sq.). *The parts must mutually determine each other as well as the whole, and the parts must be determined by a whole, but this whole must determine them in a sense before even existing as a whole as such.* Understanding the problem of purposiveness in biology is therefore, for Kant as well as for Atlan, Varela and Maturana, to understand two things: (i) the origin of living organisms, and therefore the determination by a whole which is not yet in existence — the origin of the program or the representation, (ii) the teleological functioning of certain systems — the functioning in accordance with a program or a representation.

3.2 Cybernetic Parenthesis

The Kantian definition of intrinsic purposiveness also allows us to clearly distinguish between the results of first order cybernetics and that of second order cybernetics (cf. Van de Vijver 1991, 1992).

The principal objective of first order cybernetics was to model teleological behavioral patterns in the machine. So, it was concerned with an externally defined purposiveness, to be situated in the register of control: the system itself does not develop the purposes to be attained, these purposes being imposed upon it from the outside. This applies to systems that are described in terms of feedback, but it equally applies to leading cognitivist theories that consider the cognitive and the intentional in terms of representations and programs, or, in the case of the genetic approach in biology, that characterize the living based on the notion of programme.

It is relatively easy to conceive of how to resolve the second part of the paradox we described above, proceeding from first order cybernetics. If we presume the existence of internal representations, and if we implement this 'determination by the whole' into a machine by way of a program, we are providing an answer to the second part of the problem of intrinsic purposiveness, that is, the existence of behavioral patterns that are apparently teleological. With this, however, the first part of the paradox is still insufficiently addressed, since the origin of intrinsic purposiveness remains unanswered. This is why, as Canguilhem and many others have variously put it, the concept of the machine is basically a teleological concept.

Second order cybernetics was born out of questions as to the origin of purposiveness. The theory of autopoiesis has shared essential moments with the history of second order cybernetics. Here, as with Kant, teleology was first interpreted under the sign of autonomy. The issues addressed included those of self-organization, not only with regard to the way in which a representation could be considered as an essential factor in the intentional or teleological behavior of a system, but also to the way in which this program found its origin inside the system, without external purposiveness. The principles of complexity and of self-organization that were introduced by second order cybernetics, aimed at resolving, at least in part, this issue of origin.

It can be demonstrated that the epistemological deadlocks encountered by cybernetics, are identical to the ones encountered by Kant. Here, the constructivist and the positivist positions are the two basic positions (cf. Van de Vijver 1991). But first let us see how Kant himself viewed the possibility of knowing or judging natural purposes. Let us see how he envisaged resolving the paradox.

4. THE TELEOLOGICAL PRINCIPLE

The principle for judging intrinsic purposiveness, which is at the same time the definition of natural purposes, introduces the distinction between knowing (explaining) and assessing, and immediately follows on the specifications as laid down by Kant with respect to natural purposes. This is the principle which for Kant expresses the paradox and which, simultaneously, as a principle, contains the solution. Here it is:

2c. "An organized product of nature is one in which everything is a purpose and reciprocally also a means" (CJ, § 66, 255; Ak., Bd. V, 376, italics supplied).

This principle tells us that there is nothing in this product that is useless, without purpose, or susceptible to being attributed to a mechanism that operates blindly. It is derived from experience, Kant tells us, since it finds its source in observation, but because of the necessity and the subjective universality which that principle claims for such purposiveness — we shall be returning to this aspect further on —, it cannot have a mere empirical basis, it must be based on some *a priori* principle, even if the latter is but regulative and even if these purposes are only to be found in the Idea of the one who is judging. This is why this principle is a *maxim* of the judgment of the intrinsic purposiveness of organised beings.

So Kant resolves the 'paradox' by introducing the distinction between a whole which allegedly determines the parts at the level of efficient causality, and the concept of a whole of which we serve ourselves as judges (knowing subjects), in order to understand and judge the internal purposiveness of organisms. It is this distinction between the regulative usage of concepts and their constitutive use, the distinction between *determinative judgment and reflective judgment* (knowing and thinking), which allows us to evade the paradox which manifests itself here.

The teleological principle is a *necessary and subjective* principle of reflective judgment. It is not a necessary and universally objective proposition that is *a priori* knowable. A teleological principle presents the difficulty of always being particular, being of the order of "There is purposiveness", whereas a principle which would apply universally, would for example be "Every thing has its purpose". Since teleological judgments are unable to *a priori* determine (construct) their object, they are unable to attain objective universality; their universality is a subjective one. Reflective judgment therefore requires a transcendental, subjective principle: "Hence judgment must assume, as an *a priori* principle for its own use, that what to human insight is contingent in the particular (empirical) natural laws does nevertheless contain a law-governed unity, unfathomable but still conceivable by us, in the combination of what is diverse in them to [form] an experience that is intrinsically [*an sich*] possible. Now when we find in such a combination a law-governed unity cognized by us as conforming to a necessary aim that we have (a need for understanding), but at the same time as in itself [*an sich*] contingent, then we present this unity as a purposiveness of objects (of nature, in this case). Hence, judgment which with respect to things under possible (yet to be discovered) empirical laws is merely reflective, must think of nature with regard to these laws according to a *principle of purposiveness* for our cognitive power; and that principle is then expressed in the above maxims of judgment. Now this transcendental concept of purposiveness of nature is neither a concept [we] only think of the one and only way in which we must proceed when reflecting on the objects of nature with the aim of having thoroughly coherent experience. Hence it is a subjective principle (maxim) of judgment" (CJ, V, 23; Ak., Bd. V, 183-184, italics supplied).

This is one aspect of the analogous operation we referred to earlier: it is on the basis of an analogy with a final causality within us that we are able to judge, to assess natural purposes in terms of final causality. In this way, and only in this way, do we conceive of their possibility. The sole alternative to this approach lies in adopting the position of considering the organisms as responding to a blind mechanism — which is unacceptable to Kant. For him, natural purposes are related to a "*technique of nature*" — which is to say the causality connected with the form of products as purposes (CJ, First introduction, § VII, 407 sq.; Ak., Bd. V, 219' sq.). The technique of nature consists in a relation between the things and our judgment, in which one may only find the Idea of a purposiveness of nature. It should be distinguished from a "*mechanism of nature*" which concerns the relation between scientific knowledge and the relations that exist between the particular things in nature — what Kant calls logical purposiveness.

5. THE BASIC ARGUMENT FOR THE PARTICULAR STATUS OF NATURAL PURPOSES

The essential argument for the impossibility of objectively knowing natural purposes, as upheld by Kant, is that if in nature the *nexus effectivus* was simply followed, which is to say that if one simply considered nature in mechanistic terms, one would have to admit that nature would have been able to progress in thousands of other manners without ever attaining unity, following such a principle (CJ, § 61, 236; Ak., Bd. V, 360). So it is the *essential contingency,* which Kant also refers to as the excessive diversity of nature and of the forms of things of nature that are called natural purposes — a contingency that is defined relative to efficient causality, to the *nexus effectivus* — which compels us to accept in principle that it is impossible to explain them in terms of concepts, in terms of *a priori* concepts. Hence the definition of purposiveness Kant provides us with in the first introduction to the third critique: "(...) purposiveness is a lawfulness that [something] contingent [may] have [insofar] as [it] is contingent" (CJ, First Introduction, § VI, 405; Ak., Bd. V, 217')[2]. For Kant, the form of a natural purpose is impossible following simple natural laws, if not it would be senseless to conceive of its possibility only as a purpose. "(...) the form of such a thing is, as far as reason is concerned, *contingent* in terms of all empirical laws (...) Hence that very contingency of the thing's form is a basis for regarding the product as if it had come about through a causality that only reason can have" (CJ, § 64, 248; Ak., Bd. V, 370).

It is the essential contingency of the forms of nature that implies the impossibility to know them *a priori,* setting out from concepts, and that obliges us to add *meaning,* to convert in a way the intrinsic purposiveness of living organisms into subjective purposiveness, into a kind of intentionality (*formal subjective purposiveness*) which is linked with the way in which the subject in his judgment, comports himself with regard to the forms that have

2. A much more elegant translation is the one by J. McFarland: "Purposiveness is the conformity to law of the contingent as such." For this issue, also viz. McFarland, 1970, 77 sq.

an intrinsic purposiveness. Kant did not free himself from the analogy of a clock to think about the living. He could only conceive of purposiveness in terms of a manufacturer. But at the same time, he emphasised the intrinsic relation between teleological phenomena and systems on the one hand, and the necessity of adding meaning to the description of these systems on the other hand. It is as a "logic of meaning" that the *Critique of Judgment* is firstly able to develop into a Critique of Aesthetic Judgment and subsequently into a Critique of Teleological Judgment. "The genius of Kant reveals to him that denying the possibility of a relation between the particular and the universal comes down to denying communication (...) In terms of communication and intersubjectivity, Kant formulates the issue of classical metaphysics. And here we remain his inheritors" (Philonenko 1984, 12, our translation).

5.1 On the Horizon of Contemporary Teleology: Etiological Explanation

Certain contemporary authors actualize a number of elements of this "logic of meaning", by explicitly establishing the link between teleological description and teleological explanation, or, between the identification of a behavior as teleological and its explanation in teleological terms.

One such author for example is Larry Wright, who attempted to integrate the problem of teleology into analytical philosophy and who argued for an etiological interpretation of teleology, proceeding from some sort of "linguistic reconversion" of the issue. What Wright tells us is that, if we describe a behavior as teleological, we are at the same time providing a teleological description of this behavior. "When I say the rabbit is running in order to escape from the dog, I am saying *why* the rabbit is behaving as it is" (Wright 1976, 24, italics supplied). The characterization or description of a behavior in teleological terms, therefore, logically suffices to explain the presence and the form of the behavior. Or still, the teleological 'nature' of a behavior is a *logically sufficient* condition for its teleological explanation[3].

Other authors go further still, establishing the link between the so-called teleological nature of a system and the imputation of this state of things to a selection process. Wimsatt, for example, says: "(...) the weaker claim that

3. The distinction between a teleological *system* and a teleological *behavior* here plays an essential part. Larry Wright only deals with the description and the explanation of behavior. Questions that arise with regard to teleological *systems* are, at least partially, of a different kind, and are closer to the ambitions Kant set forth in his third Critique.

a system's owing its origin and form or that of its behavior to the operation of differential selection forces is a logically sufficient condition for the appropriateness of teleological explanations and talks of purposes or ends seems both interesting and promising" (Wimsatt 1972, 15, italics supplied).

It is easy to understand why these authors adopt such a position. The basic reason is that they interpret explanation in an *etiological* manner. We saw that a teleological or even functional description of an entity (system or behavior) was considered as logically sufficient for its explanation when the function or the purpose corresponds with the raison-d'être of this entity. The explanation is etiological: it explains why the entity exists and why it has a certain form. When the function or the purpose corresponds with the raison-d'être of this entity, it is obvious that the description of this function or of this purpose implies an explanation.

But this reasoning cannot be upheld in either of two cases: when the function or the purpose do not correspond with the raison-d'être of this entity (viz. Woodfield's critique 1976), or when the explanation is not etiological. On the other hand, an etiological position of this type is highly elegant when these two objections do not hold, as for example in the case of artificial systems that were conceived with a view to a purpose or a function (cf. Cummins 1975, 746). "Etiology seems relevant only insofar as it is expected to show how the part can advance the goals we pursue with the [total system]" (Boorse 1976, 82).

This allows us to once more render explicit the reason why Kant excluded the explanation for what he labeled natural purposes. Indeed we are looking at systems for which it is difficult, as opposed to artificial systems, to affirm the existence of a purpose. Therefore, a teleological explanation that would explain a natural purpose in terms of purposes corresponding with its raison-d'être, would be an explanation ... that would require explanation.

In the same vein, the Kantian conception of purposiveness strongly underlines the necessity of distinguishing between the fact of knowing natural purposes, and the fact of identifying them as such. However, this distinction at the same time attests to the particular status Kant conferred on organisms and to his interpretation of the mechanism. Here, again, we find the dualism which we set out from: if we label mechanical laws as blind and if organisms, by definition, are the result of laws that are not blind, it is obvious that we will never be able to explain scientifically the organisms concerned (cf. McFarland 1970, 97).

This dualism allows us to discern two aspects of purposiveness in the way it is proposed by Kant: (i) the "as if" — aspect connected with the way in which we judge natural purposes and with the unity of our scientific empirical knowledge; (ii) the aspect which we might label as ontological — which relates to all that escapes the mechanistic method, and which compels us to accept something more positive, something more than this heuristic status.

We now propose to analyze what Kant calls the Idea of systematic unity of knowledge and its relation with the organism. What is the role of unity in empirical knowledge, and what connotations should we deduce from this conception of unity; what are, in other words, the Idea and the role of unity in nature ?

6. THE IDEA OF THE SYSTEMATIC UNITY OF OUR EMPIRICAL KNOWLEDGE

Kant introduces the *Ideas* in *Critique of Pure Reason.* The transcendental Ideas or the Ideas of pure reason are not deduced from experience and lie beyond our understanding. Experience holds nothing that might provide us with an illustration of an Idea. The Ideas in the first instance direct the course of science and are not deduced from science. The regulative usage of transcendental Ideas is, therefore, their legitimate usage in view of the unification of what understanding may know, of what may be known by way of concepts. This usage is not constitutive and consists in considering unity as an ideal towards which one is directed. In this way, the use of Ideas may suggest good hypotheses, but can never be considered as a knowable reality or be liable to being affirmed *a priori.*

The systematization of our empirical knowledge now presupposes two kinds of unity in nature.

In a first sense, in accordance with the Analytic of pure reason, Kant tells us of nature as a system of phenomena mutually related in a necessary way. The necessity pertains to the fact that the phenomena are determined by universal laws.

In a second sense, nature is unified in the sense that is connected with empirical diversity. This is a different type of unity than that which comes to us by way of categorical principles, since this unity is not constitutive for our experience. But Kant believes that, if science aims at necessary and systematic knowledge, we must presuppose that the empirical phenomena of nature, ultimately, comprise a unity. This is where we must seek the foundation of the analogy between the organism and the unity of knowledge (cf. the second aspect of the analogous operation mentioned in 2.).

It is said that the Idea of the purposive organization of nature does not constitute a rival conception of the mechanist conception. It would act in support of research, not as an explanatory principle. Kant assumes that we would never be able to arrive at a system of empirical laws without accepting that nature is organized in such a way that we may know it. The supposition with regard to the purposiveness of nature is linked to a purely negative mode of knowing nature: the denial of such a principle would imply the end of any form of research (cf. McFarland 1970, 87). The regulative status of the principle of teleological judgment not only brings us back, in a general sense, to the problematic status of an Idea with a regulative usage, but also to the status of mechanism.

Nonetheless, even if Kant in this way distinguishes between explaining and judging, even if the regulative usage of Ideas prohibits us from transgressing the limits of the analogous operation which we described in the preamble, we believe that the texts, subsequent to the *Critique of Pure Reason* — the *Critique of Judgment* and the *Opus Postumum* in particular — allow for a more "ontologizing" interpretation (cf. McFarland 1970; Löw 1980).

7. ONTOLOGICAL CONNOTATIONS: THE UNITY OF NATURE

The tension that exists between the conceptions of unity in nature is significant from the viewpoint of the already implied decision in the outset of the critical philosophy[4]. Distancing himself from dogmatic rationalism and from Hume's empiricism, Kant conceived of the issue of knowledge in terms of relations between form and empirical content that might come to replenish this form. It is this very conception, that underlies the whole of the transcendental construction, which is at stake here.

For the Kant of the *Critique of Pure Reason,* nature is at one and the same time a mathematical-physical system and a material system. The formal, mathematical-physical meaning refers to the essence of a given thing — which is to say the first intrinsic principle of all that pertains to the possibility of a given thing. The material meaning in turn refers to things in as much as they might be objects that may be perceived by the senses. Within the scope of natural sciences only the states of nature comprised within the material significance may constitute objects of research — and

4. A decision clearly expressed in the letter addressed to Marcus Herz, dated 21st February 1772. Zweig A., *Kant. Philosophical Correspondence (1759-1799)*, 1967, 70-76.

only as long as they meet formal requirements. Nonetheless, the unity of all phenomena, subsumption under laws, is a hypostatised notion that can only be constituted by the infinite sum of 'formal natures'. Any natural science is therefore only a science to the extent that it is apodeictic; its scientificity is proportional to the contribution of mathematics (cf. Löw 1980, 130).

Setting out from this conception of nature, Kant, in his Opus Postumum, arrives at the conclusion of the impossibility of arriving at an empirical nature; the mathematical interpretation of nature hinges on the notion of essence. Therefore, in a certain sense, no experience is possible in a science of nature. The validity of scientific knowledge is attained to the extent that one is prepared to pay the price for this: that of not being able to discuss what an experience is made of, that is, nature in the material sense.

This is why, according to Kant, reason has no other option than to make the transcendental supposition of nature as a purposive system for judgment. If nature is a significant whole, it is so only so that we humans can talk about it. Therefore, scientific knowledge implies that we presuppose this significant unity. The supposition of a nature, constituting a purposive unity, cannot be validated on the basis of possible experience, whilst, for experience as such, it is irrefutable.

With this in mind, it is interesting to look at how Kant, in order to address the problematic status of experience, in his posthumous oeuvre considers the science of nature in terms of material forces (gravity, pressure, heat...) For him a force is "the subjective possibility of being a cause" (Ak., B. XXII, 192, our translation)[5]. These forces are not dictated by reason, but by experience. Experience dictates its fundamental forces to nature. The final causes are part of these forces, and their conception must, of necessity, precede any natural science.

According to Kant, the science that might establish the link between the two conceptions of nature is physiology, viewed as the theory of general principles of moving forces. Since the notion of a force cannot be deduced from the category of cause — the possibility of being cause precedes the usage of this category — this notion must also find its origin in experience. Certain passages in the *Opus Postumum* — the interpretation of which often evokes debate — may be read in function of the Kant's increasingly

5. As far as possible, we cite from the English translation of Eckart Förster and Michael Rosen. *Opus Postumum*. Cambridge University Press, 1993 (cited from now on as: OP). Otherwise, we translate from the original edition: I. Kant, *Opus Postumum*, Ak., Bd. XXII. For the interpretation of the *Opus Postumum* we take our bearings in the first instance from R. Löw 1980.

"naturalist" preoccupation, which in the first instance expresses itself as an increase in the predominance of the body.

More concretely, the aprioristic necessity of the real experience means that the moving forces, in order to play a part in the experience, must play a part in our body. Therefore, thought itself is reduced to corporality, to the spatial extension of the body. Concrete experience is based on the fact that our organic body can be submitted to moving forces, in concordance with the aims of reason, aims which are those of the acquisition of knowledge. The dualism between body and mind blurs when it is understood as being a priori corporeal and when it is considered as being constituted by abstractions from an original experience. Teleological thought then becomes an abstraction. This leads Kant to observe, for example, that "only because the subject [is conscious] to itself of its moving forces (of agitating them) and — because in the relationship of this motion, everything is reciprocal — [is conscious] of perceiving a reaction of equal strength (a relation which is known *a priori,* independently of experience) are the counteracting moving forces of matter anticipated and its properties established" (OP, 148; Ak., B. XXII, 506)[6].

This does not imply that Kant is abandoning the critical moment. But this critical moment undergoes an alteration in the sense that the basis of the *a priori* is no longer conceived of in categorical or aesthetic terms, but in function of the role of the organic body which constitutes the connection between the physical world and the world of thought. "If we could really speak of a third 'subjectivist turn', we are not looking at connecting it with the deduction of the material unity of the thing setting out from subjective-transcendental principles but at connecting it with the deduction of moving forces setting out from the corporeal *a priori!*" (Löw 1980, 239, our translation).

Hence Reinhart Löw arrives at the following conclusion: "Just like Aristotle, Kant takes as a basic starting point of all knowledge of nature, the subject in its psycho-physiological totality, without this implying to him that

---

6. In order to shape this passage into a coherent one the required interpretation may well lead to the view postulated by Löw, who deems this passage to be essential in the 'naturalist' turn of Kant, a turn which underlines the role of experience and of reciprocity, proceeding from the organic body. We, therefore, supply the fragment in German, to present the reader with some idea of the necessity and the difficulty of making sense out of this. "Nur dadurch daß das Subject sich seiner bewegenden Kräfte (zu agiren) und da in dem Verhältnisse dieser Bewegung alles wechselseitig ist gleich stark auf sich Gegenwirkung wahrzunehemen welches Verhältnis a priori erkannt (nicht von der Erfahrung abhängig) ist werden die entgegen wirkenden bewegende Kräfte der Materie anticipirt und die Eigenschaften der Materie festgesetzt" (Kant I., *Opus Postumum*, Ak. B. XXII, 506).

he abandons the transcendental approach. Quite the contrary: 'the Aristotelian turn' contains the application of transcendental philosophy onto itself, in as much that it means to study the determinations of the possibility of the categories in a theory of the self-affectation and of the self-constitution of the knowing subject. The fact that the inescapable circularity does not lead to the abandoning of this entire approach, can be understood setting out from the fact that Kant does not reason proceeding from logic, but proceeding from hermeneutics. A transcendental philosophy in the narrow sense would lapse into psychology. Kant is able to avoid this because he is looking for the foundations in an anthropology which is essentially Aristotelian-ontological" (*Ibidem,* 238, our translation).

8. CONCLUSION

The possibility of a knowledge of nature remains an enigma in the *Critique of Pure Reason.* In the *Critique of Judgment* it becomes a supposition in the shape of a purposive unity in nature. In the *Opus Postumum,* this purposiveness has evolved into a foundation on the basis of which we are able to understand nature in general. Setting out from the necessity to expand the conception of experience in the *Critique of Judgment* and to ground it in a subjective necessity, it is the aprioristic necessity of the real experience for the constitution of the notion of force which, in the *Opus Postumum,* is seen as the foundation for possible experience. The result of this is that nature, in its formal meaning, is no longer essence and existence (Wesen plus Dasein), but that essence is the abstract nature of existence (cf. Löw 1980, 238).

Therefore, Kant illustrates the passage from the 'formalist' approach to the 'naturalist' approach of knowledge, a changeover we witnessed several times after Kant, either in phenomenology or in the approaches of purposiveness in cybernetics. We might mention, for instance, Merleau-Ponty's interest in the human body, on the subject of which Lacan observed that the conceptual "battery" of philosophy would forever be impotent in expressing that which is at stake in the reciprocal constitution of the object and the subject (cf. Merleau-Ponty 1964; Baas 1994). Or let us mention Quine's naturalized epistemology in which he proposes to make epistemology into a chapter of psychology, thus in a sense circumventing the question of ontology which has long been implicit in analytical philosophy. Or still, the contemporary cognitivist theories whose ambition it is to naturalize the phenomena of meaning and of intentionality proceeding from Husserl's phenomenology, thus taking account of, among other things,

his theory with regard to the role of perception in the constitution of knowledge.
In each of these cases, Kant's exercise is repeated in more or less identical terms, often without any great conceptual or philosophical revolution: formalism-naturalism, explanation-interpretation, mechanism-hermeneutics... these are the terms which come one after the other ... and which resemble each other. The Opus Postumum coincides with what for Kant is the time to conclude, hesitatingly, but crucial to philosophy. Indeed, all of this allows us to see that the blur that installs itself in the meaning of the transcendental, and consequently in the meaning of experience, implies a return to the abyss that separates the subject from his body and from language. Has any advance been made since those days?

REFERENCES

Atlan H. (1991). *Tout, non, peut-être.* Paris: Seuil, 340.

Baas B. (1994). Over een niet toewijsbaar object: het object *a. Psychoanalytische Perspectieven,* **25**, 75-113.

Beaune J.C. (1980). *L'automate et ses mobiles.* Paris: Flammarion.

Boorse C. (1976). Wright on Functions. *The Philosophical Review,* **85**, 1, 70-86.

Cummins R. (1975). Functional Analysis. *The Journal of Philosophy,* **72**, 20.

Dupuy J.P. (1985). L'essor de la première cybernétique. *Cahiers du CREA,* **7**, Paris, 7-140.

Kant E. (1985). *Œuvres Philosophiques.* Paris: Gallimard, Bibliothèque de la Pléiade.

- I, *Lettres et Fragments* (J. Rivelaygue, trans.). 690-697.
- II, *Critique de la Faculté de Juger* (Jean-René Ladmiral, Marc B. de Launay et Jean-Marie Vaysse, trans.). 914-1299.
- II, *Première introduction à la Critique de la Faculté de Juger* (Alexandre J.L. Delamarre, trans.). 848-912.

Kant I. *Gesammelte Schriften.* Berlin: Preussischen Akademie der Wissenschaften, B. XXII: *Opus postumum.*

Lévy P. (1985). Wittgenstein et la cybernétique, *Cahiers du CREA,* **7**: Histoires de cybernétique, nov. 1985, 257-285.

Lévy P. (1985a). L'Œuvre de Warren McCulloch, *Cahiers du CREA,* **7**: Histoires de cybernétique, nov. 1985, 211-257.

Lévy P. (1985b). Analyse de contenu des travaux du computer laboratory. *Cahiers du CREA,* **8**: Généalogies de l'auto-organisation, nov. 1985, 155-193.

Lévy P. (1985c). Le théâtre des opérations. Au sujet des travaux du BCL. *Cahiers du CREA,* **8**: Généalogies de l'auto-organisation, nov. 1985, 193-225.

Livet P. (1985). Cybernétique, auto-organisation et néo-connexionnisme. *Cahiers du CREA,* **8**: Généalogies de l'auto-organisation, nov. 1985, 105-155.

Löw R. (1980). *Philosophie des Lebendigen. Der Begriff des Organischen bei Kant, sein Grund und seine Aktualität,* Frankfurt am Main: Suhrkamp Verlag, 357.

McFarland J.D. (1970). *Kant's Concept of Teleology.* Edinburgh: Edinburgh University Press, 150.

Merleau-Ponty M. (1964). *Le Visible et l'Invisible.* Paris: Gallimard.

Pask G. (1992). Different Kinds of Cybernetics. In G. Van de Vijver (ed.), *New Perspectives on Cybernetics. Self-Organization, Autonomy and Connectionism.* Dordrecht, Kluwer Academic Publishers, 11-31.

Petitot J. (1985). *Morphogenèse du Sens I. Pour un schématisme de la structure* (René Thom, preface by). Paris: PUF.

Petitot J. (1985a). *Jugement esthétique et sémiotique du monde naturel chez Kant et Husserl,* manuscript, 9.

Petitot J. (1985b). *À propos de la querelle du déterminisme: de la théorie des catastrophes à la critique de la faculté de juger.* Paris: Centre d'Analyse et de Mathématique sociales, EHESS-CNRS, C.M.S.P. 016, nov. 1985, 52.

Petitot J. (1990). Le Physique, le Morphologique, le Symbolique: remarques sur la vision. *Revue de Synthèse,* **1-2**, Sciences cognitives: quelques aspects problématiques, 139-183.

Petitot J. (1991). *La Philosophie transcendantale et le problème de l'objectivité,* Les entretiens du Centres Sèvres, Paris: Éditions Osiris.

Petitot J. (1992). *Physique du sens. De la théorie des singularités aux structures sémio-narratives.* Paris: Éditions du Centre National de la Recherche Scientifique.

Philonenko A. (1984). Introduction à la Critique de la Faculté de Juger. In *Critique de la Faculté de Juger.* Paris: Vrin, 7-16.

Quine W. (1969). Epistemology Naturalized. (*Ontological Relativity and Other Essays,* reprinted in). New York, 69-91.

Van de Vijver G. (1991). *Van cybernetica naar connectionisme. Een epistemologische studie van doelgerichtheid,* Gent: Academia Press.

Van de Vijver G. (ed.) (1992). *New Perspectives on Cybernetics. Self-Organization, Autonomy and Connectionism,* Dordrecht: Kluwer Academic Publishers, Synthese Library, vol. 220.

Varela F.J. (1979). *Principles of Biological Autonomy.* New York (Oxford): North Holland Elsevier Company, The North Holland Series in General Systems Research, vol. 2.

Wimsatt W. (1972). Teleology and the Logical Structure of Function Statements. *Studies in the History of Philosophy of Science,* **3**, 1, 1-80.

Woodfield A. (1976). *Teleology.* Cambridge: Cambridge University Press.

Wright L. (1976). *Teleological Explanations. An Etiological Analysis of Goals and Functions.* Berkeley (Los Angeles, London): University of California Press.

LAURENCE BOUQUIAUX

ON A "MATHEMATICAL NEO-ARISTOTELISM" IN LEIBNIZ

There are many different, and even divergent, interpretations of Leibniz' natural philosophy. Some insist on his adhesion to the principles of mechanism; they evoke the dream of decomposable, totally analyzable and perfectly comprehensible nature. Others stress that Leibniz also perceived the inadequacies of mechanism and that he affirmed the necessity of reintroducing substantial forms into physical theory. Some have seen in Leibniz one of the representatives of this "new physics", for whom the world is something like a well-regulated clock, while others consider him as the first to rise up against the excesses of a paradigm which claims to reduce the world to a gigantic machine. In an article entitled *L'état actuel de la recherche leibnizienne*[1], A. Heinekamp stresses the interest aroused today by the Leibniz dynamic. He says that it is most likely linked "to the fundamental crisis of physics, where the belief in the universal value of Newtonian theories has been shaken, and to present-day scientific discussion, in which theories that could be an alternative to those of Newton are taken into consideration". This allusion unfortunately remains rather mysterious (Heinekamp mentions no name). It might be possible to see in this remark an invitation to move the spirit of Leibniz' natural philosophy closer to the ambition of those theories which today concentrate on themes such as auto-organisation or morphogenesis. In a book with the title *L'invention des formes* and sub-title *Chaos-Catastrophes-Fractales-Structures dissipatives-Attracteurs étranges*[2], A. Boutot speaks of mathematical "neo-Aristotelism". *Neo-Aristotelism* because typically Aristotelian themes that conventional science may have tended to "forget"[3]

1. *Les Études philosophiques*, 1989/2, 139-160.
2. Éditions Odile Jacob, 1993.
3. R. Thom thinks that the particularity of passing from the Aristotelian paradigm to that of conventional mechanism is that, whereas, in most cases, a theory "contains" the theory that it replaces (general relativity, for example, "contains" Newton's theory of gravitation), the

B. Feltz, M. Crommelinck and P. Goujon,(eds.), Self-Organization and Emergence in Life Sciences: 163-169.

are reconsidered. *Mathematical* neo-Aristotelism because we must not follow Aristotle in his condemnation of all mathematisation of the physical world but endeavor to give a mathematical status to the notions of form, organization, structure, etc. It seems to me not impossible to defend the idea that this project of elaborating a mathematical neo-Aristotelism was already, to a certain extent, Leibniz' own[4].

The young Leibniz adhered enthusiastically to the mechanist project of adapting geometry to physics and explaining all that happens inside bodies by means of three immediately mathematisable notions: extension, figure motion. However, as early as 1676, our philosopher began to suspect the limits of the Cartesian model. He became increasingly persuaded that the notions of extension, figure and motion were not enough, and that it was necessary to introduce into the corporal field new notions, those of force and cause (notions which, to a certain extent, correspond to the metaphysical notion of substantial form). In a famous opuscule[5], Leibniz, contradicting Descartes, establishes that physics should not be based on the principle of conservation of the quantity of motion, but on that of the equivalence between cause and effect. Like the Cartesians, Leibniz considers that force is conserved. Unlike them, he undertakes to demonstrate that force should not be estimated by the quantity of motion mv, but by the "quantity of the effect it can produce", a quantity that is, according to Leibniz, proportional to mv^2. More precisely, what Leibniz demonstrates is that, on the hypothesis that as much force is necessary to raise a one-pound body to the height h of

---

Galilean revolution completely annulled the problematics of Aristotelism (it is even, R. Thom adds, the only counter-example to the general rule). The ideas of generation and corruption, the problematic concerning the birth and destruction of forms, the general theory of change, have completely disappeared, to be replaced by a dynamic that considers only the change in place. See *Paraboles et catastrophes.* Paris: Flammarion, 1983, 125.

4. In an article in which he mentions various attempts to construct an objective theory of forms (a project that may have resulted in R. Thom's catastrophe theory), J. Petitot refers to Leibniz as a precursor, in whose work "still co-exist the Aristotelian qualitative ontology and the physical objectivity of rational mechanism", two elements exactly required by the development of an objective form theory. See article "form" in *Encyclopædia Universalis,* 9 (1990), 712. See also Structuralisme et phénoménologie: la théorie des catastrophes et la part maudite de la raison, in *Logos et Théorie des Catastrophes, À partir de l'œuvre de René Thom,* Cerisy colloquium, Minutes of 1982 international colloquium directed by J. Petitot, Geneva: Patiño, 1988, 345-376.

5. *Brevis Demonstratio erroris memorabilis Cartesii et aliorum circa Legem naturalem, secundum quam volunt a Deo eandem semper Quantitatem Motus conservari, qua et in re mechanica abutuntur.* Gerhardt (ed.), *Leibniz, mathematische Schriften* (GM below), Berlin-Halle, 1849-1863, vol. VI, 117-123. Leibniz takes up the demonstration contained in this opuscule in § 17 of *Discours de Métaphysique.*

four ells as to raise a 4-pound body to the height of 1 ell (which comes to the same thing as saying that it is supposed that force is proportional to *mh*), it must be concluded that force is proportional to mv^2. Leibniz demonstrates that the "force" of a body in motion — what enables this body to produce a certain amount of work, a certain "effect" (in this case rising to a certain height) — is proportional to mv^2 and not mv.[6] This mathematical demonstration, according to Leibniz, has a metaphysical significance of the utmost importance. The fact that it is necessary to distinguish between force and quantity of motion corresponds to the fact that "force is something different from size, figure and motion". "And from this, continues Leibniz, we can conclude that not everything which is conceived in a body consists solely in extension and its modifications, as our moderns have persuaded themselves. Thus we are compelled to restore also certain beings or forms which they have banished"[7]. Physics can no longer be constituted from geometry alone and its "axiom of the more and the less". This axiom must be replaced by a "metaphysical" axiom, that of the equivalence between full cause and entire effect. The shift is considerable. As A. Robinet[8] says, "it is no longer a question of evaluating quantitative equalities between spatio-motor conditions, but of weighing equipollences of natural effects of which one tries to determine the active causes". There is no longer an equality between sizes, but an equality between powers. It must, however, be stressed that, if there is here a return to a more " metaphysical" concept, it is however an "intelligible" concept, a measurable and calculable concept. There is no question of going back to the occult qualities of scholastics. While substantial forms have to be recalled, care must be taken to separate "the use to be made of them, and the abuse that has been made". The cause, the force has a mathematical expression, and its variations are governed by mathematical laws. Leibniz means to preserve the acquired knowledge of mechanism ; while he wishes to go further, he does not wish to banish it. If he brings back substantial forms, it is not under the influence of a naïve enthusiasm, but with full knowledge of the question, the outcome of a long

---

6. In somewhat anachronistic terms, it could be said that it is a manifestation of the energy conservation law that Leibniz discovers here, since he puts forward the conversion of potential energy mgh into cinetic energy 1/2 mv^2, (during descent) and reciprocally (during re-ascent).

7. *"Et l'on peut juger par là que tout ce qui est conçu dans le corps ne consiste pas uniquement dans l'étendue et dans ses modifications, comme nos modernes se persuadent. Ainsi nous sommes encore obligés de rétablir quelques êtres ou formes, qu'ils ont bannies."* *Discours de Métaphysique*, § 18. Translation by Leroy E. Loemker, *Philosophical Papers and Letters*, Dordrecht-London: D. Reidel Publishing Company, 1969.

8. *Architectonique disjonctive, automates systémiques et idéalité transcendantale dans l'œuvre de G.W. Leibniz*, Paris: Vrin, 1986, 203.

period of hesitation. "Perhaps I shall not be condemned so lightly", he writes[9], "when it is known that I have given much thought to the modern philosophy and that I have spent much time in physical experiments and geometric demonstrations and was for a long time convinced of the emptiness of these beings to which I am at last compelled to return in spite of myself and as by force". Nothing that tradition was able to say about the essence of the body was intelligible, so that "it is no wonder that these substantial forms have been seen by the best minds as chimerical" ; Leibniz assures us that what he will say on the subject will be "as intelligible as all that Cartesians have ever proposed in other matters".

If Leibniz definitely intends to rehabilitate the Aristotelian concepts of form or entelechy, he lends them a clarity, an intelligibility that they have never had before him. Back to Aristotle, then, but without, for that reason, abandoning the (mathematical) intelligibility requirement formulated by the "moderns". "Now, we shall," he writes, in the *Specimen dynamicum,* "reduce Peripatetic tradition of forms or entelechies, which has rightly seemed enigmatic and scarcely understood by its authors themselves, to intelligible concepts. Thus we believe that this philosophy, accepted for so many centuries, must not be discarded but be explained in a way that makes it consistent within itself (where this is possible) and clarifies and amplifies it with new truths"[10]. The body is something different from extension, motion is different from a change in place, and "mechanical principles are rather metaphysical than geometric". Because Descartes refused to recognize this, he wrote only a "fiction of physics". A purely mechanistic conception can only lead to conclusions contrary to experience: "If mechanical laws depended upon geometry alone without metaphysics, phenomena would be entirely different "[11].

Nature, infinitely varied and infinitely rich, as described by Leibniz, is beyond the reach of the principles of Cartesian mechanism. In particular,

---

9. *"Peut-être qu'on ne me condamnera pas légèrement, quand on saura que j'ai assez médité sur la philosophie moderne, que j'ai donné bien du temps aux expériences de physique et aux démonstrations de géométrie, et que j'ai longtemps été persuadé de la vanité de ces êtres, que j'ai enfin été obligé de reprendre malgré moi et comme par force", Discours de Métaphysique,* § 11. Translation by Leroy E. Loemker, *op. cit.*

10. "*[...] Peripateticorum tradita de formis sive Entelechiis (quae merito aenigmatica visa sunt vixque ipsis autoribus recte percepta) ad notiones intelligibiles revocabuntur, ut adeo receptam a tot seculis Philosophiam explicare potius, ita ut constare sibi possit (ubi hoc patitur) atque illustrare porro novisque veritatibus augere, quam abolere necessarium putemus.*" *Specimen dynamicum*, GM VI, 235. Translation by Leroy E. Loemker, *op. cit.*, 436.

11. "*Si les règles mécaniques dépendaient de la seule géométrie sans la métaphysique, les phénomènes seraient tout autres*", *Discours de Métaphysique,* § 21. Translation by Leroy E. Loemker, *op. cit.*

Cartesian mechanism is contrary to this great principle of variety which is the principle of the indiscernible, according to which there cannot be in Nature two single things which differ only *solo numero.* For Leibniz, two substances which differed only "in number", two substances which differed only quantitatively, two substances which differed only through extrinsic denominations (two substances which differed from one another only because they are not situated in the same place, for example, or because they are not the same size) would become indistinguishable, they would be in fact one and the same substance. Two substances, however similar, always differ qualitatively, they always differ by some internal denomination. There are no two identical substances, no two leaves, no two drops of water absolutely alike. For Descartes, on the other hand, bodies are only portions of an extension, an abstract extension all of whose points are identical, just as the abstract instants of time in Cartesian mechanism are identical. Deprived of any principle of diversity, Cartesian mechanism is, according to Leibniz, incapable of accounting for motion, for motion supposes a variety principle. As Y. Belaval[12] writes, the local motion Leibniz discusses "presupposes the power to become *other,* alteration, alterity, Aristotle's qualitative "ἀλλοίωσις". It is not only each monad but also each state of each monad that is unique. And it is this singularity that makes motion possible. If there is only one homogeneous undifferentiated extension, motion becomes unthinkable. In the Cartesian world, Y. Belaval makes clear, nothing allows us to say if, in a given place, is found at the instant t_1 the same portion of matter as that which was there at the instant t_0, or if another has been substituted for it. Motion, unobservable on principle, becomes something purely imaginary. "Since everything which is substituted for a prior thing must be perfectly equivalent to it, no observer, though he be omniscient, would be able to see even the slightest indication of change. And so everything would be the same as if no change or differentiation had taken place in the bodies, and no reason can be given for the diverse appearances which we experience by sense"[13]. If there is

---

12. Y. Belaval, *Leibniz critique de Descartes,* Paris: Gallimard, Coll. Tel, 401.

13. *De ipsa natura,* § 13 "(...) *cum omnia, quae prioribus substituuntur, perfecte aequipolleant, nullum vel minimum mutationis indicium a quocunque observatore, etiam omniscio, deprehendetur; ac proinde omnia perinde erunt, ac si mutatio discriminatioque nulla in corporibus contingeret: nec unquam inde reddi poterit ratio diversarum quas sentimus apparentiarum*". Gerhardt (éd.), *Die philosophischen Schriften von G.W. Leibniz* (GP below), Berlin 1875-1890, IV, 513. Translation by Leroy E. Loemker, *op. cit.*, 505.

suppression of the idea of force, of an internal determination, which would be the true cause of motion, this motion is reduced to its trajectory, to an abstract line where all points are equivalent, and which can give only an inexact idea of what motion really is. For, as Y. Belaval again stresses, motion is not for Leibniz what it is for Descartes or Galileo, a state; it remains what it was for Aristotle, a process: "It is because Aristotle had seen something of these principles, I believe, that he concluded (he being in my opinion more profound than many people think) that there is needed some alteration besides change in place, and that matter is not similar to itself everywhere and does not remain invariable"[14]. Descartes' time is made up of independent instants, his world consists of a succession of states of which, as Y. Belaval says, each one is immediately suspended from God, without owing anything to what it was itself in the preceding instant. Leibniz' time, on the other hand, is a "living" time, a time which unfolds, a continuous time where each instant is different from all the others, where the present is burdened with the past and pregnant with the future[15].

Leibniz' universe is a world where everything is in a perpetual state of flux, a world where everything is constantly transformed. There is no immobility, no rest. The world is full of souls, and the soul is perpetual restlessness. Substances are enveloped and developed, fold and unfold, are extended and drawn together, concentrated. "The body is in continuous change, like a river"[16]. "Souls continually advance and mature, like the world itself, of which they are the image, for nothing existing outside the universe can prevent it and the universe must necessarily go on advancing and developing"[17]. All this must, as it seems to me, incite us to consider circumspectly a certain interpretative tradition that makes of Leibniz the thinker of identity, of the reduction of becoming to being or the eradication of all temporality. Y. Belaval has summed up this aspect of leibnizianism very well. "In creation, each point, each instant is 'characteristic', individualised by the activity it houses and which makes a thing endure (...)

---

14. "(...) *Aristoteles, profundior mea sententia, quam multi putant, iudicavit, praeter mutationem localem opus esse alteratione, nec materiam ubique sibi esse similem, ne maneat invariabilis*", *De Ipsa Natura,* GP IV, 514. Translation by Leroy E. Loemker, *op. cit.*, 506.

15. Whereas the Cartesian world, "founded on an illusory motion, only calls for an illusory time, a dead time, where the present is absolutely not burdened with the past nor pregnant with the future". Y. Belaval, *Leibniz critique de Descartes,* 426.

16. *À Remond,* GP III, 635. Translation by Leroy E. Loemker, *op. cit.*, 658.

17. "*Les âmes avancent et mûrissent continuellement, comme le monde lui-même dont elles sont les images, car rien n'étant hors de l'univers qui le puisse empêcher, il faut bien que l'univers avance continuellement et qu'il se développe*". *À Sophie*, GP VII, 543.

everything comes into infinitely varied motion. Identity without variety would be equivalent to sleep, death, the end of consciousness, which is essentially restlessness. Bodies do not keep to a determined figure: like a river, or Theseus' ship, an organism has as its only stable element, its guiding idea, its law of organisation. In species, individuals differ : there can be found no two leaves perfectly alike. Species are varied — and may have varied — to infinity. There is no repetition. A continual flux, an unwearied temporality. A Heraclitean vision of the world!"[18].

REFERENCES

Belaval Y. (1960). *Leibniz critique de Descartes.* Paris: Gallimard, coll. Tel.

Boutot A. (1993). *L'invention des formes.* Paris: Éditions Odile Jacob.

Gerhardt (éd.) (1849-1863). *Leibniz, mathematische Schriften.* Berlin-Halle.

Gerhardt (ed.) (1875-1890). *Die philosophischen Schriften von G.W. Leibniz,* Berlin.

Heinekamp A. (1989). L'état actuel de la recherche leibnizienne. *Les Études philosophiques,* **2**, 139-160.

Leibniz (1686). Discours de Métaphysique.

Loemker E. (1969). *Philosophical Papers and Letters.* Dordrecht-London: D. Reidel Publishing Company.

Petitot J. (1988). Structuralisme et phénoménologie: la théorie des catastrophes et la part maudite de la raison. In J. Petitot (dir.), *Logos et Théorie des Catastrophes, À partir de l'œuvre de René Thom,* Cerisy colloquium, Minutes of 1982 international colloquium, Geneva: Patiño, 1988, 345-376.

Robinet A. (1986). *Architectonique disjonctive, automates systémiques et idéalité transcendantale dans l'œuvre de G.W. Leibniz.* Paris: Vrin.

Thom R. (1983). *Paraboles et catastrophes.* Paris: Flammarion.

18. Y. Belaval, *Études leibniziennes. De Leibniz à Hegel.* Paris: Gallimard, 1976, 94.

FRANÇOIS DUCHESNEAU

"ESSENTIAL FORCE" AND "FORMATIVE FORCE": MODELS FOR EPIGENESIS IN THE 18TH CENTURY

For almost a century (1672-1757) at the heart of the scientific revolution the science of the living was quasi exclusively dominated by preformationist theories of generation: these would counter any model implying self-organization of the living and discard any force accounting for the emergence of complex structures[1]. Preceded by such attempts as Buffon's and Maupertuis's, the real restoration of epigenesis came about in 1759 with the *Theoria generationis* of Caspar Friedrich Wolff (1733-1794). My first objective is to spell out the reasons that enticed Wolff into recasting embryogenesis in accordance with epigenetic concepts. In their own time, Wolff's theses had two consequences: (1) they elicited a decline of classical preformationism; (2) they themselves underwent a "vitalist", or rather "teleomechanist"[2], recasting which mitigated their drastic bent. Secondly, I shall focus on this metamorphosis as it came about when Johann Friedrich Blumenbach (1752-1840) set up his vitalist interpretation of epigenesis in *Über den Bildungstrieb und das Zeugungsgeschäfte* (1781)[3].

1. The starting point for this domination by the "preformationist paradigm" may be assigned to Marcello Malpighi's *De formatione pulli in ovo* (1672) (Adelman 1966). But numerous other epistemological and metaphysical factors have concurred in generating those abstruse versions of the paradigm based on an infinite nesting of preexisting germs (Roger 1971, Bernardi 1986).

2. This term is used by T. Lenoir to feature the blend of teleological and materialist schemes in the biology that issued from the Göttingen school, and specially from Blumenbach (Lenoir 1982). Lenoir uses also the phrase "vital materialism".

3. The present paper relies on analyses contained in Duchesneau (1982).

B. Feltz, M. Crommelinck and P. Goujon, (eds.), Self-Organization and Emergence in Life Sciences: 171-186.

1.

After Blumenbach's *Über den Bildungstrieb* was published, Wolff strove to break any ties with a potential vitalist interpretation of his *Theoria generationis*. The main point of a text like the one he published in St. Petersburg in 1789 and devoted to the "specific and essential force in vegetal as well as animal substance" (Wolff 1789)[4], was to establish that the *vis essentialis* forms so to say the essence of the organism, that it is a force analogous with those producing crystals and metallic amalgams, that it operates not by juxtaposing elements, but by assimilating inorganic material in an intimate fashion. In a word, Wolff meant a force characteristic of the organic nature, but without its depending on a preexisting organic structure. This structure is precisely an outcome of the *vis essentialis* acting on a matter more or less prone to solidify. As a material force, the *vis essentialis* selects among the material elements for the sake of organic structuring, but this "for the sake of" is only a metaphorical formula, and the discriminating function of this force should be compared with "chemical affinities" and with mechanical phenomena dependent on attraction/repulsion. However, the object to analyze consists in a complex order of functional relations and epigenetic phenomena. Because of this complexity, the conceiving of such forces boils down to describing sequences of physiological processes and morphogenetic phases. Thus, the essentially descriptive analysis deals with the phenomenal surface without necessarily reaching the underpinning mechanisms. But this type of explanation, unavoidably phenomenological, is compensated with the eliciting of laws that govern the sequence of processes as the structure-function relationships unfold in the complex organism.

In the *Præmonenda* of the 1774 edition of the *Theoria generationis*, Wolff writes:

> "As physiology considered in its entirety may be called the science that sets forth the function of the organic body, so the theory of generation may be called the very science of the organic body: the distinction between the two is evident. Physiology is to anatomy as a corollary to the theorem it is deduced from; but the theory of generation is related to anatomy as the the demonstration of the theorem to the theorem to be demonstrated" (Wolff 1774: xii-xiii note) (cf. also Wolff 1764: 12-13).

The physiological properties and their effects refer to the structures as to necessary conditions, but the more profound explanation consists in tracing

---

4. Wolff was responsible for the topic of the contest the Imperial Academy of Sciences in St. Petersburg had set up and he seized the occasion for replying to two of the memoirs which had been submitted, those of Blumenbach and Carl Friedrich Born.

the occurrence of these complex structures and the way forces intervene to produce such results. A special combination of material parts allows an activity of nutrition and vegetation to emerge, which, starting from a relatively amorphous and homogeneous structure, makes for more and more diversified and heterogeneous internal forms, and hence for emergent physiological functions. The *vis essentialis* is an intrinsic force of the *corporeal* device which produces specialized organs as derivative organic mechanisms. This force soars up from the inner and probably unanalyzable organization of the material parts joining to form "seminal" combinations. The seeds, as prime ingredients of organic matter, comprise combinations of material elements which elicit, under appropriate conditions, a dynamic disposition to achieve the complex structures required for physiological functions.

Though the *vis essentialis* seems correlative with a combination of material parts forming such a disposition, it shows up as an "emergent" force, for one cannot assign a determining reason for it beyond the phenomena if manifests itself by. As Wolff explains concerning the nutrition in plants:

> "Be this force as may, whether attractive or repulsive, or dependent on the expansion of air or composed of all these, and some more, provided it produces the aforementioned effects [the absorption and diffusion of liquids through the whole plant, and their exhalation] and provided it is supposed along with the plant and the nutritive humors received — which experience confirms — it will suffice for my present purpose and I shall call it *vis essentialis* of the vegetals" (Wolff 1759: I, § 4 , 13).

The problem with epigenesis is that of providing an adequate explanation for the complex form that befalls an originally amorphous organism[5]. One should therefore envision a "sufficient and continuous" (*perpetua*) cause for the various processes that characterize the emergence of structures and functions. From this viewpoint, the *vis essentialis* is an epigenetic force inherent in a relatively amorphous seminal structure that can mould itself into an organism. The ability to solidify (*solidescibilitas*) correlates with the essential feature of the material device wherein the *vis essentialis* acts.

The *theoria* opens up with two significant definitions: (1) that of generation: "All agree that the phrase *natural generation of the organic body* means its framing up from a manifold of parts" (Wolff 1759: Expositio et ratio instituti, § 1, 5); (2) that of the *principle* and the *laws* of generation:

5. On the experimental and theoretical reasons that moved Wolff to restore epigenesis and oppose Haller's late preformationism, cf. Roe (1981), Duchesneau (1982: 277-311) and Monti (1990).

"[...] one should consider as the principle of generation this force of the body which executes this framing up; the modes according to which it acts, form those *laws* of generation the most illustrious v. Haller is yearning for" (Wolff 1759: Expositio et ratio instituti, § 2, 5). The first definition counters the preformationists' bland rejection of the problem: it implies that the way the organism is composed, that is its mode of complex structuring for the sake of the various physiological operations which derive therefrom, may only be explained when considering the framing up of parts from one another. The notions of "principle" and "law(s)" refer to the basic methodological postulate that the explanation must relate to the necessary physical order implied in the action of a specific force.

Within this methodological framework, the *Theoria* purports to provide access to the sufficient reason of generation. The argument consists mainly in showing how this sufficient reason helps explain the sequential phenomena of epigenesis for vegetals as well as animals. The analytical process involves empirical justification once the significance and explanatory import of the *vis essentialis* have been established.

These theoretical presuppositions are elicited in the shape of definitions. Total and partial parts are thus distinguished: partial parts combine to compose organs, while total parts form the real parts of the *whole,* that is the very structure of the organism. Nutrition consists in partial parts replacing other such parts. The same scheme applies to growth and decrease, but with the addition or subtraction of similar partial parts. Vegetation, on the other hand, consists in a replacement or addition of total parts. Generation is the production of the whole body, that is of all of the embryo's total parts, by similar parent organisms. As this process may take place inside an already formed organic body, the three processes — nutrition, vegetation and generation — can be viewed as three modalities of assimilation/ disassimilation that differ only by the nature of the parts involved. So, in order to fix the main operative concepts in his theory, Wolff distinguishes between two types of parts: the ones are composed of other organic parts, the others cannot be analyzed into lower level organic parts, but they issue from a mixing of miscible inorganic parts (Wolff 1759: Expositio et ratio instituti, § 19, 6). Plain nutrition or plain or equal increasing (*æquabile*) can be so designated because it implies only the insertion of miscible parts into the organism. It contrasts with (1) organizing nutrition (*nutritio organizans*), which inserts simple parts and maintains or increases the organic volume; and with (2) vegetation proper which frames up total parts and modifies the organic structure. In principle, complex parts may be conceived through the act of vegetation. Nutrition provides the key for analyzing generative processes, since the framing up of simple parts is a necessary condition for any organic structuring. Vegetation is only a complexifying of assimilative activity. As for generation proper, it depends on "essential circumstances" determining the nutrition/vegetation process. The theoretical framework is based on the one hand on the formative sequences represented by that type of processes, on the other hand on the postulate that a force inherent in physical nature can account for epigenesis according to special laws.

This twofold aspect of the theory is set into relief in the second part of the *Theoria* devoted to the generation of animals, wherein the framing up of parts in the chick embryo after incubation is detailed, and in the third part which generalizes the implications of the theory. Rehearsing Malpighi's observations recorded in *De formatione pulli in ovo* (1672), Wolff describes the globulous and undifferentiated structure of the embryo at initial stage. If the assimilation of miscible elements to this initial structure is due to a special force, this force is to be distinguished from forces that may intervene on more accomplished structures, such as the circulatory system. Such an organic assimilation which proves independent from the resulting complex structures constrains us into postulating a notion of force analogous to the presumed concept of *vis essentialis* (Wolff 1759, II, § 168, 73). The observable phenomena correlative with the action of this force consist in the nutritive fluids penetrating between the globules of the primeval structure. In the developed organism, analogous phenomena may take place, insofar as the nutrition of parts implies absorption beyond the limits of the vessels network. The circulation of humors according to a special force is matched with the ability to solidify proper to animal organic substance, an ability considerably lesser than the equivalent property of vegetal structures. Rather, the force operates the circulation of fluids, the depositing and assimilation of nutritive matters in accordance with the ability to solidify of miscible parts. Analogy extends this process to all organic functions, starting with vegetation:

> "All this operation which achieves the growth of the embryo and the formation of circles [along the structural disposition of the embryo's parts], the dissolving of the yolk and of its matter containing an appropriate nutriment, the extracting of that nutriment, its moving about, the elimination of the residue, is not a new design of nature, singular and solely instrumental to vegetation in the embryo, but it shows up in all instances wherein animals need to feed and grow" (Wolff 1759, II, § 189, 81).

As an instance, such is the case for the separation of chyle, for the various secretions and excretions and for the transformation of blood into lymph.

We shall not follow the detailed description of the successive steps in epigenetic embryogenesis according to Wolff: the core of it resides in the progressive framing up of a venous and thereafter an arterial network. As the analysis goes on, we are left with slight justification for envisioning a preformation of the circulation that would condition the emergence of preexisting structures to sight according to Haller[6]. We may consider the *vis essentialis* as a special agent correlative with certain material dispositions providing those conditions epigenesis requires. It is indeed a force along the analogy of universal and primordial physical forces. It belongs to combinations of "physical monads", in accordance with a primordial and relatively general sort of elementary disposition, though one should not forget that the law of generation always presupposes an already formed organism, capable of priming the organogenetic process.

In view of the general properties of organic bodies, Wolff sets as a constant analogy of experience that the organic parts forming a self-preserving body must be in mutual nutrition relationship. He insists that the cohesion of organic parts correlates with their necessary nutritive dependence, the sole positive ground for a determination to unity and plurality (*determinatio unitatis vel pluralitatis*) (Wolff 1759, III, § 236, 108). The progressive interruption in the diffusion of nutriment is the principle ruling over the distinction between separable organs in the embryo's development. In individual organisms, one can readily point to a necessary condition for their functional framing up in the very dependence of distinct parts upon a common nutritive source exerting an hegemonic function. The organism will be the more complex as a greater number of distinct parts depend upon this source-organ of assimilation for their preservation and activity. The multiplication of source-parts entails a complexifying of the organic unity — as some vegetals bear witness. The formula "every organic body consists in a trunk and branches" (Wolff 1759, III, § 237, 109) expresses the principle of nutritive dependence in terms of structural relationship. On the basis of a necessary correlation between nutritive processes and integrative framing up, Wolff will set a morphological typology of parts in line with the morphogenetic phases which characterize their coming about.

Thus, a model becomes available for interpreting the various phases of embryological developement according to modes of assimilation. Whereas Haller postulated a predisposed determining structural reason without which an adequate organogenetic analysis might not be conceived, Wolff relies on modifying factors that affect a basic assimilation-disassimilation function: this function would depend on the *vis essentialis* operating in and on the

6. That was the purpose of Haller's series of observations on the emergence of vascular structures in the chick embryo (Haller 1758 and 1762-1768, II).

minimal embryonic structure. But then, how are we to conceive models for analyzing vital activity? The major difficulty with the Wolffian position is thus epitomized:

> "It seems that the mechanism (machina) provides animal life with the same advantage it provided vegetal life, namely: it modifies the instrumental conditions for the function — and so function can proceed more easily — or it assists in letting function come to existence through its determining causes" (Wolff 1759, III, § 255, schol. 3, 129).

This boils down to saying that the vital operations derive basically from the *vis essentialis* and the ability to solidify which depends on the minimal organic components. Apart from special limiting or targetting conditions, the whole organic activity flows from assimilation and its modalities. And so, the explanation of complex emerging functions should be deducible from the basic organic operations.

But would not this be only possible because the *vis essentialis* assumes an architectonic role and is responsible for the ability to complexify that results in the organism and its integrative functions? Indeed, this seems to foreshadow the notion of a *vital principle.* But, looking more closely, it appears that those bodies on which vegetation acts are inorganic in their ingredients; and the end products of this process are devices of a complex mechanical type on which the functions of the resulting organism depend:

> "Evidently, the bodies capable of vegetation are not machines, but plain inorganic substance. This vegetalizing substance must be absolutely distinguished from the machine it is wrapped in and adheres to. And this machine should be considered as its product" (Wolff 1759, III, § 253, 123).

Thus, the vegetation process links a so-called inorganic state of corporeal parts with a state wherein the complex structure provides an instrumental disposition for specialized functions. The sufficient reason for this process is not to be found in the degree of composition and integration of the developed organism: witness the fact that vegetal and animal organisms develop considerably at the very time they resemble inorganized masses more. The potential for growth itself seems to regress as individual organization becomes more complex. And even in the adult organism, nutrition operates not so much as a result of complex organization, but rather because of the simpler elements whose combination forms the integrative organic whole[7]. The architectonic role of the *vis essentialis* gets

7. Wolff (1759, III, § 253, schol., 123-124): "Et hisce igitur non sequi videtur organicam compositionem perfectam ad vegetationem valde necessariam esse [...] Et si nutriuntur saltem in adulto animale etiam vasa, aliæ organicæ partes, id non fit quatenus illa vasa, et quatenus hæ partes organicæ sunt, sed quatenus ex inorganica substantia, suis qualitatibus prædita, componuntur."

ultimately to this: though analogous with forces in the physical universe, it differs for its capacity to determine through epigenesis the emerging of complex organic structures on which integrative functional processes depend. For sure, the *Theoria* does not admit of any preformationism of the «virtual form». The *vis essentialis* belongs to the general order of forces in physical nature. What is daring in Wolff's analysis is his conceiving a necessary order of epigenetic development based on nutritive assimilation. He takes this order as conditioning the emergence of structures. Wolff's *Theoria generationis* may be held responsible for the systematic criticism of Haller's structural preformationism. But it is not equally true that Wolff succeeded fully in imposing his epigenetic doctrine concerning the force of vegetation. In his time, what seemed more promising was his sequential description of embryological transformations. In particular, the theory of layers at the core of later embryological descriptions is already foreshadowed in Wolff (1769 and 1773). This said, the problematic stance of *vis essentialis* will open the field for analyses that will attribute the specific causality of biological phenomena to vital principles. The paradigmatic case to be considered in this instance is Blumenbach's.

2.

The main influences that exerted initially on Johann Friedrich Blumenbach were Haller's and C.F. Wolff's. Haller's research program provided Blumenbach with the methodological framework of a physiology that aimed at unveiling the structural microdispositions and those emerging functional properties which could be analytically linked with given sets of microdispositions[8]. On the other hand, as we have just seen, Wolff had restored epigenesis in his *Theoria generationis* by resorting to a *vis essentialis* that was presumed to act in the midst of organic matter so as to promote the framing up of a progressively more complex organism. This *sui*

---

8. In general, Haller's physiology is characterized by his notion of *anatome animata*. His methodology combines a subtile description of structures (in accordance with a notion of fibre as the physiological unit) and the experimental identification of functional properties correlative with such elementary structures or with integrative sets of structures. Properties like irritability and sensibility make for sets of dynamical phenomena linked with distinct structural bases. However, their ontological status and causal foundation seem to escape empirical determination. Haller is content with relating them to what I have termed a "special mechanism": hence a considerable latitude for theoretical reinterpretation among the following generations (Duchesneau 1982).

generis force for growth and nutrition was also supposed to emerge from the primordial inorganic composition of organic fluids. Now, Blumenbach revises this concept as well as the Hallerian notion of functional properties inherent in organic microstructures. He presumes that the structuring of the organism requires as its *explanans* a force capable of foreshadowing the structural/functional organization to be achieved, a force that would embody a kind of immanent plan and actualize it while adapting to external and internal circumstances affecting organic development. By analogical extension, the functional properties of the developed organism, starting with Hallerian irritability and sensibility, will be represented by special forces: the correlated action of these will depend on the generative force which, as it unfolds, manages to integrate the organism and its various functions.

The starting point for the studies on "formative tendency" (*Bildungstrieb, nisus formativus*) is afforded by experiments on the reconstitution of structures after various mutilations in *Conferva fontinalis* and fresh water polyp. This is the basis for Blumenbach's treatises *Über den Bildungstrieb und das Zeugungsgeschäfte* (1781), *De nisu formativo et generationis negotio nuperœ observationes* (1787), and *Über den Bildungstrieb* (1789a). But, for the sake of comparing functional properties, Blumenbach relies on further observations concerning the fetal production of abnormal membranes, bones and vessels: though these cannot be attributed to preexisting germs, they manifest a structuring activity bending towards a certain functional combination of parts. In abnormal ossifications which guide Blumenbach's analysis, one cannot suppose a structural predetermination for a development that shows up as strictly contingent, but one should admit the equivalent of a design for functional preservation in the structure issuing from abnormal development: such phenomena as would be assigned to a kind of *natura medicatrix,* require in fact a formative force that implies a plan for functional compensation.

As Newton's force of attraction interpreted in skeptical terms, that is to say without any deductive link with the general mechanical order of causes, the *Bildungstrieb* means a force whose determination as an occult quality consists in correlating a certain set of empirical effects, independently of any assumption concerning their causal derivation[9]. The whole set of

---

9. Cf. Blumenbach (1799, § 9, n. 2, 18): "Hoffentlich ist für die mehresten Leser die Erinnerung überflüssig, daß das Wort Bildungstrieb selbst so gut wie die Benennungen aller andern Arten von Lebenskräften an sich weiter nichts erklären, sondern eine besondere (das Mechanische mit den zweckmässig Modificirbaren in sich vereinende) Kraft unterscheidend bezeichnen soll, deren constante Wirkung aus der Erfahrung anerkannt worden, deren Ursache aber so gut, wie die Ursache aller andern noch so allgemein anerkannten Naturkräfte für uns hienieden im eigentlichen Wortverstande 'qualitas occulta' bleibt."

correlated phenomena consists here in the sequence of epigenetic processes from an initial amorphous state to the achievement of an organism comprising integrative dispositions and a complex functioning. In the treatises influenced by his reading of Kant after 1785 (Lenoir 1980), Blumenbach underlines the specificity of this force in contrast with the properties inherent in the material dispositions of the organism, but also in contrast with the vital forces responsible for determinate organic functions:

> "It is a propension which thus belongs to the vital forces, but which clearly differs from the other sorts of vital forces in the organized bodies (contractility, irritability, sensibility, etc.), as it does from the universal physical forces in bodies in general; this seems to be the first most important force for all generation, growth and reproduction; and, so as to distinguish it from the other vital forces, one can designate it by the name *Bildungstrieb* (*nisus formativus*)" (Blumenbach 1789a, 24-25).

The question is raised concerning the connexion of the *Bildungstrieb* with more specialized forces which nonetheless cannot be reduced to the order of properties resulting from the structure's mechanical dispositions. A fundamental observation introduces the tentative answer. Blumenbach notes that the building force embodying the organic plan exerts its architectonic action in the framing up of the organism, but also in the partial regeneration of structures, if needs be. The more complex the organism gets, the more it develops subordinate structures, and the lesser intervening of the formative force is allowed by the specialized ones; conversely, these specialized vital forces manifest themselves the more markedly, eliciting what might be termed a physiological division of work.

In his analysis of Blumenbach's theory, T.S. Hall (1969, II, 100-105), insists on the plurality of vital forces that are considered part of the *explanans;* he even suggests that a certain *ad hoc* procedure entails the development of such concepts for the sake of fostering hypotheses. In fact, the typology boils down to the following categories: (1) the *Bildungstrieb* as the force of generation and development; (2) contractility or cellular force which resides in the various membranes and rules therefore almost on the whole body — a force not so different from Stahl's *tonus;* (3) (Hallerian) irritability which belongs properly to muscular fibres and is excited by specific stimuli; (4) sensibility or nervous force whose seat is in the nerve marrow: it brings the impressions affecting sense organs to sensibility centres. Contractility, irritability and sensibility are termed common forces, for being associated with typical structures in whichever part they are to be found. But one must add: (5) the proper life (*vita propria*) affecting this or that particular organic structure because of the functions it is endowed with: think, for instance, of the motions of the iris or Fallopian tubes, and of the phenomena related with the several secretions. Specially according to the later texts, the *Bildungstrieb* may be considered as controlling a sequence of unfolding for those forces and as manifesting an architectonic order in the entailment of functions[10]. The cellular force and the proper life have no equivalents in Haller. The former belongs to the elementary constitution of all organic parts: it serves as a base for more complex functions by producing a global functional regulation for the whole organism and maintaining the latter's dynamic disposition. Proper lives own nothing Hallerian, for they involve a manifold of processes and structures (circulatory, lymphatic, nervous). Blumenbach's idea is that these forces intervene as principles of determination and integrate the lower level organic operations required for this or that effect: for instance in the secretion of bile. As for Hallerian properties, they change meaning in the new Blumenbachian synthesis. Thus, Blumenbach admits that sensibility may operate in a way functionally independent from the activity of the brain centre. Some motions can be determined and guided by an integrative action that takes place at the level of the spinal marrow, and even of more localized centres — this is probably a lesson drawn from Robert Whytt rather than Haller (Whytt 1751 and 1755, French 1969). Besides, from now on sensibility comprises a reactivity function which varies with the integration level it operates at Blumenbach accepts certain theses of Johann Gottfried Zinn (1727-1759) and Johann August Unzer (1727-1799) according to which ganglions and plexuses may serve in channeling nervous actions and organizing a vital reaction controlled by sensibility (Zinn 1749, Unzer 1771, Canguilhem 1955). Irritability conceived as a vital force, and not any more as a property inherent in a given structure, reveals itself through a network of complementary effects in such systems of organs as that of blood circulation: the phenomena of irritability show up as functional sequences, due to the function's integrative character which appears irreducible to its particular operative conditions.

Since vital forces according to Blumenbach seem relatively detached from the requirement that they inhere in specific organic devices, how is one to develop a coherent representation within the framework of physiological

10. Cf. Blumenbach (1798, § 44, 35-36): "Ordo quem in enarrandis hisce variis viribus servavi, idem est, quo in homine nascendo et nato *alia post aliam* manifestant. Primo equidem loco nisum formativum efficacem fuisse oportet, antequam de ipsa novi conceptus existentia certiores reddi possimus. Proxima tunc in gelatinoso tenelli embryonis corpusculo agit contractilitas. Post, ubi iam musculosæ carnes effictæ fuerunt, in ipsis earum fibris motricibus irritabilitas. Tum in paucis iis organis quorum motus neque ad contractilitatem neque ad irritabilitatem commode referri potest, vita propria. Denique in homine nato præter eas vires quoque sensilitas."

theory? In his *Commentatio de vi vitali sanguinis* (1788), Blumenbach asserts that no vital force may be postulated in whichever case the effect does not seem to relate to a process maintaining or producing a functionally adapted organic activity. This functionality manifests itself in phenomena elicited by the various vital forces; and it is corroborated whenever we consider pathological alterations of the corresponding functions. In all cases — this is neatly set into relief by deviances — the various forces keep acting and reacting on each other; and one can draw therefrom the notion of a potential for adaptating their respective effects. The relative integration between these categories of closely correlated phenomena refers back to the architectonic connexion between the corresponding specific forces: these are essentially defined as faculties or occult properties that emerging special effects would justify. The relative harmony in the effects gets reflected in the integrative system of vital forces. And we interpret this theoretical construction as implying a hierarchy of functional reasons embodied in the *Bildungstrieb* and in the principles or forces that derive from it as the organic construction goes on. Actualizing the architectonic plan, the *Bildungstrieb* comprises potentially the various forces which will unfold later in operations correlative with the emerging organic structures. The concepts we use in accounting for the formative force involve a teleology of functional organization and they throw light accordingly on the functions that express the vitality of the organism[11].

In my opinion, Blumenbach's physiology features well a certain type of vitalism: it implies that the analysis of physiological phenomena could not be fully achieved if we rested satisfied with locating their *explanans* in the dispositions of the smaller organic machines and structural dispositions, as well as in the forces emerging therefrom. Haller, for instance, had tried to accord the analysis of vital phenomena with theoretical constructions about organic microdispositions and their resulting effects (Haller 1757-1766). Along the new trend, theoretical representation aims at translating the teleological order elicited by phenomena more directly: organic activity reveals a functional integration of processes and reflects a certain type of architectonic disposition. Thus we get concepts of *vital principle* and *Bildungstrieb;* but these need be further detailed into series of models, concerning synergies and sympathies, derivative forces and properties

11. Cf. Blumenbach (1798, § 587, note h, 465): "[...] Contra vero ipsum cardinem in quo universa hæc de nisu formativo doctrina versatur, et qui vel solus sufficit ad distinguendam eam a veterum vi plastica aut Wolffii desideratissimi vi quam vocabat essentiali, aliisque id generis hypothesibus, in connubio consistere binorum principiorum explicationis naturæ corporum organicorum, physico-mechanica inquam cum mere teleologico."

apparently supervenient on complex organic structures (contractility, irritability, proper life, sensibility), modes of the self-preservation, and special powers of acting[12]. This architecture of principles and subordinate concepts serves a twofold purpose: (1) determining a system of sufficient reasons for classifying and ordering physiological data: (2) representing at the theoretical level the *architectonic competence of nature* in forming and animating organic structures. This second objective outrides the first and provokes a move away from both iatromechanism and animism, as well as from the analytic system of physiological properties Haller had sketched. Indeed, this does not mean that the new system avoids or discards speculative constructions. On the contrary, this theory shows its speculative purport in integrating teleological concepts of a metaphorical or reflexive origin, concepts which appeal to an immanent finality in order to interpret functional effects and processes. But the physiologist's intent was to use them for analytic and heuristic purposes. Blumenbach will combine an analysis of complex structures and derivative forces in Hallerian style with an epigenetist hypothesis concerning organic formation and regeneration. The *Bildungstrieb* concept inserts architectonic dispositions as a virtuality within a special force, in presumed analogy with Newtonian properties. And this *Bildungstrieb* determines what interpretation shall be given for the subordinate and more specialized forces which are viewed as combining functional self-regulation with inherence in complex organic structures.

* * *

In a neo-mechanist and reductionist perspective, C.F. Wolff supposed the organism frames up by processes which would prove analogous with those of inorganic nature, but caused and regulated by an "essential force" (*vis essentialis*). Drawing his means from embryological observations and experiments, he would define this specific vegetative force by reference to such phenomena as would illustrate its effects. This undertaking was methodologically complex, but it has served to prime a program of embryological morphogenesis (*Entwicklungsgeschichte*), which develops in the first decades of the 19th century with Heinrich Christian Pander, Carl Ernst von Baer and Johannes Müller (Lenoir 1982, Duchesneau 1987). One can even link the still later program of *Entwicklingsmechanik* with the Wolffian tradition[13]. However, Wolff's *vis essentialis* entailed a fundamental

12. On the theoretical styles among the principal vitalist physiologists at the end of the 18th century, cf. Duchesneau (1985).

13. This methodological trend intends to do away with historical (= developmental) or phenomenological pseudo-explanations for the benefit of eliciting the physico-chemical

ambiguity: as a principle, it was held indirectly responsible for the architectonics of the complex organism to be achieved and for the specialized functions it would possess. As a consequence, it seemed difficult to reject the tendency to refer to the *vis essentialis* as the virtual form of all vital activities including the more complex. Therefore, could not a preformationism of the virtual conform with the order of things if the analysis of vital phenomena as they unfold were pursued, starting with the initial phases of embryological development? It is thus easy to understand the vitalist turn Blumenbach initiated in analyzing the forces acting and the processes emerging in embryogenesis. The concept of "formative force" (*Bildungstrieb, nisus formativus*) which he frames up and which inspires Kant in his *Kritik der Urteilskraft* (1790)[14], means a specific force embodying the architectonic plan for the organism to be developed. The emergence of complex structures would depend on *sui generis* laws under the ægis of specialized forces, and would prove irreducible to material self-organization of the type Wolff had envisioned. The Blumenbachian program will set a deep mark on 19th century biology, even when biologists will try and emancipate themselves from its vitalist underpinnings. As a consequence, the tensions between the two rival and correlative programs will be found surfacing in more recent doctrines about the generation of vital forms and the emergence of integrative structures and functions. A question got to the fore in the last decades of the 18th century to which Wolff's and Blumenbach's theses can be viewed as having afforded tentative answers. It concerned the nature of such an inherent principle in organic matter as may produce and mould an organism whose essential features are architectonic integration and self-regulation. It was my purpose to show that the physiologies of Wolff and Blumenbach provided an initial, highly polarized, framework for some upcoming debates.

REFERENCES

Adelman Howard B. (1966). *Marcello Malpighi and the Evolution of Embryology.* Ithaca (N.Y.): Cornell University Press, 6 vol.

Bernardi Walter (1986). *Le metafisiche dell'embrione. Scienza della vita e filosofia da Malpighi a Spallanzani.* Firenze: L.S. Olschki.

mechanisms that would explain concretely embryogenetic organization and differentiation sequences: cf. in particular Wilhelm His (1874) and Wilhelm Roux (1881).

14. Cf. in particular Kant (1790, § 81: Von der Beigesellung des Mechanismus zum teleologischen Prinzip in Erklärung eines Naturzweckes als Naturproduktes, 292).

Blumenbach Johann Friedrich (1781). *Über den Bildungstrieb und das Zeugungsgeschäfte.* Göttingen: J.C. Dieterich.

Blumenbach Johann Friedrich (1787). *De nisu formativo et generationis negotio nuperæ observations.* Gottingæ: apud J.C. Dieterich.

Blumenbach Johann Friedrich (1788). *Commentatio de vi vitali sanguinis.* Gottingæ: apud J.C. Dieterich.

Blumenbach Johann Friedrich (1789a). *Über den Bildungstrieb.* Göttingen: J.C. Dieterich.

Blumenbach Johann Friedrich (1789b). *Zwo Abhandlungen über die Nutritionskraft... Nebst einer fernern Erläuterung eben derselben Materie von C.F. Wolff,* St. Petersburg: Kayserliche Akademie der Wissenschaften.

Blumenbach Johann Friedrich (1798). *Institutiones physiologicæ, editio nova auctior et emendatior.* Gottingæ: apud J.C. Dieterich.

Blumenbach Johann Friedrich (1799). *Handbuch der Naturgeschichte, 6. Ausgabe.* Göttingen: J.C. Dieterich.

Canguilhem Georges (1955). *La formation du concept de réflexe aux XVIIe et XVIIIe siècles.* Paris: Presses Universitaires de France.

Duchesneau François (1982). *La physiologie des Lumières. Empirisme, modèles et theories.* La Haye-Londres-Boston: M. Nijhoff.

Duchesneau François (1985). Vitalism in Late Eighteenth-century Physiology: The Cases of Barthez, Blumenbach and John Hunter. In W.F. Bynum and R. Porter (eds), *William Hunter and the Eighteenth-century Medical World.* Cambridge: Cambridge University Press.

Duchesneau François (1987). *Genèse de la théorie cellulaire.* Paris: Vrin; Montréal: Bellarmin.

French Roger K. (1969). *Robert Whytt, The Soul, and Medicine.* London: Wellcome Institute of the History of Medicine.

Hall Thomas S. (1969). I*deas of Life and Matter. Studies in the History of General Physiology 600 B.C. – 1900 A.D.* Chicago: University of Chicago Press, 2 vol.

Haller Albrecht von (1757-1766). *Elementa physiologiæ corporis humani.* Lausannæ: Sumptibus M.-M. Bousquet (F. Grasset; Bernæ, Sumptibus Societatis typographiæ), 8 vol.

Haller Albrecht von (1758). *Sur la formation du cœur dans le poulet; sur l'œil; sur la structure du jaune.* Lausanne: M.-M. Bousquet, 2 vol.

Haller Albrecht von (1762-1768). *Opera minora emendata, aucta, renovata.* Lausannæ: Sumptibus F. Grasset, 2 vol. His Wilhem (1874). Unsere Körperform und das physiologische Problem ihrer Entstehung, Leipzig, F.C.W. Vogel.

Kant Immanuel (1790). *Kritik der Urteilskraft.* Hamburg: Felix Meiner Verlag, 1974.

Lenoir Timothy (1980). Kant, Blumenbach, and Vital Materialism in German Biology. *Isis,* **71**, 77-108.

Lenoir Timothy (1982). *The Strategy of Life. Teleology and Mechanics in Nineteenth Century German Biology.* Dordrecht: D. Reidel.

Monti Maria Teresa (1990). *Congettura ed esperienza nella fisiologia di Haller. La riforma dell'anatomia animata e il sistema della generazione.* Firenze: L.S. Olschki.

Roe Shirley A. (1981). *Matter, Life, and Generation. Eighteenth-century Embryology and the Haller-Wolff Debate.* Cambridge: Cambridge University Press.

Roger Jacques (1971). *Les sciences de la vie dans la pensée française du XVIIIe siècle,* 2e éd. Paris: A. Colin.

Roux Wilhelm (1881). *Der Kampf der Theile im Organismus. Ein Beitrag zur Vervollständigung der mechanischen Zweckmässigkeitslehre.* Leipzig: W. Engelmann.

Unzer Johann August (1771). *Erste Gründe einer Physiogie der eigentlichen Natur der thierischen Körper.* Leipzig: Weidmanns, Erben & Reich.

Whytt Robert (1751). *An Essay on Vital and Other Involuntary Motions of Animals.* Edinburgh: Printed for John Balfour, 1763.

Whytt Robert (1755). *Physiological Essays.* Edinburgh: Hamilton, Balfour & Neill.

Wolff Caspar Friedrich (1759). *Theoria generationis.* Hildesheim: G. Olms, 1966.

Wolff Caspar Friedrich (1764). *Theorie von der Generation.* Hildesheim: G. Olms, 1966.

Wolff Caspar Friedrich (1769 et 1773). De formatione intestinorum. *Novi commentarii Academiæ Scientiarum Petropolitanæ,* **13**, 478-530; **17**, 540-575.

Wolff Caspar Friedrich (1774). *Theoria generationis.* Editio nova aucta et emendate. Halæ ad Salam: Typis et sumtu J.C. Hendeli.

Wolff Caspar Friedrich (1789). *Von der eigenthümlichen und wesentlichen Kraft der vegetabilischen sowohl als auch der animalischen Substanz.* St. Petersburg: Kayserliche Akademie der Wissenschaften.

Zinn Johann Gottfried (1749). *Dissertatio inauguralis medica sistens experimenta quœdam circa corpus callosum, cerebellum, duram meningem, in vivis animalibus instituta.* Gottingæ: apud A. Vandenhoeck.

PHILIPPE GOUJON

FROM LOGIC TO SELF-ORGANIZATION – LEARNING ABOUT COMPLEXITY

ABSTRACT

This article traces how Second-Order cybernetics came into being. It emphasizes the objections raised against first-order cybernetics and, in doing so, describes the process whereby a new type of epistemology — an epistemology of the observer — appeared in the United States at the end of the 1940s. At the same time it explains the implications of this epistemology.

1. AN OVERVIEW OF THE LOGICAL FORM OF MACHINES: FROM LOGIC TO THE UNIVERSAL MACHINE

From Plato to Hilbert via Aristotle, Descartes, Leibniz, Morgan, Bool, Peano, Russell, Whitehead and Frege, it becomes gradually more and more obvious that mathematical logic implies manipulating symbols according to rules which have been clearly defined in advance. This line of thought opens the way for a calculatory and computational paradigm. Church, Kleen, Gödel, Turing and Post's work formalized the notion of a logical sequence of stages which led them to recognize that the essential element in any mechanical process which regulates its own dynamic behavior is an abstract control structure or program. It was discovered that the essential characteristic of any machine did in fact come from its logical form and not from its material structure. Alan Turing recognized that new, abstract machines, programmed by putting their instructions into code and operating via a series of codes, provided a new vision of what he called the human

B. Feltz, M. Crommelinck and P. Goujon, (eds.), Self-Organization and Emergence in Life Sciences: 187-214.

calculator[1]. In order to tackle problems connected with thought and the brain, the most important thing for Turing was not the physical structure of the brain, but rather its logical organization which could supposedly be replicated by another type of physical mechanism. His thesis was that the mind could be accurately described as a Turing machine because it described the world with the same degree of complexity, that of discreet logical systems. The process was therefore not one of reduction, but rather an attempt to transfer natural systems to an artificial brain.

After the last war such a hypothesis took on a very concrete significance. A large number of decoding machines, which were essentially the same as Turing machines, had already been produced.

In this way many quite unheard of possibilities opened up which remained to be explored experimentally by creating a universal machine, a Turing machine, which, once it had been built, could imitate the behavior of any other machine. As he thought, with the help of his assistant Don Bailey, about possible ways of building his machine, Turing imagined what he called an A.C.E., or Automatic Computing Engine, the prototype of what would later become the computer.

With his abstract representation, Turing had not only invented the computer, he had also played his part in the birth of a new type of mechanism and a symbolic, logical, operational and even information-based understanding of nature itself.

Turing's challenge — to build a bridge between the symbolic and the physical — would give rise to much interest and influence a large number of cyberneticians as well as leading to an understanding of life itself as a formal system.

Once machines have been thought of in terms of their abstract specifications, their organization, it becomes possible to understand living systems in the same way. Also, if one has a universal Turing machine on hand, it becomes possible to simulate them. This is what enabled the computer, that genuinely universal machine, to come into being.

The amazing capacity for imitation present in computers allowed them to explore the behavior patterns of a considerable number of potential machines. It is hardly surprising that, during the war and then during the fifties and sixties, there was a growing interest in electromechanical and computational models of artificial life.

---

1. Turing A. (1936). On Computable Numbers with an Application to the Entscheidungs Problem. *Proceedings of the London Mathematical Society,* 2e serie, Vol. **42**, 3e part, November 12, 230-265 and (1937) 2e serie, Vol. **43**, 7e part, May 20, 1937, 546-550; (1950) Computing Machinery and Intelligence. *Mind*, Vol. LIX, **236**.

2. CYBERNETICS OR A NEW WAY OF REPRESENTING PHENOMENA

In the 1940s the logico-mathematical traditions and the data processing tradition developed in the specific area of telecommunications had already converged and played their part in making information and communication the key concepts in a whole new way of representing phenomena.

Cybernetics, defined by Norbert Wiener in 1948 as the study of "control and communication theory, whether in the machine or in the animal", came out of the work done by Wiener and Julian Bigelow during the early months of 1940 in order to design a machine which could automatically control anti-aircraft gun fire. The very practical problem of predicting the future position of the aircraft revealed, from a mathematical point of view, the need for extrapolation. While they were working on this problem, Wiener and Bigelow became aware of the importance of feedback. At the same time they saw that inappropriate feedback (too much or not enough) could be harmful. With insufficient feedback it was impossible to make accurate adjustments. On the other hand, too powerful feedback led to the line of fire being adjusted too far, which made another adjustment in the other direction necessary and so on and so forth in interminable oscillations. In the first case, the situation resembles that of a man or an animal suffering from ataxia, a nervous complaint which leads to uncoordinated body movements. Wiener and Bigelow asked Arturo Rosenbluth if the second type of problem also occurred in humans or animals. Rosenbluth replied immediately that one example of this was the involuntary shaking which sometimes occurs in patients suffering from brain injuries.

Wiener, Bigelow and Rosenbluth were therefore led to realize that feedback played a similar role in a wide variety of natural and artificial systems, and that an interdisciplinary research project into the way teleological machines function, both when they are working properly or when they malfunction, could well reveal important information about similar mechanisms in living organisms. A new conception of theoretical biology, of the art of model-making, and of science itself would grow out of this discovery which undoubtedly led to the appearance of Second-Order cybernetics and of speculations about the possibility of creating artificial life.

Two articles were published in 1943 which may be considered as the birth certificate of cybernetics and of a new way of perceiving both Man and the world where logic and information would progressively gain in importance.

The first article was Wiener, Bigelow and Rosenbluth's Behaviour, Purpose and Teleology[2]. Essential to any understanding of the subject, it provided a new vision of the world. Reality can be interpreted entirely in terms of information. The only thing that matters is the logic behind events and behavior. The behavioral method of study would lead Wiener to emphasize the notions of information and communication.

The other article, Walter Pitts and Warren McCulloch's *A Logical Calculus of the Ideas Immanent in Nervous Activity*[3], presented for the first time how the behavioral method could be applied to the study of the brain. It demonstrates a type of Turing machine which may be considered, on account of its structure and behavior, as an ideal representation of the anatomy and physiology of the brain. *We have become machines.*

Recognizing the organisational role of communication and information (which belong to the realm of physics in Wiener's opinion), which is what gives cybernetics its identity as the science of control and communication, led the way to the creation of a vast movement. In particular, it formed the basis of a complex theory of organization and, more generally, of a triple revolution: an epistemological revolution (the rehabilitation of the analogical method and a new epistemology of the machine), an ontological revolution and a transdisciplinary revolution. Seeing as cybernetics presupposes, without specifically stating it, the physical nature of all systems, and seeing as it sets itself up as being founded on the principle of organizational communication, there can no longer be, in cybernetics, any barriers between physics, biology, sociology and anthropology — at least at a certain organizational level. This explains how cybernetics can encompass at one and the same time both the world of machines and that of natural automata.

Wiener, Rosenbluth, McCulloch and Pitts, in addition to Von Neumann and Morgenstern[4] with their logical representation of actors in the field of economics, joined together in their manner of seeing Man through the lens of logic, information and communication theory as transparent, with no hidden depths. What matters is no longer the physical structure but rather the information contained within. The appearance of the first computers at this period reinforced this way of looking at Man.

2. Rosenblueth A., Wiener N. and Bigelow J. (1943). Behavior, Purpose and Teleology. *Philosophy of Science*, Vol. **10**, 18-24.

3. McCulloch W. and Pitts W. (1943). A Logical Calculus of the Ideas Immanent in the Nervous Activity. *Bulletin of Mathematical Biophysics,* **5**, 115-133.

4. Von Neumann J. and Morgenstern O. (1980). *Theory of games and economic behavior,* First edition, Princeton University Press, 1944, Sec. ed., 1947, Third ed., 1953, Princeton University Press.

The formal, information-based view of life is at the very center of cybernetics. It would also be at the heart of the theory of self-reproducing automata developed by John Von Neumann in the last years of his life.

The technology of control systems which, in its progressive form, had led to the creation of cybernetics, would, in its discreet form, lead to Von Neumann's robots.

2.1 Von Neumann and the Theory of Self-Reproducing Automata

The first truly computer based attempt to solve the problem of how to generate behavior which imitates that of natural automata was the work of a genius, the Hungarian mathematician John Von Neumann. According to Arthur W. Burks, at the end of the 1940s Von Neumann was looking into the following problems: "What kind of logical organization is needed to make automata self-replicating? This question is quite vague and includes both trivial aspects of the question and those which are really interesting."

As he asked this question, Von Neumann was thinking of the phenomenon of self-reproducing organisms. However, he did not attempt to copy a natural self-reproducing system on the genetic or biochemical level. He hoped to isolate the logical aspect of the problem of how a system can reproduce itself[5]. For Van Neumann it was not a question of analyzing the internal structure of such organisms, but rather of examining their behavior when faced with certain unambiguous stimuli. The organisms in question were therefore perceived as functioning like black boxes in aircraft. As Van Neumann intended to consider at the same time how these organisms were constructed from elements which in themselves resembled black boxes, his perspective gave rise to a theory about how these black boxes were produced and then programmed, in other words a theory of automata.

Von Neumann's cellular automata are good examples of this kind of computational paradigm which would later occur in the context of artificial intelligence: local determination of behavior combined with a parallel, upward approach.

Von Neumann demonstrated that it was possible to build 29-state cellular automata which contained a Turing machine, were self-replicating and could reproduce any other type of Turing machine.

5. Burks A.W. (ed.) (1970). *Essays on Cellular Automata.* Urbana: University of Illinois Press.

Whether we consider Von Neumann's approach or that of the cyberneticians, we see that a new form of mechanical representation of living organisms came into being at that time. These two approaches resemble each other insomuch as neither analyses the internal structure of organisms, but both examine their behavior when faced with certain unambiguous stimuli. In both cases, the material composition of the systems is of no importance compared with the logic underlying events and behavior.

2.2 The Post-war Period

During the years which followed the publication of Von Neumann's book and Wiener, Bigelow, Rosenblueth, McCulloch and Pitts's work, other academics latched onto and followed the same intuitions and basic ideas, extending them, simplifying them and suggesting alternative models to explain and classify the behavior of living organisms. In the fifties the mathematical approach took on greater importance. The founders of artificial intelligence — Herbert Simon, John McCarthy and Marvin Minsky — supported the notion that intelligence was a mechanism in Turing's sense of the word. For such orthodox cognitivists as Simon and Minsky any rigorously accurate description potentially had its equivalent computer programme. Seymour Papert considered that computer science was above all concerned with describing complex forms of behavior. The connectionists, following Warren McCulloch's lead, defined the brain as a machine for processing information, and neurons as data processors. The processes devised from this research remained in equation form, but the success of computer-based ontology was as spectacular as that of molecular biology.

A cybernetic revolution did then truly take place. Yet, it had its limits and its lacks.

3. THE LIMITATIONS OF FIRST-ORDER CYBERNETICS

Cybernetics, in its way of conceiving machines, after having gone beyond the reductionism which broke everything down into its constituent parts, developed a new kind of reductionism which equated every living machine/being and in fact just any machine or being at all, with the artificial machine model.

By reducing every information program to its basic substance, cybernetics tended to overlook the limitations of the reified model where everything is simply artificial, having no existential, ecological or organizational understanding of being open to outside influences because the information it contains is always pre-existant and directionless.

Cybernetics, insomuch as it conceived of machines as being autonomous (only an artificial autonomy resulting from the fact that society had been brushed aside), certainly showed the need for a theory of the essential nature of machines, but it omitted to actually elaborate the theory itself, sticking instead to artificial machines which, for all their complexity, are simple when your compare them with those natural machines which can program themselves.

Cybernetics in general, or rather the mechanistic wing of cybernetics, did not seem in the slightest bit interested in the problem of evolution. On the contrary, it was only interested in regulating and stabilizing machines or correcting technical errors. It had completely brushed aside one of the two faces of that basic concept *feedback:* the positive face, which should naturally be considered in partnership with the negative face. It had completely brushed aside the part played by noise in the growth of heterogeneity and in transforming systems. Noise can in fact be a source of morphogenesis or of those instances of evolution which occur to overcome a flaw.

This bias against 'chance' in the first cyberneticians can be seen, as J.P. Dupuy has pointed out, in the proceedings of the 8th conference of the Josiah Macy Jr Foundation.

"The fact that chance may, in certain circumstances, bring meaning is not however, over all, a theme which cyberneticians readily welcome. The British researcher Donald Mackay, who was invited to the 8th conference, learnt this to his cost. After having presented his idea of an automaton which would be able to make inductive inferences by means of various random strategies, he was severely criticized by Leonard Savage, himself a statistician. Savage went on at length on his theme that including a element of chance in the working of a machine could in no way help it to imitate human behavior and, in any case, would definitely not make it more effective in solving problems"[6].

In fact, cybernetics had ignored the limits of artificial machines and, by doing so, deliberately disregarded the epistemological division between living and artificial organisms and the variations and multiple dimensions

6. Dupuy Jean-Pierre (1985). L'essor de la première cybernétique (1943-1953). *Histoires de cybernétique.* Cahier du CREA, **7**, Paris: CREA, 67.

which exist in reality. It took upon itself the right to decide the difference between a signal and a noise and, considering the things it was studying as no more than means of transforming input into output. This way of thinking, if the leading lights of Second-Order cybernetics from Von Foerster to Varela are to be believed, led to a certain confusion between signals and information. This mechanistic and ideological reductionism came in the long run from the limitations of the model based on artificial machines. It prevented first-order cybernetics from rising to the level of complexity found in living organisms and, in the same way, blinded it to those purely mechanistic aspects of its work which led it to disregard anything which seemed to be less than strictly rational.

At this stage of our analysis, we are perhaps better equipped to grasp the significance of Simondon's criticism[7], which underlined the fact that the initial theory — that living creatures and self-regulating machines were analogous — could act as a brake to progress. If researchers stick to this theory, they will also be sticking to an artificial, over-simplified concept of reality which cannot grasp systems whose unity derives from their substance but only those, and these only too well, whose unity derives from their accidents, a result of precisely determined elements and the relationship between them.

Cybernetics lacked a complexity principle so that it could fully develop the epistemological revolution inherent in the idea of organizing communication channels. Cybernetics lacked a complexity principle which would allow it to include the idea of disorder. That is why it was, at least at the beginning, incapable of understanding the concept of systems which are continually reorganizing themselves or of that conflict which brings about the reorganization of natural machine/beings or of the existential, ecological and organizational meaning behind the idea of openness to input from outside.

3.1 Towards an Organization theory

Both cybernetics and information theory have used a complex organization theory which takes further the relationship between information, organization and noise perceived by Shannon and Wiener.

The first move towards a greater complexity was introduced via the problem of model-making. The model, which is an image of reality beyond the academic's control, must however represent reality as accurately as

7. Simondon (1969). *Du mode d'existence des objets techniques.* Aubier-Montaigne, 137.

possible and, in the most extreme cases which were envisaged by Wiener and Rosenbluth in their article entitled *The Role of Models in Science*[8], may even be a clone of the original object and therefore exactly equivalent to it.

In this way the model progressively moved further away from reality. Instead of the model being clearly subordinate to its original, it gradually became identical and therefore equal to it, which resulted in the notion of a model undergoing certain modifications. Cybernetic models are already post-structuralist models. They are only models of themselves or of other models, reflecting no real objects but just mirrors. The only scholar to seriously attempt to justify this gradual emancipation from reality was Van Neumann.

This move towards complexity had already appeared in the U-turn accomplished by Van Neumann. In his opinion, and Wiener and Rosenbluth had also supported this point of view in 1945, a model should try to reach the same level of complexity as the original in order to become not merely a model of the object in question, but also a model of itself or of its own behavior, no longer referring back solely to itself or focussing all the attention on itself. Unlike the technocentric approach, this type of model brought back reality as the main reference. As the model frees itself, becoming more autonomous and more complex, the scientist seeks to grasp the principles behind that autonomy and complexity.

In this way Van Neumann presented the problem of complexity and, as a consequence of this, revolutionized the philosophy of model-making. Although I would not like to minimize the obvious importance of his work, in general cyberneticians did not encounter the problem of complexity as a result of his influence. They mainly encountered it in the work of certain scientists whom they invited to the Macy conferences, in particular Weiss, Lashley, Bavelas and above all W. Ross Ashby, whose work provided the springboard for a genuine cybernetic revolution.

4. CHALLENGING CYBERNETICS

Shannon had constructed an information theory which excluded any mention of meaning. However, it is fascinating to observe the way in which, as the Macy conferences progressed, the discussion gradually started to bypass or to challenge this exclusion and reintroduce the question of meaning. If attempts were made at the Macy conferences to introduce

8. Rosenblueth A. and Wiener N. (1945). The Role of Models in Science. *Philosophy of Science*, Vol. **12**, 4, 316-321.

complementary theories to Shannon's, theories concerning semantic information — and in particular by Donald Mackay at the eighth conference and by Bar-Hillel and Carnap at the tenth — we can affirm that the problem arose with the greatest clarity after presentations made by psychologists. The experimental psychologists present, although in many ways close to the cyberneticians, preferred to take information as their frame of reference but, in doing so, were criticized by their gestaltist or holistic colleagues, who never failed to underline the fact that, in spite of their precautions, the meanings they thought they had quashed kept slipping back into their experiments.

For instance, as early as the sixth conference, John Stroud summarized the experiments he was making for the navy in order to calculate the maximum quantity of information (in Shannon's sense of the word) a man could assimilate in a given time unit. Kluver had no trouble proving that these results were neither useful nor particularly interesting because the man's capacity would vary enormously according to whether he had to assimilate a series of meaningless symbols or sequences which made some sort of sense. A conflict broke out between those delegates who thought they could ignore the question of meaning (information conceived as merely physical) and those who thought that this was impossible. This conflict did not only concern the nature of information, but also two opposing conceptions of the nature of rationality.

4.1 Lashley, Weiss, Bavelas and Birch: Introducing the Complexity Debate

One of the characteristics of first-order cybernetics was, to quote Dupuy, its artificiality: that approach which consists of taking a function and then trying to find the structure which will make this function possible, moving from the whole to the constituent parts and what joins them together, while supposing that the functions determine the structure. McCulloch was the first to support this approach, although he was also criticized for his atomistic view of neural function.

Very early on; this over-simplified, reductionist, artifical and mechanical, pseudo-holistic approach (this type of holism is always artifical) was, from the Hixon Symposium in 1948, confronted with another way of understanding those organized forces which opposed it.

First of all, Lashley[9] criticized McCulloch from a purely neurophysiological point of view. He contradicted the dominant theory, which supposed that signals entering the sense organs went into a system where the majority of the neurons were at rest and followed a predetermined route in order to produce output. Lashley underlined that the brain should be seen as a huge network of reverberating circuits in perpetual motion, a network which, in the absence of external stimuli, settles down into regular patterns of activity fulfilling several functions. Sensory imput acts on these circuits causing them to reorganize themselves in order to produce output. Output is therefore the result of interaction between a trigger event and a global structure which is made up of a whole system of spontaneously interacting neurons.

What Lashley was criticizing, in addition to McCulloch's atomistic approach, was above all that his conception of cybernetics neglected the fact that natural systems are autonomous, and that he saw the nervous system as a kind of machine for transforming incoming messages into outgoing messages, in fact as a cybernetic machine. This criticism is important because it includes, although not explicitly, the criticism which the Second-Order cyberneticians would later formulate concerning the confusion existing in first-order cybernetics between signals and information.

Information produced cannot be reduced to a structure, nor even a collection of structures. It is rather a variety of structures. The point of view of the engineer who tended to put information on a pedestal had to be put into perspective. In the same way, and this was stressed by the Second-Order cyberneticians, the difference between noise and signals also needed modification because, if it was to be seen from the point of view of its organization or its host organism, there was nothing to prevent a trigger event acquiring meaning.

This was stated explicitly at the seventh Macy conference where, for example, Lawrence Franck explained that "every culture creates a world by selecting from the background noise of events, certain signals which it treats as messages by giving them a meaning"[10].

---

9. Lloyd A. Jeffres (ed.) (1951). Hixon Symposium. *Cerebral Mechanisms in Behavior.* California Institute of Technology, September 1948, New York: John Wiley and Sons, 70-71 et 112-133, *op. cit.,* 74-75.

10. Von Foerster H., Mead Margared and Hans Lukas Teuber (eds) (1950). Cybernetics-circular Causal and Feed-back Mechanisms in Biological and Social Systems. *Transactions of the Seventh Conference,* March 23-24, New York: Josiah Macy; New York: Jr. Foundation, 1951, 153-154.

Returning to the Hixon Symposium, Lashley's criticism was picked up and enlarged by Weiss[11] using language much closer to that used to describe self-organization theory.

In short, he reproached McCulloch with having neglected the main characteristic of the nervous system, it's basic autonomy. Unlike McCulloch, who saw the nervous system as a mere machine which transformed incoming messages into outgoing messages, Weiss emphasized the fact that the nervous system is a system with its own internal coherence. Stimuli, or input, could choose to trigger different autonomous modes in the brain pattern and could possibly even modify them. Weiss, speaking as an embryologist, in his book *Science of Life: The Living System,* underlines that "we still need to know how a mass of molecular activity can turn into an all-embracing, integrated system and how the varied, imprecise behavior of individual cells can bring about organs which, in any one species, resemble each other far more than do the detailed processes of morphogenesis which bring them into being… When we face up to these problems, the concept of "transferring information" falls apart like a train which, before reaching its destination, hits a broken rail, is derailed and gets stuck in a sandy desert. It could only finish its journey if a previously installed automatic pilot took over automatically when the rails failed. What may appear to us as an unstructured void is not necessarily a desert, but may be a genuine system in which an overall dynamic process has replaced the mechanical type of guidelines."

The reductionist, cybernetic point of view was therefore seen to contradict another kind of logic. These two opposing conceptions, or ways of reasoning, were seen to be contradictory at the eighth Macy conference in March 1951, during the discussion which followed the social psychologist, Alex Bavelas's, paper. He had talked about the psychology of small groups. During his research he had conducted several experiments both in his laboratory and in factories. These experiments involved giving a limited number of people a task to perform together which would require them to exchange information among themselves, while at the same time limiting the possibles means of communication. He described some of his experiments to the conference and in particular this experiment, which was one of the simpler ones: "Five people, completely isolated from one another, each have to write on a piece of paper a number between 0 and 5 and the total of the five numbers must be 17. First method: after each attempt the research assistant announces the total and they start again until they hit on the right

11. Weiss, Hixon Symposium, *op. cit.,* 72-74 et 140-142.

total. Second method: the research assistant just says 'No good. Start again' until they hit on the right total. Bavelas' experiments showed with no possible ambiguity that all the groups reached the correct total faster with the second method than with the first."

The cyberneticians, imprisoned in their overly technical point of view which prevented them from perceiving any information as negative, by their engineer's mentality which led them to pronounce authoritively on what is information and what is noise, and by their hope, nourished by Turing's 'theorem', that it will always be possible to create a machine which can reproduce everything that men can do, saw this apparent paradox as completely irrational.

On the other hand, Mackay's reaction was completely different. At the eighth Macy conference he presented his idea for an automaton which would be able to make inductive inferences using random strategies. He too met with strenuous opposition. Together with Kubie, Kluver and in particular Bavelas, he sought to understand the reason behind Bavelas' results. Leaving the aim of the experiments to one side, he focussed on the structure of the group and its behavior — in other words, on the complexity issue.

"Bavelas suggests a form of reasoning based on speculation. When the research assistant gives the total, everyone starts to put himself in the others' shoes and notices that the others are doing the same for him. Therefore the uncertainty increases and the groups studied tried to compensate for that by working out theories about role distribution in the adjustment process"[12].

The importance of organization and complexity is also underlined by Birch who, at the same conference, supported the view that the most complex forms of behavior observed in the animal kingdom ultimately result from their dependence on a certain organization of the senses which, if their circumstances change, can provoke most unacceptable behavior. "The intelligence of the whole does not, he concluded, lead us to infer that the constituent parts are also intelligent, as these do not have collective behavior as their aim"[13]. Birch's hypothesis only met with a limited amount of resistance from the cyberneticians who only protested because the concepts he used were, from their perspective, meaningless, as they were based on such notions as intelligence, conscience, memory and the learning process.

We can therefore surmise that the cyberneticians were beginning to learn about complexity from their contact with concrete examples of experimental science. We should not forget that this confrontation between the artificial

12. Dupuy J.P., L'essor de la première cybernétique, 1943-1953, *op. cit.,* 69.
13. *Ibidem,* 87.

conception of organized systems favored by the cyberneticians and the other view supported by a group of academics including neurophysiologists like Lashley and Gerard, psychologists like Kohler and Kluver and, above all, the embryologist Weiss, introduced a certain number of theses and hypotheses into the original arguments. In particular we should mention the idea that chance and meaning are two sides of the same coin and that, if we face the problem from the point of view of the way the organism being studied is organized, outside stimuli can give birth to meaning. They can also help us recognize the main characteristic of natural systems : their autonomy. These are the theses, hypotheses and ideas which we find at the heart of Second-Order cybernetics and which, even if they would still be severely criticized by some cyberneticians, would open new perspectives for cybernetics.

The cyberneticians' reticence in accepting these new ideas, and in particular the revolutionary idea that chance can produce meaning, above all the strongly contested theory that introducing random processes into a mechanism helps it to imitate human behavior more accurately or, at the very least, as the advocates of the gestaltist and holist schools maintain, increases its efficiency when trying to solve problems, would be revealed in dramatic fashion when, as Dupuy put it, "the Ashby tornado swept away the ninth Macy conference".

4.2 Ashby's Homeostat: a New Mechanism and the Beginning of a New Era

W. Ross Ashby only took part in one Macy conference, the ninth. At this conference he presented two papers: one on his not yet famous homeostat and the other on the following problem. Can a chess playing machine beat his creator in a match? Ashby was convinced that by playing at random, this should be possible. With a specially arranged chessboard which looks inoffensive, but where one particular move would lead to certain victory, the random player will certainly find it one day because, for him, no move is logically excluded. The machine does not know which is the best move, it just tries out, randomly or systematically, all the possible combinations[14]. The homeostat is the mechanical 'incarnation' of this principle. This cybernetic automaton which was not designed to perform any particular task and which claims to reproduce the way the brain interacts with its environment was supposed to illustrate a thesis which Ashby considered

14. de Latil Pierre (1953). *La pensée artificielle.* Éd. Gallimard.

was of universal validity: life and intelligence will necessarily develop in any isolated system. Being alive means being able to maintain a small number of basic variables within certain physiological limits in a wide range of environments by means of internal adaptation. Ashby considered that he managed to recreate this teleological ability (maintaining basic variables), previously believed to be found exclusively in living organisms.

Grey Walter's machine, called Lora, whose creator had intended, from the start, to endow it with the ability to acquire conditional reflexes, did indeed perform in an extraordinary way, even if Walter apparently failed to make it react to sound in the same way as it reacted to light; he did however succeed in linking sound to shock. Lora not only shied away from shocks, but also from the sound of a whistle blowing once the two stimulants had occurred simultaneously a certain number of times. We can therefore imagine, as a first hypothesis, that it had the ability to acquire a conditional reflex.

The fact remains that Walter's machine was nothing more than an electronic structure, via which he could determine a delayed reaction in advance, this reaction being linked to a predetermined signal. Unlike the flexibility we see in the way animals acquire conditional reflexes, Lora's reflexes were characterized by their rigidity. Lora's 'conditional' reflex could not adapt as it had no internally coherent aim, unlike the reflexes acquired by living creatures which may change as they are the way these creatures continuously adapt to unpredictable conditions in their environment in pursuit of their own specific aims. Ashby's homeostat was a form of research into this kind of internal teleology.

Ashby's homeostat, or the way it functioned, produced an appearance of purposeful behavior. The homeostat appeared, at the very least, to be markovian, although without seeing the graph in question, we can say no more than 'appeared'. It was capable of regaining homeostatic stability (its purpose), but not by following a fixed programme. On the contrary, it reached this point by random reconfigurations, seeming to be tentatively feeling its way towards its goal and only stopping when the desired stability was reached. To Ashby, among others, this appeared to demonstrate that the machine was capable of learning. Therefore the homeostat appeared to prove Ashby's thesis.

However, Ashby received a lot of opposition from those present at the ninth Macy conference. His audience criticized what appeared to them to be completely irrational, the use of random processes to simulate thought and adaptation to unpredictable situations.

Quastler, Pitts and Bigelow asked him whether he really thought that Brown's movement was the best way for an organism to invent new solutions. Ashby replied that he did not know any other way for a machine to do so. Bigelow insisted: "How on earth can you say that your homeostat is learning when all it is doing is feeling its way gradually to a position of stability? Would you say that marbles rolling around inside a box which finally reach the only exit, have learned how to find the way out ?" Ashby said that this did not bother him seeing as learning, in his understanding of the term, was an objective process having nothing to do with introspection. Bigelow then lost his temper and said that "Shannon's rat could learn something but not your contraption".

The cyberneticians were completely incapable of understanding Ashby. As advocates of mechanistic mindlessness, they were reduced to quoting the positive and negative points of mechanisms which were in some way "mindful". Ashby therefore was dishing out to them the very arguments they had used to refute Kubie and Birch. In the same way, Ashby was showing them that the basic characteristics of life were in no way unique. While they were gradually learning to face problems of complexity and complexification, Ashby had provided an embodiment for them in his experiments with elementary and self-contradicting thought forms. Once their pedagogical simplicity had been revealed, nothing remained of the illusion they had created. Ashby was therefore the precursor of a new era, one where the problem of complexity would come to the fore the start of the new cybernetics, Second-Order cybernetics.

Ashby's homeostat played in important role in this change. It introduced a new type of machine and clearly demonstrated that such a machine could have extremely long and complex chains of cause and effect. As regards retroactive systems, the homeostat presented the possibility of a circular process of cause and effect. The homeostat also brought out the importance of regulating sound intensity and of the need for a complex structure as a basic condition for a system's survival in an aggressive environment full of random sources of perturbation. All these factors had been scientifically established by Ashby's experiments and allowed him to define his *law of indispensable variety*[15], an essential law for understanding the minimal structural conditions needed for a system to survive.

This law states that a wide variety of available responses is indispensable in order to ensure that a system which aims to maintain itself in a limited number of states, can actually adapt satisfactorily when confronted with a

---

15. Ashby W.R. (1958). Requisite Variety and its Implications for the Control of Complexe Systems. *Cybernetica,* Vol. **1**, 2, 83-99.

wide variety of perturbations from the outside. Or, more directly, for a system to be autonomous it must be able to function and to be structured in a variety of ways.

Ashby, via his law of indispensable variety, was able to define the minimal conditions necessary for the survival of an autonomous system. For there to be genuine autonomy, there must be a variety of possibilities both in the system's basic structure and in the ways it can function. On the other hand he found that, in complex systems, the organization of the system will inevitably consist of a compromise between variety and redundancy. These conditions, important as they undoubtedly are, do not explain all the underlying mechanisms which give natural systems their autonomy.

4.3 The Search for Principles of Autonomy

In order to discover these principles, a methodological and epistemological revolution was necessary in order to go beyond the prevailing predominance of the command and information theory whose basic paradigm was the giving of commands or instructions.

The way in which we recognize or characterize a system, by interacting with it, is inseparable from the way in which we understand its results and its cognitive activities. Therefore the command theory remains closely linked to a conception of information as a means of instruction and representation.

This way of perceiving commands and forms of representation is perfectly applicable to allonomic systems like computers. In such systems, information is reduced to a preprogrammed instruction to the system which may therefore be considered, according to the classic point of view in first-order cybernetics, as a machine which transforms input into output. When considering natural systems, this viewpoint is problematic. First of all, if we consider the question of representation, in the brain there is no internal entity we can consult or examine in order to calculate the number of direct links between the brain and the outside world. In addition, applying the command theory to a living being could be interpreted as reducing it to an allonomic system, a conclusion which originates in the mind of the observer. Our external view of the system under observation is also what enables us to remark the regular features in its behavior, whether these be symbolic or cognitive. We can notice the regularity or irregularity of these because we have access at the same time to way the system functions and to interaction with the outside, so we can refer to system we are studying as a single entity. From the system's point of view, these links have no reality, because we are the ones who define their existence from a point of view which is not that of the system itself.

5. FROM OBSERVED TO OBSERVER: THE CREATION OF SECOND-ORDER CYBERNETICS

For this reason, certain thinkers felt led to contest the restrictive point of view which made pontifical declarations about the difference between information and noise and which confused information and signals. This was the dominant viewpoint at the time, but was incomplete and inadequate when attempting to reach a genuine understanding of living systems. This was also the reason why the same researchers decided to reconsider their understanding of the nature of information, seeing it no longer as merely a series of instructions or as nominal representations, but as a phenomenon formed and constructed (literally *in-formati*) within the system itself. While they were aware that the obstacles encountered when applying the artificial, mechanistic viewpoint typical of first-order cybernetics called into question the implicit epistemology and ontology of classical scientific method, they aimed to create a Second-Order cybernetics, which would be more reflective, a cybernetics concerned with observer as much as with the observed. What they were seeking was a cybernetics where the descriptions used would reveal rather than hide characteristics of the observer, and where the conception of knowledge and of reality would take into consideration the fact that we play an active part in formulating these conceptions.

The mere fact of talking about a Second-Order cybernetics allows us to suppose that a gap had formed between those who spoke in such a way and the proponents of first-order cybernetics. This gap becomes clearer still when we consider the name given to the laboratory where Second-Order cybernetics came into being : the Biological Computer Laboratory (BCL).

While the Macy conferences had seen a dialogue established between cybernetics specialists and the thinking being, the question of the nature of living beings had been presented solely by physiologists and was considered as already more or less answered. The BCL group, on the other hand, were now going to try and understand living beings as active in unstable surroundings.

The key concepts here were distance, of course, but also proximity. Treating a living creature as a biological computer supposed that the BCL group, like the traditional cyberneticians, would try to understand the behavior of living creatures, and their interaction with the world, as mathematical problems. Von Foerster, the former secretary of the Macy conferences and director of the BCL which he founded at the University of Illinois in 1958, was interested in trying to define what was meant by a biological computer — *i.e* a computer which can work out by itself and for itself which information is relevant and for whom, unlike a manufactured computer, the problem of which calculations to make and the problem of its own survival are indissociable.

This is where we can see the originality of the BCL group. The problems confronting a living being whose priority is to survive in a hostile and unpredictable environment, and not just to reason logically so as to solve a predetermined problem, do indeed take us into another dimension from that supposed by the problems presented under McCulloch's influence at the Macy conferences.

A new way of working was inaugurated: considering the object or the natural automaton, or even the collection of natural automata, not as a simple black box but as an autonomous, self-regulating system. This involved grasping the internal complexity specific to each natural automaton without overlooking the complexity of its relationship with the outside world, which in fact had permitted its internal complexity to develop. More precisely, the aim was to try and find the organizational principles were underlying these self-regulatory or self-reproducing properties which had been defined as the typical characteristics of living organisms.

From this point, new concepts starting with the prefix self-sprung up all over the place. Among these were 'Self-Organizing' which became the buzz word at conferences for academics working in the field of self-regulatory systems between 1960 and 1962[16]. These conferences were organized by Yovits, Cameron, Zopf, Jacobi, Goldstein and, above all, by the BCL. Together with research scientists like Ashby, McCulloch, Günther, Löfgren, Weston, Varela, Pask and Maturana, Von Foerster devoted himself to research projects where his taste for paradox could be indulged to the full: circular casuality, the regulatory properties of chance etc.

As the BCL was trying to understand what constituted a biological computer, internal organization became the central problem. This implied distancing themselves somewhat from McCulloch and first-order cybernetics. The very nature of information was called into question. The origin of information was to be found in the organism itself, as a whole, and

16. Yovits M.C. and Cameron S. (eds) (1960). *Self-Organizing systems.* Proceedings of an Interdisciplinary Conference, 5 and 6 May, 1959, Oxford, London, New York, Paris: Pergamon Press; Foerster H. von and Zopf Jr. G.W. (eds) (1962). *Principles of Self-Organization.* Oxford, London, New York, Paris: Pergamon Press; Yovits M.C., Jacobi G.T. and Goldstein G.D. (eds) (1962). *Self-Organizing systems.* Washington: Spartan books.

no longer merely in the organizational abilities of a third-party who would make a distinction between noise, signals and information within that organism. The complexity of the automaton had to be fully understood. This was the principal message of Ashby's homeostat. This led to a more reflective kind of cybernetics which confronted the issue of how the observer system organized itself.

Although the work done to develop Second-Order cybernetics could be accused of a certain eclecticism, they all lay claim to a certain hidden identity which can be discerned by the following characteristics:

- They refused to accept any concept of reality which could be discovered without taking the observer's own perception into consideration. (Information is constructed).
  - They considered that the observer was by definition implicated in the systems he observed.
- They opposed any reductionist understanding of cybernetics.
- Above all, they advocated the use of the epistemological, experimental method where the experiments are carried out by a process of formal logic without worrying about how the results can be realized in practice.

Second-Order cybernetics, by its attempts to understand self-regulatory systems which appear as meta-organizations in comparison with artificial, preprogrammed systems whose organizing principle is always exterior to themselves, completely revolutionized first-order cybernetics. The main problem tackled by the Second-Order cyberneticians was to discover the underlying logic which enabled these systems to function. It is therefore not surprising that the greater part of the research undertaken by the BCL group was devoted to providing an accurate definition of self-organization.

5.1 Defining Self-Organisation

The inventers of the term, Gordon Pask[17], George Zopf[18] and Heinz Von Foerster[19] considered self-organization as a property which makes a system capable of observing the person who is observing it and of relating to them. This implied that it had at its disposal a mode of interaction which went beyond the simple discovery of the logic behind its own system. As far as

17. Pask G. (1958). *Organic Control and the Cybernetic Method,* Vol. **1**, 3, 155-173.
18. Pask G. (1962). Attitude and Context? In H. Von Foerster and G. Zopf (eds) (1962). *Principle of Self-Organisation.* Oxford, London, New York, Paris: Pergamon Press, 325-346.
19. Pask G. and Von Foerster H. (1960). A Predictive Model for Self-Organizing Systems. *Cybernetica,* Vol. **3**, 4, 268-300.

we can see, this definition actively criticized the explanatory schemas produced through first-order cybernetic research and, at the same time, criticized the proponents of Artificial Intelligence.

Despite the revolutionary nature of this conception, it did not attain the same popularity as those propounded by Ashby[20] and Von Foerster[21], more paradoxical conceptions which demonstrated that self-organization, in the strictest sense of the word, was impossible and which underlined the importance of noise. Information (organization) is created by the system out of noise. Self-organization can therefore only exist via the means of hetero-organisation.

For Ashby, the only rational definition of the forms of behavior possible for a given system had to be related to a function which, given the state the system is in at the time and the surrounding environment, would determine the following state to be attained. Understood in this way, self-organization was merely a way of explaining that the function which appeared to be regulating itself had been badly defined. The only properties which had any real meaning were those which resisted the observer's point of view.

As far as cybernetics was concerned, Ashby made a U-turn. The aim of the system was not predetermined. It was a consequence of an evolutionary process in which an organism had found a way of surviving in a given environment. Self-organization was therefore, in Ashby's opinion, an illusion born of a misunderstanding of the system's true function. Dialogue was no longer necessary and any problem which may arise was obliged to conform to the conditions defined by Ashby. The heuristic approach was rejected. From an epistemological point of view, using this method, Ashby cut down to size any ambition cybernetics may have had to find a solution to the controversy between the vital and the mechanistic approaches. He removed the basic presupposition behind this ambition: that an analogy exists between living systems and teleological machines. For the cyberneticians, he made a decisive break with all interdisciplinary preoccupations. Dialogue was no longer possible.

In the same way Von Foerster demonstrated that self-organization was impossible from the physicist's point of view. He showed that, in accordance with the second law of thermodynamics, a Self-Organizing

---

20. Ashby W.R. (1962). Principles of the Self-Organizing System. In H. Foerster, G.W. von Zopf Jr. (eds) (1962). *Principles of Self-Organization,* Oxford, London, New York, Paris: Pergamon Press, 255-278.

21. von Foerster H. (1960). On Self-Organizing Systems and their Environments. In M.C. Yovits and S. Cameron (eds) (1960). *Self-Organizing systems.* Oxford: Pergamon Press, 31-50.

system cannot not organize itself entirely by itself, but needs help from its environment.

According to him, such systems are characterized by increased redundancy and by their ability to transform noise into order, into information for its own sake. This position differed greatly from the one accepted by first-order cybernetics and would lead to cybernetics turning back upon itself.

If self-organization could not come into being by the system's own devices, it could do so for the system's own ends. Let us take the example of Von Foerster's magnets. The noise allowed them to be organized: there were two possible opinions about this organization. Seen from the outside, the system seemed to transform noise into information. For the inside observer however, the information was already potentially present. We can see that this thought experiment caused a problem. In fact, the conditions needed for Von Foerster's definition of a Self-Organizing system were not fulfilled – despite all our efforts, we remain uncertain how we should describe the system. The observer who was surprised about this is merely the one who chose not to take the lid off. The choice appears to be a unilateral one and not, as Pask and Zopf and Pask and Von Foerster's articles claimed, a heuristic choice essential to any intelligible interaction. Two themes appeared to coexist in an ambiguous way: that of the unilateral decision, which may be called Self-Organizing, in which 'I' decide not to take the lid off, and that of a property in relationship with realities beyond itself with which I could interact in a productive and relevant manner.

For both Von Foerster and Ashby self-organization lost some of its meaning. This may explain why the term disappeared from their articles around this time. For Ashby, it no longer had any meaning, while for Von Foerster it merely referred to a system's ability to transform noise into information. Self-organization was, using that definition, possible for its own ends.

Von Foerster and his colleagues retained this understanding of the term, which lead them to recognize the existence of circular processes.

By trying to understand the behavior of living systems as a mathematical equation, the members of the BCL group adopted an ontological position: living beings are machines, life is an immense cognitive process of self-understanding and all physical functions may legitimately be described in terms of information, mathematical equations, treatment of input and data creation which, from this point of view, is not pre-programmed but constructed. The problem of recognizing which properties belong exclusively to the domain of the living is reduced to a matter of arithmetic.

Von Foerster's non-trivial machines are a case in point. Although these machines were quite definitely deterministic, they were not predictable. The relationship between input and output was not invariable, and the output at a given point depended on the system's history and on previous input. This distinction made it possible to envisage forms of collective behavior which would resist attempts to take the lid off, to introduce an infinite number of possibilities.

There was thus no further point in Von Foerster and Pask's attempt to clarify and predict the ways in which a system may transform by itself the framework in which it can be questioned. This type of problematic definition had the advantage of being able to resist Ashby's demonstration, based as it was on the epistemology of observed systems: for Ashby the only properties which had any meaning were those which could resist an omniscient observer capable of defining his object as a machine, in Ashby's sense of the word.

The nature of the criticism addressed to the first-order cyberneticians was changing. Researchers no longer used the notion of self-organization focussed around the distinction between trivial and non-trivial machines. The epistemology of observed systems sought mechanisms which could treat their environment like a trivial machine. This would lead to the realization that anything an organism does is done as if that organism were autonomous, as if it were treating its environment according to conditions established by itself, as if it could itself produce its own information. This distinction underlined the importance of both reflexivity and arithmetic.

As non-trivial machines are a sub-group of the category of mechanisms traditionally represented by the Turing machine, and as Turing machines are logically equivalent to calculators, we can conclude that the new type of mechanism invented by Ashby and developed by Von Foerster took the calculator as the basic model for all types of machines. As these machines use symbols, and as arithmetic can be reduced to data processing, we can say that for the BCL group, explaining a phenomenon was the same thing as producing a model of it by data processing. The complete break brought about by the new type of mechanism now became clear.

Life was a cognitive process and reproduction, memory and recognizing forms could be described like non-trivial machines in terms of recursive equations. It was in fact their reflexivity which enabled these machines to be non-trivial and, for Von Foerster and the Chilian school of biology, reflexivity was the basic principle of life.

Reflexive equations could be defined in the following way : they were an economic principle which included the concept of history without reducing memory to an information storehouse. The function was alive at each moment. These equations provided an explanation for the circular characteristics of cognitive activity and tended to converge with real values. Considered in this way, the being organized and defined its own world. It calculated one possible reality from the constraints which made up its system.

From an ontological point of view, all this meant that the being got lost in an infinite mass of calculations and that organisms were perceived as observers about whom researchers were still trying to establish a theory and, as a consequence, a "How do we know?" rather than a "What do we know?" epistemology. The meaning of the materialism/idealism debate had thus moved on, the object could no longer be studied independently of its subject and vice versa.

The meaning of materialism could not therefore be found in an ecological vision which always needs something beyond the subject itself. Representations and information were always the result of an arithmetical calculation and everything remained within the system. The brain was not a computer, but functioned as a closed system. It did not receive information from outside but imposed information on its surroundings by a mathematical maneuver.

Knowledge did not solely come from Hermes but also from Turing.

5.2 A Conclusion with Regard to the BCL Group

Von Foerster's and Ashby's articles taken together succeeded in confusing the issues at stake in the self-organization debate as Pask, Zopf and Von Foerster himself had originally understood it. Uncertainty about the rules of the game no longer appeared to be adequate differentiation criteria, even with regard to the distinction to be made between homeostatic adaptation and active learning. In his 1969 article "The Meaning of Cybernetics in the Behavioral Sciences" (*The Progress of Cybernetics,* Gordon and Breach, New York), Pask no longer talked about self-organization. For the BCL group, self-organization did not survive its association with the themes of omniscience and original production. The problems raised were redistributed to other areas of research : the distinction between trivial and non-trivial machines, the natural language issue etc. On the other hand, and we could even say that he reinvented self-organization, Atlan rediscovered Ashby's and Von Foerster's texts and saw them as the precursors of his own research into self-organization seen not heuristically but how it relates to points of view.

We must emphasize the fact that the notion of an unexpected proliferation of possibilities, associated with Von Foerster's non-trivial machines, was also rediscovered in the context of Conway's Life Game on computer.

The final irony comes from the fact that it was physicists specializing in spin glasses and not cyberneticians who took the initiative which made neo-connectionism a reputable pioneering science. The importance of automata, possessing a defined form of energy, was in fact typical. These automata were at the heart of the new discipline and were not unlike Von Foerster and Pask's competing automata, which were in fact only in competition with them on account of a global constraint which gave them some strategic importance.

5.3 General Conclusion about Self-Organization

As far as cybernetics was concerned, the BCL remained a minority movement and finished with a failure. It did not succeed in imposing its point of view. We could suggest from the texts that Von Foerster's attempt to make some sort of plausible and convincing sense of his dilemma, before the eyes of those who did not share his interest in interaction with living systems, opened the way for an increasing formalization which in fact took him further away from the original question and sent him scrambling up an epistemological cliff face. Ashby reinforced this tendency when, speaking from an omniscient point of view, he took the point away from the interaction issue and merely associated it with a lack of knowledge.

In the area of embryology, the results were not much more positive. For the embryologists, the notion of self-organization implied a change in strategy and in the way of defining the objects being studied. This change took place at a time when embryology was losing its reputation as a pioneering science to genetics and bacteriological biochemistry. From this point of view, self-organization became a stumbling block which those who put their trust in molecular biology or reductionist strategies tripped over. In the eyes of the proponents of this strategy, self-organization became the the emblem of a backward science.

Finally, with regard to the Prigogine group, their project only concerned a small minority. However, their starting point was firmly anchored in an established science to such an extent that it no longer attracted the best scientists. Only other thermodynamics specialists are capable of judging the originality of their project from within their own tradition.

5.4 An Epistemological Synthesis

In spite of all these factors, the appearance of this concept was not without a certain profound significance. In the hands of those who defined the concept, it brought into question one of the dominant representations of scientific rationality with regard to the opposition between subjects and objects, observers and observed. It relativized the manichean conception of science according to which the question of how living organisms are organized had, by its very nature, nothing to do with scientific experimental methods. We should underline the fact that this relativisation was itself relativized by Ashby, who restored the dichotomy by adopting an omniscient viewpoint and condemning, on principle, the heuristic approach. We shall see what will happen next with the research into neuron networks. If we read Atlan's work, we may wonder if, at last, the question of self-organization has not been solved.

If self-organization has returned to prominence today, ignoring Ashby's impossibility theorem which had logically condemned it, this is because today's scientists are, in practice, concentrating on networks, and in consequence are no longer interested in the functions and the laws governing their behavior, but rather in what they are capable of doing. This is because neo-connectionism does not treat automata like logical machines but like networks of interconnected elements. This point of view means that Ashby's condemnation is no longer relevant. The important point here is that the questions asked of random networks consider the predetermined connection between the initial situation and subsequent performance as of no interest, which involves a complete change in the ways networks are understood. They are no longer seen as logical machines, but as networks of interconnected elements. Without this new understanding, we find ourselves, as Atlan has explained in his articles, unavoidably faced with Ashby, his omniscience and the perception of self-organization as meaningless which this entails.

REFERENCES

Ashby W.R. (1962). Principles of the Self-Organizing System. In H. von Foerster and G.W. Zopf Jr. (eds) (1962). *Principles of Self-Organization,* Oxford, London, New York, Paris: Pergamon Press, 255-278.

Ashby W.R. (1958). Requisite Variety and its Implications for the Control of Complexe Systems. *Cybernetica,* Vol. **1**, 2, 83-99.

Burks A.W. (ed.) (1970). *Essays on Cellular Automata.* Urbana: University of Illinois Press.

de Latil Pierre (1953). *La pensée artificielle.* Éd. Gallimard.

Dupuy Jean-Pierre (1985). L'essor de la première cybernétique (1943-1953). *Histoires de cybernétique.* Cahier du CREA, **7**, Paris: CREA, 67.

Lloyd A. Jeffres (ed.) (1951). Hixon Symposium. *Cerebral Mechanisms in Behavior.* California Institute of Technology, September 1948, New York: John Wiley and Sons, 70-71, 74-75 et 112-133.

McCulloch W. and Pitts W. (1943). A Logical Calculus of the Ideas Immanent in the Nervous Activity. *Bulletin of Mathematical Biophysics,* **5**, 115-133.

Pask G. (1958). *Organic Control and the Cybernetic Method,* Vol. **1**, 3, 155-173.

Pask G. (1962). Attitude and Context? In H. Von Foerster and G. Zopf (eds) (1962). *Principle of Self-Organisation.* Oxford, London, New York, Paris: Pergamon Press, 325-346.

Pask G. and Von Foerster H. (1960). A Predictive Model for Self-Organizing Systems. *Cybernetica,* Vol. **3**, 4, 268-300.

Turing A. (1936). On Computable Numbers with an Application to the Entscheidungs Problem. *Proceedings of the London Mathematical Society,* 2e serie, Vol. **42**, 3e part, November 12, 230-265 and (1937) 2e serie, Vol. **43**, 7e part, May 20, 1937, 546-550; (1950) Computing Machinery and Intelligence. *Mind,* Vol. LIX, **236**.

Rosenblueth A., Wiener N. and Bigelow J. (1943). Behavior, Purpose and Teleology. *Philosophy of Science*, Vol. **10**, 18-24.

Rosenblueth A. and Wiener N. (1945). The Role of Models in Science. *Philosophy of Science*, Vol. **12**, 4, 316-321.

Simondon (1969). *Du mode d'existence des objets techniques.* Aubier-Montaigne.

Von Foerster H., Mead Margared and Hans Lukas Teuber (eds) (1950). Cybernetics-circular Causal and Feed-back Mechanisms in Biological and Social Systems. *Transactions of the Seventh Conference,* March 23-24, New York: Josiah Macy; New York: Jr. Foundation, 1951, 153-154.

von Foerster H. (1960). On Self-Organizing Systems and their Environments. In M.C. Yovits and S. Cameron (eds) (1960). *Self-Organizing systems.* Oxford: Pergamon Press, 31-50.

von Forester H. and Zopf Jr. G.W. (eds) (1962). *Principle of Self-organization,* Oxford, London, New-York, Paris, Pergamon Press.

Von Neumann J. and Morgenstern O. (1980). *Theory of games and economic behavior,* First edition, Princeton University Press, 1944, Sec. ed., 1947, Third ed., 1953, Princeton University Press.

Weiss, Hixon Symposium. *Cerebral Mechanisms in Behavior.* California Institute of Technology, September 1948, New York: John Wiley and Sons, 72-74 et 140-142.

Yovits M.C. and Cameron S. (eds) (1960). *Self-Organizing systems.* Proceedings of an Interdisciplinary Conference, 5 and 6 May, 1959, Oxford, London, New York, Paris: Pergamon Press.

Yovits M.C., Jacobi G.T. and Goldstein G.D. (eds) (1962). *Self-Organizing systems.* Washington: Spartan books.

PAUL MENGAL

THE CONCEPT OF EMERGENCE IN THE XIXth CENTURY: FROM NATURAL THEOLOGY TO BIOLOGY

1. INTRODUCTION

During the first half of the XIXth century, the theologians who developed the standpoint of natural theology for zoology were divided into two camps concerning the delicate question of animal instinct. This question, a direct descendant of the dispute over the souls of beasts, opposed those who lent to the animal some reason to those who argued a total break between animality and humanity. The latter argued that animal behavior was directed by instinct enacting a project of which the animal is ignorant. The extraordinary variety of animal behavior and its marvelous efficiency could therefore only be the expression of the all powerfulness of the world creating divinity. In this conception one encounters one of the main uses of the physico-theological proof of the existence of God proposed by the theists. The physico-theological argument was the basis of the discourse of natural theology, and was adopted time and time again by all naturalist ecclesiastics of the XIXth century. Natural theology belongs to religious immanentism since it was developed in opposition to revealed theology that is thoroughly transcendental. We would like to show how the immanentist conception, played — and still does today — a decisive part in understanding the concepts of emergence and self-organization that life and human sciences use to explain development.

1.1 Natural Theology and Immanentism

Since John Ray and his book *The Wisdom of God Manifested in the Works of the Creation* (1691) the arguments of physico-theology have been well

B. Feltz, M. Crommelinck and P. Goujon, (eds.), Self-Organization and Emergence in Life Sciences: 215-224.

known. The exemplary organization of the world is a witness to the creator's *design,* the remarkable efficiency of the mechanisms installed are consequence of the *benevolence* of God. These two qualities also reveal the *goodness* of God, in particular with regards to man. The three aspects of the *design, benevolence,* and *goodness* of God were tirelessly illustrated and commented upon by the adepts of natural theology.

These three aspects appeared in *Natural Theology*[1] by William Paley, published in 1802, and were reproduced in identical form in the *Bridgewater Treatises on the Power, Wisdom and Goodness of God as manifested in the Creation*[2] published in London between 1833 and 1836. According to C. Blanckaert, these texts were conceived in order to: "conjure the French menace, to refute materialist science advocate of cerebral determinism of thought, spontaneous generation, transformism and polygenism"[3].

When Kant had opposed revealed theology to rational theology he had divided the latter into *deist* and *theist.* When theology becomes theist it proves the existence of the author of the world through the order and unity found within it. It is natural theology, indeed, that is referred to. In the *Critique of Pure Reason,* Kant developed this argument:

> "This arrangement of means and ends is entirely foreign to the things existing in the world — it belongs to them merely as a contingent attribute; in other words, the nature of different things could not of itself, whatever means were employed, harmoniously tend towards certain purposes, were they not chosen and directed for these purposes by a rational and disposing principle, in accordance with certain fundamental ideas"[4].

We know, at least since Spinoza, that a short distance separates divine intelligence from fecund intelligence. Immanentism is the best way of reconciling this apparent opposition. In order to truly understand its effect one must go back to the Aristotelian source and to the distinction between the two types of action in the Metaphysics. Aristotle opposed the action in

---

1. Paley William (1802). *Natural Theology or, Evidences of the Existence and Attributes of the Deity collected from the Appearances of Nature.* London: Hamilton. Translated into French by C. Pictet, under the title of: *Théologie naturelle ou Preuves de l'existence et des attributs de la Divinité tirées des apparences de la nature,* Genève: Imp. de la Bibliothèque britannique, An XII-1804.

2. The reverend Francis Henry Egerton, count of Bridgewater had bequeathed a certain amount to the Royal Society for the production of the books. Amongst the authors the names of the philosopher W. Whewel, of the geologist W. Buckland, the physiologist Charles Bell, the naturalist W. Kirby and of the doctor and chemist W. Prout are to be found.

3. Blanckaert C. (1900). La "Théologie naturelle" de Louis-François Jéhan (1803-1871). Science, apologétique, vulgarisation, *Nuncius. Annali di storia delle scienza.* **2**, 167-204.

4. Kant I. (1924). *Critique of Pure Reason.* Trad. J.M. Meiklejohn, London: G. Bell & Sons, 384.

which the end is exterior to the subject to the one where the end is within the acting subject:

> "Now whereas in some cases the ultimate thing is the use of the faculty, as, e.g., in the case of sight seeing is the ultimate thing, and sight produces nothing else beside this; but in other cases something is produced, e.g. the art of building produces not only the act of building but a house" 1050a, 459-461[5].

In his commentary of Aristotle, Thomas Aquinas also distinguished an action that aims to produce a new object, such as a house, from one which ends in the subject and which carries him to a higher level of perfection. It is this distinction that the new scholastics, such as Cajetan, referred to as *transitive action* and *immanent action.*

Adriaan Heereboord, an influence on Spinoza, recommended in his *Meletemata Philosophica* of 1654, not to open the book of Aristotle but the book of nature and to found the knowledge of God on natural light. With Spinoza, immanentism became more radical in that in the *Short Treatise* (around 1654), it is written that God

> "... is an immanent and non transitive cause in that he acts within himself and not outside of himself, since nothing exists outside of him"[6].

In this sense, Spinoza's immanence[7] is an activity that finds in the subject that is its seat, the principle and the meaning of its development. Spinoza's immanentism is coupled with monism in which "the soul and the body, thought and being cease to be discrete things each for itself"[8]. If Spinoza could only reach his aim at the cost of the elimination of contingency and historicity, Hegel saved historicity by substituting Spinoza's Absolute, or God considered as a unique Substance, for the absolute Spirit that only reaches its full self consciousness by constructing itself in time and history.

Naturphilosophy, inspired by Herder, made God's design into an "idea of God" enacted by the organic force that is externalized in the project of creation. This organic force is the endogenous source of the dynamism of all

---

5. Aristotle (1956), *The Metaphysics,* transl. Tredennick H.M.A., Harvard University Press, 1050a.

6. Spinoza (1963), *Short Treatise on God, Man and His Well Being.* New York: ed. A. Wolf.

7. The notion of *natural light* found in A. Heereboord (1613-1631), Spinoza's individual piety and philosophy of immanence are, in fact, traits that already characterize the *Collegiant* movement developing itself in the Netherlands following the defeat of the Arminian current after the national meeting of Dordrecht in 1618. On the *Collegiant* movement and the notion of *natural light,* see the study by Andrew Fix (1989). Angels, Devils, and Evil Spirits in Seventeenth Century Thought: Balthasar Bekker and the Collegiants. *Journal of the History of Ideas*, Vol. L, **4**, 527-547.

8. Hegel G.W.F. (1955). *Lectures on the History of Philosophy* (transl. E.S. Haldane and F.H. Simpson, 3 vols, London, 1892-1896). London: Routledge and Kegan Paul, New York: Humanities Press.

development. This immanentism that leads so easily to pantheism was severely criticised by the Catholic church in that man and world themselves containing the reasons of "divine" effects produced within them, God and the world did not then amount to separate beings.

2. IMMANENTISM AND EMERGENCE

It is in this immanentist perspective that G.H. Lewes developed his conception of emergence. George Henry Lewes, born in London in 1817, received a rare form of education for the period. He first of all studied physiology but did not graduate. As early as 1836, he planned to write a treatise that would rebuild Scottish philosophy on a physiological basis but gave it up immediately to visit Germany. When he returned, he attempted to be an actor and in parallel published articles including *The Modern Philosophy in France* which appeared in 1843 in the *British and Foreign Quarterly.* He expressed the view that Victor Cousin was a charlatan and that Auguste Comte's positivism was the ultimate aim of philosophy. From 1854 to 1857 he again went to stay in Germany and coming back he returned to his former interest in physiology. He visited the marine zoology centre of Ilfracombe and published as a result *Seaside Studies* (1858). He also wrote *Physiology of Common Life* (1859) and *Studies in Animal Life* (1862) before returning to his initial project the first volume of which was published in 1874 under the general title *Problems of Life and Mind.* This work presented itself as a step beyond the confrontation of materialism and spiritualism and dismissed both Condillac and Kant. The former for having mistaken ideation for sensation, the latter for not having used the biological method of objective analysis. To approach the relation between the mind and the body one must drop the classical opposition of object and subject

> "We know ourselves as Body-Mind; we do not know ourselves as Body and Mind, if by that be meant two coexistent independent Existents; and the illusion by which the two Aspects appear as two Reals may be made intelligible by the analysis of any ordinary proposition"[9].

Contrary to Auguste Comte, Lewes believed in a possible science of the mind based on the one hand upon physiology and on the other upon sociology. For the mind as for the body there is not, he stated, preformation or pre-existence but, evolution and epigenesis. Kant's error, he insisted, was to have mistaken anatomy for morphology and logic for psychology. By

9. Lewes G.H. (1877). *The Physical Basis of Mind.* London: Täbner and Co, Ludgate Hill, 350.

considering the adult mind only, philosophers took these built up forms for original conditions. The procedure is perfect for logic the function of which is to show the manners of thought and not its origin. This interest in the epigenesis of intelligence lead him to formulate a definition of instinct that integrated perfectly well this new dimension:

> "Instinct, which, because it is so frequently cited to prove the doctrine of Innate Ideas, may best serve to illustrate the doctrine of evolution. The marvel and mystery of Instinct naturally render it a favourite topic in the writings of those who oppose the experiential School. (Instinct is often regarded as so superior to Intelligence in the certainty of its action, that nothing except Creative Wisdom is admitted in explanation of it; while from other sides it is regarded as so removed from all community with Intelligence, that is declared to be the blind action of a mechanism, not the operation of a rational soul.) Psychogenesis seems to me to teach the direct contrary to all this. It teaches that Instinct is organised Experience/ *i.e.* undiscursive intelligence; that is to say, while the neural and logical processes are the same in both, the operations in what is specially termed Intelligence are facultative, and involve the element of choice in the selection of means to ends: Intelligence is therefore discursive; whereas in Instinct the operations are fixed, uniform, with no hesitation in the selection of means"[10].

Instinct, for a given generation, is therefore only the result of ancestral experiences the organization of which the present individual has inherited, and biological science must trace its evolution path or genesis. If what is meant by organization is the totality of necessary conditions, then life, for Lewes, is proportional to the organization. And if there is a unity and consensus in the organization, it must not be attributed to a life principal independent of the organism. How are the variations in evolution and psycho genesis to be accounted for in this case? Lewes's thesis is immanentist in that it is in nature or in the individual himself that we must look for the principle of betterment, and its functioning is simply the result of the manner in which the necessary conditions to produce an organization come together.

In *Problems of Life and Mind,* Lewes opposed resultants and emergents. This distinction takes place in a development on theories of causality in which Lewes refuted the difference between cause and effect claiming that it is only a case of distinct expressions referring to identical processes considered from different angles. *Resultants* are reached by adding up their components, whereas emergents are the product of *coalescence* or fusion of components at the end of a process beyond description. Thus, Lewes commented, adding heat to heat gives a measurable result but adding heat to different substances produces varied effects: expansion, in one case,

10. Lewes G.H. (1874). *Problems of Life and Mind,* Vol. I, London: Träbner and Co., Ludgate Hill, 226-227.

liquefaction, in another case, crystallization, in yet another case, and decomposition, in a final case. If there are different emergents it is because, in each case, there is a distinct mode of co-operation. Lewes quoted Hegel to reinforce his position recalling a few words from *Logic*

> "The effect is necessary just because it is the manifestation of the cause, or is this necessity which the cause is"[11].

Applying this distinction to resultant and emergent for the development of psychical processes, Lewes indicated the mode

> "The great problem of Psychology as a section of biology is, in pursuance of this conception, to develop all the psychical phenomena from one fundamental process in one vital tissue. The tissue is the nervous: the process is a Grouping of neural units. A neural unit is a tremor. Several units are grouped into a higher unity, or neural process, which is a fusion of tremors, as a sound is a fusion of aerial pulses; and each process may in turn be grouped with others, and thus, from this grouping of groups, all the varieties emerge"[12].

This way of conceiving how psychological processes emerge from the organisation of the nervous tissue is a simple illustration of Lewes's theoretical positions as expressed in his experimental philosophy. Two of these rules are the basis for the principle of emergence:

> "Rule VIII. — Because the significance of a phenomenon lies wholly in its relation to other phenomena we must never isolate it from this relativity, and draw conclusions respecting it *per se.*"
> "Rule IX. — We are not to conclude the properties of elements from the properties of the groups they form; nor *vice-versa*"[13].

Whereas today biological epistemology opposes emergence and reductionism, it is remarkable that Lewes included in a methodological reductionism his own conception of emergence. Lewes used the resolutive-recompositive method criticized by holist epistemologists but distanced himself simply from the traditional interpretation by accepting two distinct modes of recomposition that lead one to the *resultants* and the other to the *emergents.*

3. PHILOSOPHICAL IMMANENTISM AND DEVELOPMENTAL MODEL

It is probably in the liberal Protestant movement of the end of the XIXth and beginning of the XXth centuries that philosophical immanentism was developed. The most well built synthesis between this philosophical leaning

11. Hegel G.W.F. (1975), *Hegel's Logic,* (transl. W. Wallace), Oxford Clarendon Press, II, 218.
12. Lewes G., *op. cit.,* I, 135.
13. Lewes G., *op. cit.,* I, 96-97.

and the domain of biology is without doubt to be found in Jean Piaget's genetic epistemology. This biologist by training conducted an in depth reflection on the relations between religious immanentism and biological epistemology. As a militant at the *Swiss Association of Christian Students,* Piaget gave a few conferences to the members of this association[14]. In 1928 and 1929 in particular Piaget gave two conferences on the question of religious immanentism. It is above all the second conference, published in 1930, which allows us to understand how his genetic epistemology is embedded in this theological conception[15]. Inspired by both the reading of Bergson's *Evolution créatrice* and the works of the Protestant theologian Auguste Sabatier[16] Piaget distanced himself with his immanentism of interiority from Bergson's immanentism of exteriority and Sabatier's transcendence of interiority. For Piaget, the organism or knowledge both made-up, in their own way, systems enclosed within their own immanence whose development, vital or intellectual, was pure efference or *enaction* to use a more fashionable word. The immanent God proposed by Piaget is a Value-God as oppose to the Cause-God of transcendence. The value is at the root of truths and moral obligations not of events or facts. Values are not the result of experience but determine the conditions of possibility:

> "The immanent God does not solve in causal terms but in implications. (...) The immanent God is therefore not the source of physical or psychological realities, whatever their type, but the principle of moral and intellectual conscience, that is to say hearth of all values necessary for the functioning of this conscience"[17].

This theological conception is also present in Piaget's genetic psychology where he used it to surpass the psycho-physiological parallelism which dominated XIXth century psychology. Returning later to a lay version of this opposition, Piaget rejected parallelism showing that the physical or

---

14. On Piaget's formative years the best study is by F. Vidal (1994). *Piaget before Piaget.* Cambridge: Harvard University Press.

15. Piaget J. and de la Harpe J. (1930). *Deux types d'attitude religieuse: Immanence et Transcendance.* Genève: Robert. These conferences lead to a debate with A. Reymond, *cf.* F. Vidal, *op. cit.*

16. The protestant theologian Auguste Sabatier (1839-1901) proposed *symbolo-fideism* or symbolic character of dogmatic expressions that leads to the analysis of the religious phenomenon in the most general framework of psychological experience. Inspired by Schleiermacher, he maintained none the less the transcendence of the interiority of conscience. The text to refer to is A. Sabatier (1897). *Esquisse d'une philosophie de la religion d'après la psychologie et l'histoire.* Paris: Fischbacher.

17. Piaget J., *Immanentisme et foi religieuse, op. cit.,* 9-10.

physiological domain was governed by causality whereas the psychological domain was governed by implication "between values the construction of which is the function of conscious activity"[18]. The values mentioned by Piaget on this occasion are logical values such as those found in truth tables. Causality and implication are, however, the two sides of the same reality as, for Piaget, rationality emerged from biological organization and intelligence was a specific form of biological adaptation. No parallelism can therefore exist between two processes that follow each other and have identical laws of composition. This is the reason why Piaget referred to isomorphism of structure relating back, in so doing, to *Gestaltpsychology.* This "isomorphism of conscious implication and organic or material causality"[19] kept firmly the unity of the biological and the psychological, and allowed Piaget to maintain that the principles of biological development were the same as the ones underlying psychological development. A few years later Piaget reformulated his position with the new terms of cybernetics and cognitivism erasing therefore all traces of references to theology:

> "Cognitive processes seem, then, to be at one and the same time the outcome of organic autoregulation, reflecting its essential mechanisms, and the most highly differentiated organs of this regulation at the core of interactions with the environment"[20].

Associating to his immanentist choice the predominance of the interiority over the outside world, Piaget confirmed his theological position by openly declaring himself party to a scientific immanentism that gave predominance to the endogenic dynamism at the expense of the influence of the environment. Novelty on both the ontogenetic and phylogenetic scale could therefore only be the result of emergence considered as the end of internal reorganisations. Piaget referred under the term of *équilibration majorante* to the processes of reorganisation of cognitive structures and proposed the mechanism of phénocopie[21] to explain the reorganisation of the genome, thus maintaining structural identity between the cognitive and biological processes.

---

18. Beth W.E., Mays W. and Piaget J. (1957). *Épistémologie génétique et recherche psychologique. Études d'épistémologie génétique,* Vol. 1, Paris: PUF, 76.
19. Beth W.E., Mays W. and Piaget J. (1957), *op. cit.,* 82.
20. Piaget J. (1971). *Biology and Knowledge,* Chicago: The University of Chicago Press, 26. First french edition, *Biologie et connaissance,* Paris: Gallimard, 1967, 38.
21. For a more detailed account of these processes see J. Gayon and P. Mengal (1992). *Théorie de l'évolution et psychologie génétique chez Jean Piaget.* In D. Andler *et al., Épistémologie et cognition,* Liège: Mardaga, 41-58.

4. CONCLUSION

Immanent action, to refer back to the Aristotelian distinction, that carries the subject of which it is the seat to a superior level of perfection still has today the leading part in developmental models. Its scientific reformulation is done in embryology where William Harvey introduced the notion of embryogenesis, an idea that owed a lot to the philosophy of Aristotle as shown by Walter Pagel. Epigenesis is part of a monist perspective:

> "There is only matter that is also alive, functioning and perfected by virtue of the immanent vital impulses that are inseparable from it"[22].

Aristotelian hylomorphism described development as the action of form upon a passive nature that enacts change.

If contemporary science has rid itself of the concepts of form and entelechy, Wilhem Roux and Hans Driesch use the word *Selbstregulation* to illustrate the idea that the embryo draws from its own organization the possibility of surpassing it. And Piaget used the same language to describe cognitive progress to which he referred to as an "embryology of reason". Religious immanentism has disappeared behind an epistemological immanentism, self-organization has replaced immanent action and emergence taken the place of the productive faculty which held the vital substance.

REFERENCES

Aristotle (1956), *The Metaphysics,* transl. Tredennick H.M.A., Harvard University Press, 1050a.

Beth W.E., Mays W. and Piaget J. (1957). Épistémologie génétique et recherche psychologique. *Études d'épistémologie génétique.* Paris: Presses Universitaires de France, vol. 1.

Blanckaert Claude (1990). La "Théologie naturelle" de Louis-François Jéhan (1803-1871). Science, apologétique, vulgarisation. *Nuncius, Annali di storia delle scienza,* **2**, 167-204.

Fix Andrew (1989). Angels, Devils and Evil Spirits in Seventeenth-century Thought: Balthasar Bekker and the Collegiants. *Journal of the History of Ideas,* vol. L, **4**, 527-547.

22. Walter Pagel, William Harvey revisited, *History of Science,* Vol. **8**, 1-31 and Vol. **9**, 1-41.

Gayon Jean et Mengal Paul (1992). Théorie de l'évolution et psychologie génétique chez Jean Piaget. In D. Andler *et al., Épistémologie et cognition.* Liège: Mardaga.

Hegel G.W.F. (1955). *Lectures on the History of Philosophy* (transl. E.S. Haldane and F.H. Simpson, 3 vols, London, 1892-1896). London: Routledge and Kegan Paul, New York: Humanities Press.

Hegel G.W.F. (1975), *Hegel's Logic,* (transl. W. Wallace), Oxford Clarendon Press, II, 218.

Kant I. (1924). *Critique of Pure Reason.* Trad. J.M. Meiklejohn, London: G. Bell & Sons.

Lewes G.H. (1874). *Problems of Life and Mind,* vol. I. London: Träbner and Co, Ludgate Hill.

Lewes G.H. (1877). *The Physical Basis of Mind.* London: Täbner and Co, Ludgate Hill.

Pagel Walter, Harvey William (revisited). *History of Science,* vol. 8, 1-31 and vol. 9, 1-41.

Paley William (1802). *Natural Theology or Evidences of the Existence and Attributes of the Deity collected from the Appearances of Nature.* London: Hamilton.

Piaget Jean (1930). *Immanentisme et foi religieuse.* Genève: Robert.

Piaget J. (1971). *Biology and Knowledge,* Chicago: The University of Chicago Press.

Piaget Jean et de La Harpe Jean (1928). *Deux types d'attitude religieuse: Immanence et Transcendance.* Genève: Labor.

Sabatier Auguste (1897). *Esquisse d'une philosophie de la religion d'après la psychologie et l'histoire.* Paris: Fischbacher.

Spinoza (1963), *Short Treatise on God, Man and His Well Being.* New York: ed. A. Wolf.

Vidal Fernando (1994). *Piaget before Piaget.* Cambridge: Harvard University Press.

II.HISTORIC APPROACH

B. Contemporary Origins

JEAN-CLAUDE HEUDIN

ARTIFICIAL LIFE AND THE SCIENCES OF COMPLEXITY: HISTORY AND FUTURE

1. INTRODUCTION

The field of Artificial Life (ALife) has recently emerged through the interaction of Biology and Computer Sciences, but also with important contributions from Physics, Mathematics, Cognitive Sciences and Philosophy. Many researchers from diverse backgrounds share the ALife approach and apply it in their own discipline. They seek to understand, through synthetic experiments, the organizational principles underlying the dynamics of living organisms. Then, these principles are used for synthesizing models or artificial systems with lifelike properties. This paper introduces ALife from an historical point of view in four parts. Firstly, it describes its historical roots. The second part gives its foundational principles and emphasizes the ALife approach. The third part gives a methodological-oriented classification of the main research trends. The fourth part introduces emergence as the core concept of ALife and replaces it in the framework of the sciences of complexity.

2. HISTORICAL FOUNDATIONS

2.1 Los Alamos

During the second world war, many of the best scientists joined the Manhattan project at Los Alamos. It was a pretty remarkable team that one observer at the time called the greatest gathering of intellects since ancient Athens: Robbert Oppenheimer, Enrico Fermi, Niels Bohr, Hans Bethe, Richard Feynman, Eugene Wigner, John von Neumann and many others. The project started with a very specific research challenge: building the

B. Feltz, M. Crommelinck and P. Goujon, (eds.), Self-Organization and Emergence in Life Sciences: 227-247.

bomb in a race against the Nazis. At the same time, some of these researchers were also beginning to think about complexity. This was the time computers have begun to be used for simulations, standing halfway between theory and experiment, making thinking about complex systems possible. In this framework, John von Neumann supervised the design of unprecedentely powerful computers. While wrestling with practical problems, he became interested in the abilities of Cellular Automata and machine self-reproduction.

2.2 Von Neumann's Self-Reproduction Theory

Even though the formalization of the concept of Cellular Automata was initiated by John von Neumann in the beginning of the 1950's, the true founder of research in artificial growth and evolution was the mathematician Stanislas Ulam, who designed the first experiments on one of the first stored program computers at Los Alamos. Ulam was interested in the growth patterns of two- and tree-dimensional geometrical figures generated from very simple recursive rules. This idea of complexity resulting from the combination of simple rules is one of the key notion in ALife. The works done by Ulam inspired John von Neumann and allowed him to design the first model of Cellular Automata. Von Neumann was interested in the process of reproduction and searched for the logical conditions *sufficient* for a non-trivial self-reproduction. He had formulated first a *kinematic* model consisting of a robot floating in a lake with all the components needed to built other robots. He pictured the robot collecting components and assembling them into a copy of itself. Von Neumann essentially succeeded in showing how the floating robot could reproduce, but unfortunately, much of his analysis was bogged down with the problem of motion in the lake. Thus, von Neumann took the abstraction one step further and adopted Ulam's approach. His idea was not to simulate self-reproduction at the genetic level, but to *abstract its logical form*: if self-reproduction is describable as a logical sequence of steps, then it exists a universal Turing machine which can perform its own reproduction. John von Neumann defined a two-dimensional cellular automaton with 29 possible states. The state of each cell is the result of a transition rule applied on the current cell and its four orthogonal neighbors. The non-trivial self-reproduction principle of von Neumann can be summarized as follow:

1. The system encapsulates a *description* of itself. Infinite regression is avoided because the self-description does not include itself. Instead, the description serves a dual role: it is an uninterpreted model of the system

and, at the same time, it is a coded description of the system (excluding the description itself).
2. The system includes a *supersory unit* which is able to perform any computation (simulate any Turing machine). It knows about the dual role of the description and makes sure that it is interpreted both ways during reproduction.
3. The system includes a *universal constructor* which can build any of a large class of objects given its description, in an empty region of the cellular space.
4. Self-reproduction occurs when the supervisory unit instructs the universal constructor to build a new copy of the system, including a description.

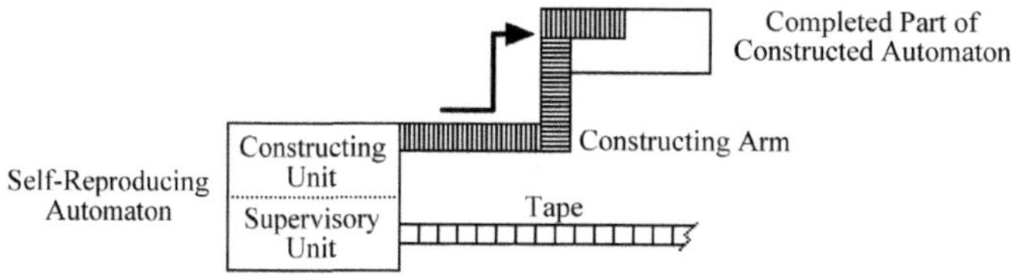

Figure 1. Simplified diagram of von Neumann's self-reproducing automata (not drawn to scale)

Without any specific assertion about biology, von Neumann showed that one of the main feature of life could be explained by means of logical principles instead of some magic property of matter. Unfortunately, he died in 1957 and did not finish his proof. Arthur Burks, who worked with him on the logical design of the EDVAC (one of the first computers) completed and edited his works[1]. By abstracting from the natural self-reproduction its logical (computational) form, John von Neumann is now recognized as the pioneer of the ALife approach.

2.3 Conway's Game of Life

The Game of Life is certainly the best example of the idea that complex worlds could emerge from simple rules. Life was designed in the 1970's by John Horton Conway, a young mathematician at Gonville and Caius College of the Cambridge University. It was introduced to the world at large via

1. von Neumann J. (1966). *Theory of Self-Reproducing Automata.* Urbana: University Illinois Press.

Martin Gardner's columns in Scientific American[2]. Then, the advent of home computers has opened Life to a much wider audience. Conway adapted Ulam and von Neumann's approach based on Cellular Automata. The state of each cell (alive/dead) is the result of two rules applied on the cell and its eight neighbors. Life's rules are marvelously simple: if the number of "alive" cells is exactly three, the current cell will be "alive" in the next generation; if the number of "alive" cells is zero, one, four, five, six, seven or eight, the cell will be "dead" in the next generation. Life has been experimented with extensively. Many of the configurations which emerge seem to have a "life" of their own. One of the most remarkable example of life's structures is the glider, a configuration of period four which displaces itself diagonally.

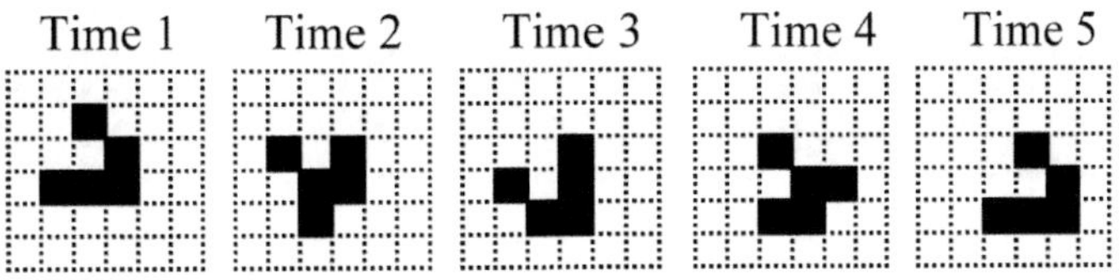

Figure 2. The glider displaces itself in four steps

2.4 Langton's Self-Reproducing Automata

In 1965, Edgar Codd, a student of Burks at the University of Michigan, was able to simplify von Neumann's cellular model[3]. In 1984, Christopher Langton, another student of Burks, designed a self-reproducing pattern based on an extremely simple configuration of Codd's automaton called the periodic emitter, itself derived from the periodic pulser organ in von Neumann's 29-state automaton[4]. Christopher Langton demonstrated that the capacity of universal construction was not a necessary condition for self-reproduction. The automaton is based on eight-state cells which are used (1) as data to copy in the cellular space causing the generation of an offspring and (2) as instructions to execute according to the transition rule. After 151 time steps, the initial structure has succeeded in reproducing itself. Then, each of these "loops" go on to reproduce itself in a similar

2. Gardner M. (1970). Mathematical Games: The Fantastic Combinations of John Conway's New Solitaire Game "Life". *Scientific American,* **223**.
3. Codd E. (1968). *Cellular Automata.* Academic Press.
4. Langton C.G. (1984). Self-Reproduction on a Cellular Automata. *Physica D,* **10**.

manner, giving rise to an expanding colony of "loops". This experiment captures the flavor of what goes on in natural development: the genotype codes for the constituents of a dynamic process in the cell, and it is this dynamic process that is primarily responsible for "computing" the expression of genotype in the course of development.

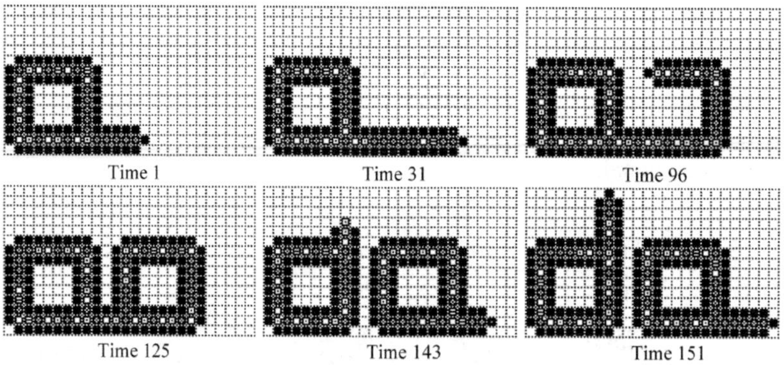

Figure 3. Langton's self-reproducing automaton

2.5 From Los Alamos to Santa Fe

From the day it was founded, Los Alamos had been a leader in advanced computing and non-linear research. By the early 70's, it seems clear that many nonlinear problems were the same kind of problem in the sense of having a similar mathematical structure. Thus, the result was a vigorous program for research in nonlinear sciences and the creation of the Centre for Nonlinear Studies. However, many interesting problems were not part of the laboratory's basic mission. George Cowan, who has worked with Enrico Fermi during the Manhattan project and then be involved into management responsibility at Los Alamos, began to imagine a new and independent institute. In the spring of 1983, he decided to take the idea to his companions, the Los Alamos senior fellows: Pete Carruther, Stirling Colgate, Nick Metropolis, David Pines and others. In May 1984, the Santa Fe Institute was incorporated, but with no location or staff. However, drawing from diverse field, scientists such as Nobel Laureates Murray Gell-Mann and Kenneth Arrow loved the project and joined the institute. In the autumn of 1986, Philip Anderson and Kenneth Arrow organized the first economics meeting, while George Cowan was making a deal with the

Archdiocese of Santa Fe for leasing the Christo Rey Convent: in February 1987, the institute staff moved in[5].

2.6. The Birth of Artificial Life

Christopher Langton was greatly inspired by the works of von Neumann, Ulam and Conway. He invented the term "Artificial Life" in late 1971, during a night he was hacking with the game of Life on a PDP-9 computer of the psychology department at the Massachusetts General Hospital in Boston. Several years after, he worked as a teaching assistant for Burk's history of computing class at the University of Michigan and worked also with John Holland, the pioneer of evolutionary computing. In June 1984 he went to a conference on cellular automata at MIT and meet Doyne Farmer. This was the same period when Doyne Farmer, Norman Packard and Stuart Kauffman worked on autocalytic networks and helped to get the Santa Fe Institute up and running. In August 1986, Christopher Langton arrived at Los Alamos for a postdoctoral appointment in the Center for Nonlinear Studies. In September 1987, he organized the first workshop on ALife at the Los Alamos National Laboratory. The workshop was sponsored by the Center for Nonlinear Studies, the Santa Fe Institute and Apple Computer Inc. It brought together 160 computer scientists, biologists, physicists, anthropologists and others, all of whom sharing a common interest in the simulation and synthesis of living systems.

3. WHAT IS ARTIFICIAL LIFE?

3.1 Foundational Principles

The foundations of ALife have been proposed by Christopher Langton in the paper introducing the proceedings of the first ALife workshop[6]. Biology is the scientific study of life, but it employs an *analytical* approach which is largely concerned with the material basis of life. ALife is the study of possible lifes: it contributes to theoretical biology by locating *life-as-we-know-it* within the larger picture of *life-as-it-could-be*. Thus, ALife gives a

5. Waldrop M.M. (1992). *Complexity: The Emerging Science at the Edge of Order and Chaos*. Touchstone Book.
6. Langton C.G. (1989). Artificial life. In C.G. Langton (ed.), *Artificial Life, SFI Studies in the Sciences of Complexity*, Vol. VI. Addison-Wesley.

new framework which complements the traditional biological sciences. ALife is based on a *synthetic* approach in two phases: (1) the first phase abstracts the logical principles of living organisms; (2) the second phase implements these principles through synthesis on another media like a computer. The results are *models* for studying the living (modeling approach) or *artificial systems* with lifelike properties (engineering approach). The ALife approach is also based on two hypotheses: (1) *life is a property of the organization of matter rather than a property of the matter which is so organized;* in other words, life is a property of form rather than something that inheres the matter itself; (2) *complex behaviors or properties (like life itself) can emerge from interactions of collections of simple processes.*

3.2 A Taxonomy of Possible Artificial Lifes

Recently, Claus Emmeche has stated the different possible trivial and non-trivial forms of artificial living systems[7]. We extend it here by pointing out the possible media:

1. The first trivial form consists of all artificially *modified living organisms.* This can be done, for example, by genetic engineering or cell fusion. We can see most of our new food crops has artificial systems produced for the purpose of man.
2. The second trivial form includes mathematical, conceptual and physical *models.* Everybody agrees upon the possibility of modeling living phenomena. This is the most classical version of the Artificial Life research program.
3. The first non-trivial form attempts to make real material systems with lifelike characteristics using *biochemical synthesis.* Examples of such approaches are *in vitro* experiments of prebiotic processes, hypercyclic systems, primitive metabolisms, replicating micelles.
4. The second non-trivial form can be seen as a new approach of *robotics.* In this group we find animates and robots designed for technical purposes which are seen to behave in a lifelike manner.
5. The third non-trivial form of artificial living systems is *virtual (computational) life:* computer programs with emergent properties that in the eye of the beholder seems to approach real life organisms.

---

7. Emmeche C. (1994). Is Life a Multiverse Phenomenon? In C.G. Langton (ed.), *Artificial Life III, SFI Studies in the Sciences of Complexity,* Vol. XVII. Addison-Wesley.

Media	Trivial Forms	Non-trivial Forms
Carbon-based Matter	*Modified Organisms*	*Biochemical Synthesis*
Non-Organic Matter	-	*Robots*
Computer Programs	*Models*	*Virtual Life*

Figure 4. Artificial Life Forms

3.3 A Moderated Functionalism

Many philosophers have explored the analogy between the mind/brain problem (weak or strong AI) and the life/body problem (weak/strong ALife), such as E. Sober[8]. Functionalists claim that psychological theories can be formulated by abstracting away from the physical details that distinguish one thinking system from another. This position has attracted a great deal of attention, both in the form of advocacy and in the form of attack. ALife leads us to view natural life as an example of life and enables us to abstract away from physical details. The idea is that life properties are multiply realizable. The problem is to be able to do this without going too far. This problem is especially pressing when the mathematical structure of a phenomena is confused with its empirical content, since it can lead one to say that a system is alive when it does not. On one hand, reducing the number of possible realizations to only one leads to something close to the identity theory. On the other hand, abstracting too far from physical details leads to an overly liberal conception of life and results in a dualist or vitalist claim. However, a moderated functionalist approach is a liberating doctrine, which enable researcher to explore important questions about the possibility of ALife forms. Even if some answers will be negative, it will represent a significant progress in the sciences of the living.

4. RESEARCH TRENDS

The taxonomy of artificial living systems (cf. section 3.2) shows that all these interesting disciplines cannot be easily integrated in a single and coherent research framework: ALife is not a unitary field, but includes several different trends. The first classification of these trends was proposed

8. Sober E. (1991). Learning from functionalism - Prospects for Strong Artificial Life. In C.G. Langton, C.E. Taylor, J.D. Farmer and S. Rasmussen (eds), *Artificial Life II, SFI Studies in the Sciences of Complexity,* Vol. X. Addison-Wesley.

by C.G. Langton[9]. We propose here a more methodologically-oriented classification close to the one of C. Taylor[10].

Cellular Automata: Cellular Automata are generally used for modeling complexity. Cellular automata are array of cells, each of which assumes a discrete state. The state of a cell may change through discrete time according to a well-defined transition rule which takes into account the current state of the cell together with the states of its immediate neighbors. In most cases, all cells are updated simultaneously using a parallel and synchronous iteration algorithm, or sequentially using a stochastic and asynchronous iteration algorithm. One of the most important research trend concerns self-reproduction, which is one of the main property of living organisms. The studies of von Neumann[11], C. Langton[12] and those of S. Wolfram[13] are typical examples of this trend.

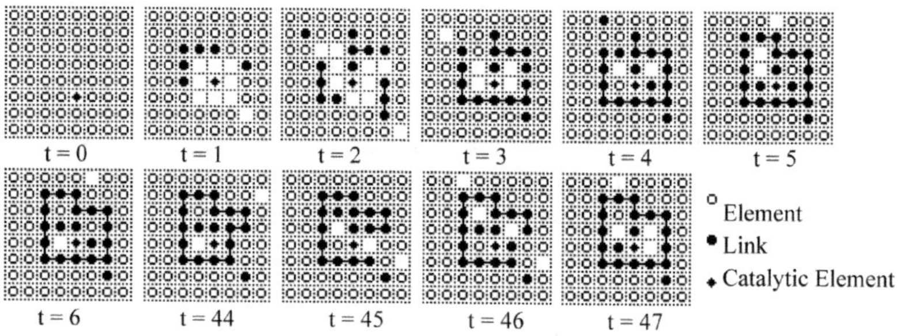

Figure 5. Simulation of Autopoiesis using 2D Cellular Automata is another example of their appropriateness for modeling complexity (after Varela 82)

Artificial Embryologies: The ability of a living system to develop itself from a single cell to a complete organism is another important property of Life. The study of artificial embryologies, often based on fractal geometry, shows that complex and beautiful lifelike forms can emerge from simple

9. Langton C.G. (1989). Artificial life. In C.G. Langton (ed.), *Artificial Life, SFI Studies in the Sciences of Complexity,* Vol. VI. Addison-Wesley.

10. Taylor C.E. (1991). "Fleshing Out" Artificial Life II. In C.G. Langton, C.E. Taylor, J.D. Farmer and S. Rasmussen (eds), *Artificial Life II, SFI Studies in the Sciences of Complexity,* Vol. X. Addison-Wesley.

11. von Neumann J. (1966). *Theory of Self-Reproducing Automata.* Urbana: University Illinois Press.

12. Langton C.G. (1984). Self-Reproduction on a Cellular Automata. *Physica D,* **10.**

13. Wolfram S. (1986). *Theory and Application of Cellular Automata.* Singapore: World Scientific.

recursive procedures. Examples in this category include the L-Systems of A. Lindenmayer and P. Prusinkiewicz[14] and the Biomorphs of R. Dawkins[15].

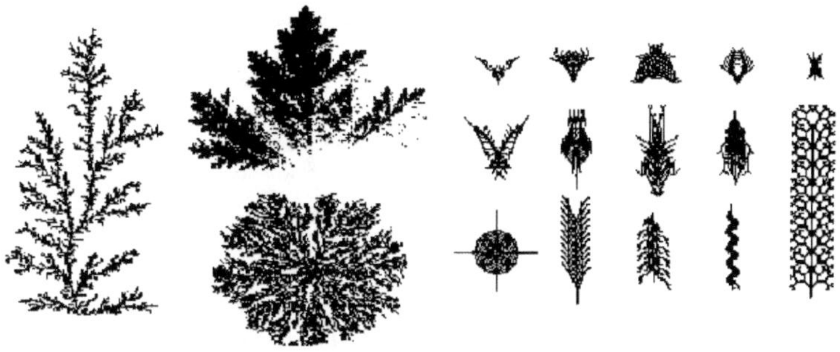

Figure 6. Fractal-based embryological studies (left) and a selection of Dawkins's biomorphs (right)

Evolutionary Computing: Evolutionary Computing is a set of optimization methods and algorithms based on the principle of evolution by natural selection. Most of the research projects use Genetic Algorithms as proposed first by John Holland[16]. A Genetic Algorithm generates a set of offspring from a parent population, and is primarily concerned with producing variants having a higher success in the environment. The variants are generated by applying genetic operators, such as mutation and crossing-over, on the genotypes of the most successful phenotypes in the population. John Holland has also pioneered the application of Genetic Algorithms to the problem of machine learning in the form of the Genetic Classifiers[17]. More recently, J.R. Koza has demonstrated the emergence of evolutionary self-improving computer programs using Genetic Programming: a method for programming computers by means of natural selection[18].

---

14. Lindenmayer A. and Prusinkiewicz P. (1989). Developmental Models of Multicellular Organisms. In C.G. Langton (ed.), *Artificial Life, SFI Studies in the Sciences of Complexity,* vol. VI. Addison-Wesley.
15. Dawkins R. (1989). In C.G. Langton (ed.), *Artificial Life, SFI Studies in the Sciences of Complexity,* Vol. VI. Addison-Wesley.
16. Holland J. (1975). *Adaptation in Natural and Artificial Systems.* Ann Arbor: University of Michigan Press.
17. Goldberg D.E. (1989). *Genetic Algorithms in Search, Optimization and Machine Learning.* Addison-Wesley.
18. Koza J.R. (1991). Genetic Evolution and Co-Evolution of Computer Programs. In C.G. Langton, C.E. Taylor, J.D. Farmer and S. Rasmussen (eds), *Artificial Life II, SFI Studies in the Sciences of Complexity,* Vol. X. Addison-Wesley.

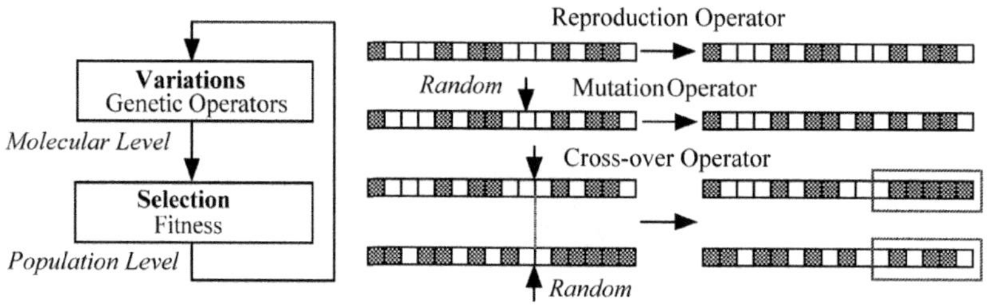

Figure 7. Principles of a simple Genetic Algorithm

Autocatalytic Networks: Another important trend concerns the study of the possible origins of life and models for prebiotic evolution. Most of these works use autocatalytic networks, where nodes are interpreted as specific types of RNA sequences and oriented arcs as catalytic interactions. An especially well-developed example of this approach is the theory of hypercycles, pioneered by M. Eigen and P. Schuster[19]. Examples of autocatalytic networks are those developed by J.D. Farmer[20], S. Kauffman [21] and S. Rasmussen[22].

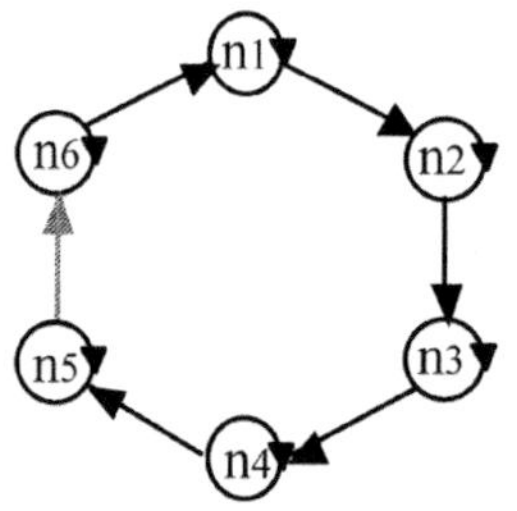

Figure 8. Example of an hypercyclic reaction

Computational Life: A growing research trend concerns the design of computer programs that exhibit lifelike behaviours but which are not explicit

19. Eigen M. and Schuster P. (1979). *The Hypercycle: A Principle of Natural Self-Organization.* New York: Springer Verlag.
20. Bagley R. and Farmer J.D. (1991). Emergence of Robust Autocatalytic Networks. In C.G. Langton, C.E. Taylor, J.D. Farmer and S. Rasmussen (eds), *Artificial Life II, SFI Studies in the Sciences of Complexity,* Vol. X. Addison-Wesley.
21. Kauffman S.A. (1986). Autocatalytic Replication of Polymers. *Journal of Theoretical Biology,* **119**.
22. Rasmussen S. (1989). Toward a Quantitative Theory of Life. In C.G. Langton (ed.), *Artificial Life, SFI Studies in the Sciences of Complexity,* Vol. VI. Addison-Wesley.

simulations or models of any particular known biological organisms. Even if they are fundamentally different from natural life, computer processes are capable of reproducing, interactions with the environment, and evolution. The Tierra program of T. Ray is probably the best example of such an approach to synthesis of life within computers[23]. Another example is computer virus, which are more a problem than a real research trend[24].

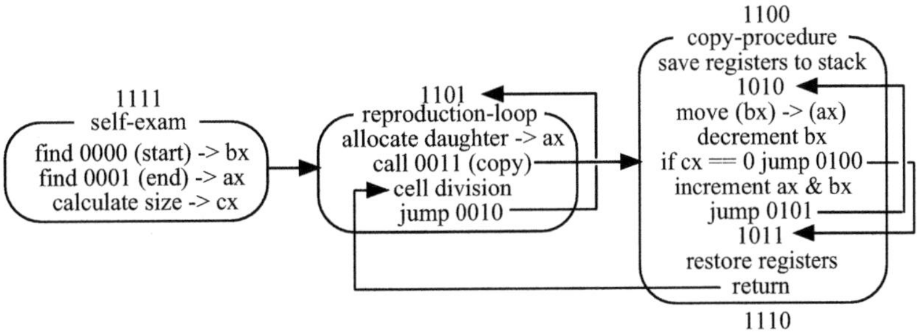

Figure 9. The "ancestor" creature in Tierra (after T. Ray 91)

Collective Intelligence: This trend includes research programs which are close to those of Distributed Artificial Intelligence and Multi-Agent Systems. However, the approach differs since ALife is mainly bottom-up and is explicitly based on biological models. Examples in this category are the swarm networks of M. Millonas[25], the study of ant colonies by J.L. Deneubourg[26], and many implementations of evolving neural networks like[27].

---

23. Ray T.S. (1991). An Approach to the Synthesis of Life. In C.G. Langton, C.E. Taylor, J.D. Farmer and S. Rasmussen (eds), *Artificial Life II, SFI Studies in the Sciences of Complexity,* Vol. X. Addison-Wesley.
24. Spafford E.H. (1991). Computer Viruses - A Form of Artificial Life? In C.G. Langton, C.E. Taylor, J.D. Farmer and S. Rasmussen (eds), *Artificial Life II, SFI Studies in the Sciences of Complexity,* Vol. X. Addison-Wesley.
25. Millonas M.M. (1994). Swarms, Phase Transitions, and Collective Intelligence. In C.G. Langton (ed.), *Artificial Life III, SFI Studies in the Sciences of Complexity,* Vol. XVII. Addison-Wesley.
26. Deneubourg J.L. and Gross S. (1989). Collective Patterns and Decision making. *Ethology Ecology and Evolution,* **1**.
27. Belew R.K., McInerney J. and Scraudolph N.N. (1991). Evolving networks: Using the Genetic Algorithm with Connectionist Learning. In C.G. Langton, C.E. Taylor, J.D. Farmer and S. Rasmussen (eds), *Artificial Life II, SFI Studies in the Sciences of Complexity,* Vol. X. Addison-Wesley.

Evolutionary Robotics: Robotics represents the "hardware" end of ALife. Many researchers concentrate on the design of autonomous individual robots like R. Brooks and his Insectoïds based on a reactive layered architecture[28]. Other research programs concentrate on collective behaviors of populations of simpler and smaller robots[29]. A complementary approach analyzes robot control structures which can take advantage of the process of evolution, like I. Harvey's SAGA[30].

Evolvable Hardware Devices: A promising research trend concerns the design of evolvable hardware devices. De Garis has first pointed out the need to achieve hardware evolution[31]. Recently, we have described different approaches for the implementation of a Genetic Microprocessor based on hardware genetic classifiers[32]. D. Mange and his colleagues have set up a research group for studying self-repairing and self-reproducing hardware based on biological-like properties[33]. Many studies are also done for designing new sensors such as artificial retinae[34]. The main problem for implementing evolvable hardware is that silicon lacks certain basic characteristics that are essential to the evolutionary capacity. Thus, many researchers concentrate on using FPGA-like (Field-Programmable Gate Array) technology in order to obtain the required level of adaptiveness[35].

---

28. Brooks R. (1986). *A Robuts Layered Control System for a Mobile Robot.* IEEE J. Robot. and Automation.
29. Nolfi S., Floreano D., Miglino O. and Mondada F. (1994). How to Evolve Autonomous Robots: Different Approaches in Evolutionary Robotics. In R.A. Brooks and P. Maes (eds), *Artificial Life IV, Proceedings.* Bradford Book: MIT Press.
30. Harvey I. (1994). Evolutionary Robotics and SAGA: The Case for Hill Crawling and Tournament Selection. In C.G. Langton (ed.), *Artificial Life III, SFI Studies in the Sciences of Complexity,* Vol. XVII. Addison-Wesley.
31. de Garis H. (1993). *Evolvable Hardware: Genetic Programming of Darwin Machines.* International Conference on Neural Networks and Genetic Algorithms, lecture notes in computer sciences, Springer-Verlag.
32. Heudin J.C. (1994). *Towards Genetic Microprocessors for Symbolic Control,* SODIMA Technical Report.
33. Marchal P., Piguet C., Mange D., Stauffer A. and Durand S. (1994). Embryological Development on Silicon. In R.A. Brooks and P. Maes (eds), *Artificial Life IV.* Bradford Book: MIT Press.
34. Carnapete L.S., Nguyen P.E., Nguyen R.G. and Bernard T.M. (1995). *A Miniature Retinae-Based AGV Called Vampire.* IEEE Computer Architectures for Machine Perception, **95**, Como, Italy.
35. Hemmi H., Mizoguchi J. and Shimora K. (1994). Development and Evolution of Hardware behaviours. In R.A. Brooks and P. Maes (eds), *Artificial Life IV.* Bradford Book: MIT Press.

We have proposed recently an artificial evolution paradigm for hardware devices which follows the Genetic Programming approach[36].

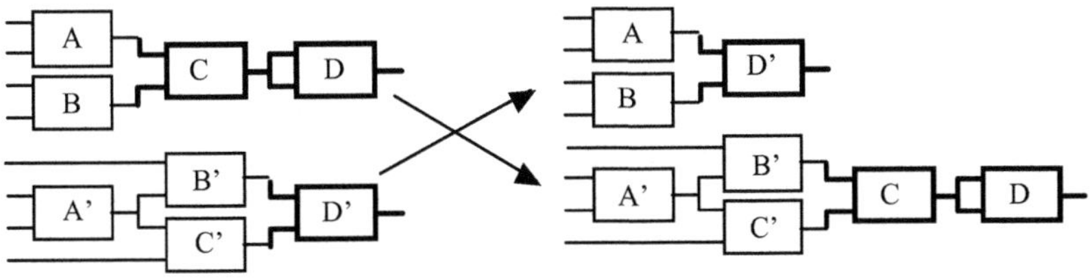

Figure 10. Crossover operator in evolvable hardware devices as proposed by J.C. Heudin

Nanotechnologies: ALife involves the synthesis of processes associated with natural life on new media, new scales or with new organizations. One possible trend for creating ALife is based on Richard Feynman's suggestions in 1959 for ultraminiaturization and extension of our industrial manufacturing capabilities all the way down to the molecular level. Many reseachers have developed related concepts and further generalized them[37].

Biochemical Synthesis: ALife is not confined to computers and an important trend concerns in vitro experiments of RNA reproduction, prebiotic artificial life forms, synthesis and evolution of RNA chains[38], autocatalytic reactions, and osmotic growths[39].

5. ARTIFICIAL LIFE AND THE SCIENCES OF COMPLEXITY

5.1 The Science of Emergence

The first workshop gave birth to the field of Artificial Life. Since that workshop, a large number of people have become interested in the field, its methodological approaches, and have initiated new research projects. Many

36. Heudin, J.C. (1995). *Artificial Life and Evolutionary Computing for Machine* Perception. IEEE Computer Architectures for Machine Perception, **95**, Como, Italy.
37. Schneiker C. (1988), NanoTechnology with Feynman Machines: Scanning Tunneling Engineering and Artificial Life. In C.G. Langton (ed.), *Artificial Life, SFI Studies in the Sciences of Complexity,* Vol. VI. Addison-Wesley.
38. North G. (1990). Expanding the RNA Repertoire. *Nature,* **345**.
39. Zeleny M., Klir G.J. and Hufford K.D. (1989). Precipitation Membranes, Osmotic Growths and Synthetic Biology. In C.G. Langton (ed.), *Artificial Life, SFI Studies in the Sciences of Complexity,* Vol. VI. Addison-Wesley.

of them were reported in the following ALife workshops. The second one was held in February 1990 and the third one in June 1992, both in Santa Fe and organized by C. Langton and his colleagues. Now, the field has grown and matured to the point where it no longer belongs just to Santa Fe, but becomes a truly international field. The first European Conference on ALife was held in December 1991 and organized by F. Varela and P. Bourgine. The fourth ALife workshop was held at MIT in July 1994, and organized by R. Brooks and P. Maes. In all the research projects reported, the "key" concept is emergence. Thus, ALife could be seen as the science of *emergence.* An emergent property is a global behavior or structure which appears through interactions of a collection of elements, with no global controller responsible for the behavior or organization of these elements. The idea with emergence is that it is not reductible to the properties of the elements. In other words, we can say "the all is more than the sum of its parts". Traditional sciences are based on a reductionist analytical approach which analyzes a system as a structure composed of simpler elements, and continues this process, breaking things down as far as possible. It has proven its efficiency, where the reductionist ontology of western physics is probably the most successful example. However, this approach is intrinsically limited in the case of studying complex systems which exhibits emergent properties, simply because when one breaks such a system into pieces, the emergent properties disappear. Therefore, a synthetic approach, which brings together parts rather than disassembling them, gives a promising complementary research framework. Emergent phenomena have been found everywhere in nature and in every domains of science. However, since an arbitrary system could be viewed as emergent simply by properly choosing a level of abstraction at which to consider it, we must define it more clearly.

5.2. Complexity and emergence

Many systems can be observed at different scales. These scales can be described as a set of hierarchical abstract levels. Then, given a particular level of abstraction, the system is describable as a network of structures which interact and give rise to the next level of complexity. At lower levels, there is a great number of structures, but in few categories. At higher levels, structures are more complex and in a greater number of categories. All levels form together a pyramid of complexity[40]. It is an abstract model

40. Heudin J.C. (1994). *La Vie Artificielle.* Paris: Éditions Hermès.

because structures can be interpreted as particles, molecules, living organisms, information, symbols, etc. However, all levels are combined together and form a tangled hierarchy where all levels cross. Structures have properties in the form of their potential links with other structures of the same level. Thus, the interactions of these structures with others become the starting point for the formation of new dynamic structures. These new structures are characterized by new properties in the form of their potential new interactions with other structures of the new level. In this context, we can define an emergent property in terms of the *non-existence* of a sub-level over which this property can be reduced to a simple and local combination of its structures. A more formalized definition can be found in[41].

A large range of systems could be described using this model. As an extreme example, our universe can be seen as a pyramid of complexity[42]. Quarks, particles, atoms, molecules, bio-molecules, cells, living organisms, etc : elements of a given level combine to form new elements of the upper level. At the atomic level, the atom of helium is very stable and has no "links". The atom of helium cannot forms new structures with other atoms. In contrast, the atom of carbon has four "links", which enable a large number of possible interactions with other atoms. Thanks to its interactions abilities, carbon is included in every large molecular structures and represents with water the material basis of life.

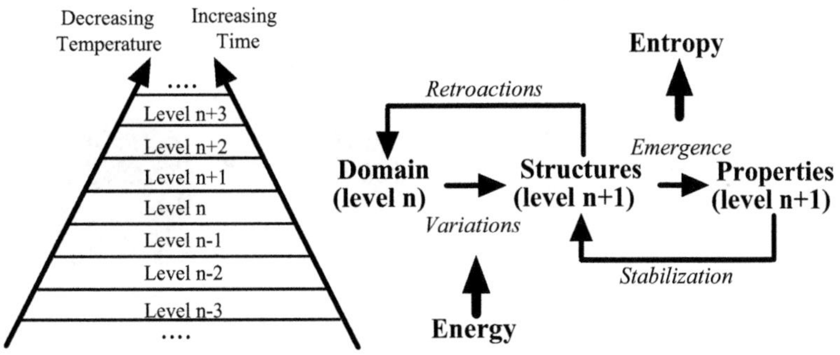

Figure 11. Pyramids of complexity

41. Heudin J.C. (1995). *Évolution de la complexité et Vie Artificielle.* University of Paris XI.
42. Reeves H. (1981), *Patience dans l'azur - L'évolution cosmique.* Paris: Éditions du Seuil.

5.3 The Evolution of Complexity

What is the principle which enables new structures with new properties to emerge from a pyramid of complexity? The principle involves two main processes: *variation and stabilization.* These processes are *continuous* and occur in *parallel* at all levels of the pyramid. Considering an arbitrary level of a pyramid of complexity, the structural units of this level are potentially able to take a large number of different configurations. When some thermodynamical fluctuations occur, the deterministic description breaks down and far-from-equilibrium processes began. These fluctuations create random *variations* of structural units configurations. These variations allow the formation of a large number of transient structures which relate to a higher level of organization. Some of these transient structures stabilized themselves due to their "fitness" to the environment. This *stabilization* process involves structural properties of the units combined with those of the environment. These new structures form a new level of complexity. Stabilization processes evolve to four classes of behaviors which refer to those defined by S. Wolfram[43] and C. Langton in his phase transition work[44]: (1) fixed and homogenous states, (2) simple periodic structures, (3) chaotic aperiodic structures, and (4) complex structures. Langton has suggested that complex structures such as living systems need to avoid either of these ultimate outcomes by learning to maintain themselves near a "critical" transition between order and chaos.

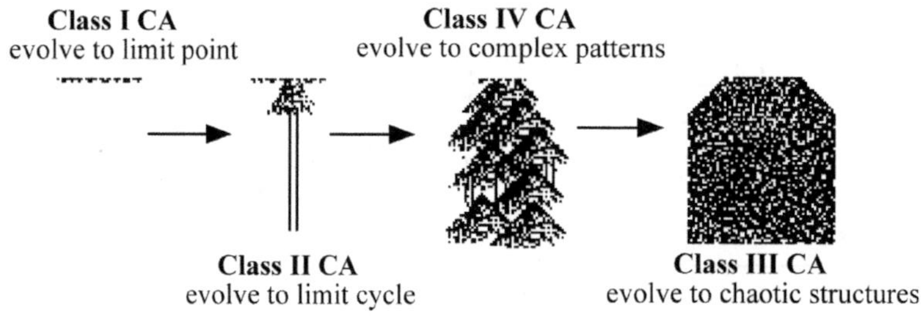

Figure 12. Schematic drawing of 1D Cellular Automata rule space indicating the progression through the spectrum of dynamical behaviors (after Langton 91)

---

43. Wolfram S. (1986). *Theory and Application of Cellular Automata.* Singapore: World Scientific.
44. Langton C.G. (1991). Life at the Edge of Chaos. In C.G. Langton, C.E. Taylor, J.D. Farmer and S. Rasmussen (eds), *Artificial Life II, SFI Studies in the Sciences of Complexity,* Vol. X. Addison-Wesley.

A particular example of a stabilization process based on self-organization in the biological cellular domain is *autopoiesis*[45]. We have recently proposed an organization-oriented definition of "minimal" life based on autopoiesis[46]. We argue that the Darwinian "natural selection", based on a *mutation-selection* principle, is a particular example of the *variation-stabilization* principle. One of the main differences is that results of the stabilization process are not optimal solutions. Selection occurs only when there is a competition for limited resources. In contrast with the natural selection principle which optimizes a fitness function, the structural stabilization process chooses "good" solutions that globally satisfy the environment constraints. It is a satisfaction process rather than an optimization process which relates to the natural drift proposition of F. Varela and his colleagues[47]. A second important difference is that all levels are tangled in a hierarchical mode. The environment is composed of all the levels including all structures. Recently C. Langton suggested that such local to global back to local inter-level feedback loops are essential to life[48].

6. CONCLUSION

We have introduced the foundational principles of Artificial Life from an historical point of view. We have first described the historical roots of ALife and introduced its foundational principles and its approach in more details. Then, we have given a methodological-oriented classification of its main research trends. We have emphasized emergence as the core concept of ALife. The future of ALife will be to become the synthetic end of the sciences of complexity. To be successful, ALife must keep its multi-disciplinary approach and focus first on theoretical contributions about emergence and self-organization. However, this must be done without sacrifying practical experiments and applications which will certainly influence this new and promising discipline.

---

45. Varela F.J., Maturana H. and Uribe R. (1974). Autopoiesis: The Organization of Living Systems, Its Characterization and a Model. *Biosystems,* vol. 5.
46. Heudin, J.C. (1995). *Artificial Life and Evolutionary Computing for Machine* Perception. IEEE Computer Architectures for Machine Perception, **95**, Como, Italy.
47. F. Varela, E. Thompson and E. Rosch (1993), L'inscription corporelle de l'esprit, Editions du seuil, Paris
48. Langton C.G. (1991). Life at the Edge of Chaos. In C.G. Langton, C.E. Taylor, J.D. Farmer and S. Rasmussen (eds), *Artificial Life II, SFI Studies in the Sciences of Complexity,* Vol. X. Addison-Wesley.

REFERENCES

Bagley R. and Farmer J.D. (1991). Emergence of Robust Autocatalytic Networks. In C.G. Langton, C.E. Taylor, J.D. Farmer and S. Rasmussen (eds), *Artificial Life II, SFI Studies in the Sciences of Complexity,* Vol. X. Addison-Wesley.

Belew R.K., McInerney J. and Scraudolph N.N. (1991). Evolving networks: Using the Genetic Algorithm with Connectionist Learning. In C.G. Langton, C.E. Taylor, J.D. Farmer and S. Rasmussen (eds), *Artificial Life II, SFI Studies in the Sciences of Complexity,* Vol. X. Addison-Wesley.

Brooks R. (1986). *A Robuts Layered Control System for a Mobile Robot.* IEEE J. Robot. and Automation.

Carnapete L.S., Nguyen P.E., Nguyen R.G. and Bernard T.M. (1995). *A Miniature Retinae-Based AGV Called Vampire.* IEEE Computer Architectures for Machine Perception, **95**, Como, Italy.

Codd E. (1968). *Cellular Automata.* Academic Press.

Dawkins R. (1989). In C.G. Langton (ed.), *Artificial Life, SFI Studies in the Sciences of Complexity,* Vol. VI. Addison-Wesley.

de Garis H. (1993). *Evolvable Hardware: Genetic Programming of Darwin Machines.* International Conference on Neural Networks and Genetic Algorithms, lecture notes in computer sciences, Springer-Verlag.

Deneubourg J.L. and Gross S. (1989). Collective Patterns and Decision making. *Ethology Ecology and Evolution,* **1**.

Eigen M. and Schuster P. (1979). *The Hypercycle: A Principle of Natural Self-Organization.* New York: Springer Verlag.

Emmeche C. (1994). Is Life a Multiverse Phenomenon? In C.G. Langton (ed.), *Artificial Life III, SFI Studies in the Sciences of Complexity,* Vol. XVII. Addison-Wesley.

Gardner M. (1970). Mathematical Games: The Fantastic Combinations of John Conway's New Solitaire Game "Life". *Scientific American,* **223**.

Goldberg D.E. (1989). *Genetic Algorithms in Search, Optimization and Machine Learning.* Addison-Wesley.

Harvey I. (1994). Evolutionary Robotics and SAGA: The Case for Hill Crawling and Tournament Selection. In C.G. Langton (ed.), *Artificial Life III, SFI Studies in the Sciences of Complexity,* Vol. XVII. Addison-Wesley.

Heudin J.C. (1994). *La Vie Artificielle.* Paris: Éditions Hermès.

Heudin J.C. (1994). *Towards Genetic Microprocessors for Symbolic Control,* SODIMA Technical Report.

Heudin J.C. (1995). *Évolution de la complexité et Vie Artificielle.* University of Paris XI.

Heudin, J.C. (1995). *Artificial Life and Evolutionary Computing for Machine* Perception. IEEE Computer Architectures for Machine Perception, **95**, Como, Italy.

Hemmi H., Mizoguchi J. and Shimora K. (1994). Development and Evolution of Hardware behaviors. In R.A. Brooks and P. Maes (eds), *Artificial Life IV.* Bradford Book: MIT Press.

Holland J. (1975). *Adaptation in Natural and Artificial Systems.* Ann Arbor: University of Michigan Press.

Kauffman S.A. (1986). Autocatalytic Replication of Polymers. *Journal of Theoretical Biology,* **119**.

Koza J.R. (1991). Genetic Evolution and Co-Evolution of Computer Programs. In C.G. Langton, C.E. Taylor, J.D. Farmer and S. Rasmussen (eds), *Artificial Life II, SFI Studies in the Sciences of Complexity,* Vol. X. Addison-Wesley.

Langton C.G. (1984). Self-Reproduction on a Cellular Automata. *Physica D,* **10**.

Langton C.G. (1989). Artificial life. In C.G. Langton (ed.), *Artificial Life, SFI Studies in the Sciences of Complexity,* Vol. VI. Addison-Wesley.

Langton C.G. (1991). Life at the Edge of Chaos. In C.G. Langton, C.E. Taylor, J.D. Farmer and S. Rasmussen (eds), *Artificial Life II, SFI Studies in the Sciences of Complexity,* Vol. X. Addison-Wesley.

Lindenmayer A. and Prusinkiewicz P. (1989). Developmental Models of Multicellular Organisms. In C.G. Langton (ed.), *Artificial Life, SFI Studies in the Sciences of Complexity,* Vol. VI. Addison-Wesley.

Marchal P., Piguet C., Mange D., Stauffer A. and Durand S. (1994). Embryological Development on Silicon. In R.A. Brooks and P. Maes (eds), *Artificial Life IV.* Bradford Book: MIT Press.

Millonas M.M. (1994). Swarms, Phase Transitions, and Collective Intelligence. In C.G. Langton (ed.), *Artificial Life III, SFI Studies in the Sciences of Complexity,* Vol. XVII. Addison-Wesley.

Nolfi S., Floreano D., Miglino O. and Mondada F. (1994). How to Evolve Autonomous Robots: Different Approaches in Evolutionary Robotics. In R.A. Brooks and P. Maes (eds), *Artificial Life IV, Proceedings.* Bradford Book: MIT Press.

North G. (1990). Expanding the RNA Repertoire. *Nature,* **345**.

Rasmussen S. (1989). Toward a Quantitative Theory of Life. In C.G. Langton (ed.), *Artificial Life, SFI Studies in the Sciences of Complexity,* Vol. VI. Addison-Wesley.

Ray T.S. (1991). An Approach to the Synthesis of Life. In C.G. Langton, C.E. Taylor, J.D. Farmer and S. Rasmussen (eds), *Artificial Life II, SFI Studies in the Sciences of Complexity,* Vol. X. Addison-Wesley.

Reeves H. (1981), *Patience dans l'azur - L'évolution cosmique.* Paris: Éditions du Seuil.

Schneiker C. (1988), NanoTechnology with Feynman Machines: Scanning Tunneling Engineering and Artificial Life. In C.G. Langton (ed.), *Artificial Life, SFI Studies in the Sciences of Complexity,* Vol. VI. Addison-Wesley.

Spafford E.H. (1991). Computer Viruses - A Form of Artificial Life? In C.G. Langton, C.E. Taylor, J.D. Farmer and S. Rasmussen (eds), *Artificial Life II, SFI Studies in the Sciences of Complexity,* Vol. X. Addison-Wesley.

Sober E. (1991). Learning from functionalism - Prospects for Strong Artificial Life. In C.G. Langton, C.E. Taylor, J.D. Farmer and S. Rasmussen (eds), *Artificial Life II, SFI Studies in the Sciences of Complexity,* Vol. X. Addison-Wesley.

Taylor C.E. (1991). "Fleshing Out" Artificial Life II. In C.G. Langton, C.E. Taylor, J.D. Farmer and S. Rasmussen (eds), *Artificial Life II, SFI Studies in the Sciences of Complexity,* Vol. X. Addison-Wesley.

Varela F.J., Maturana H. and Uribe R. (1974). Autopoiesis: The Organization of Living Systems, Its Characterization and a Model. *Biosystems,* Vol. 5.

Varela F., Thompson E. and Rosch E. (1993), *L'inscription corporelle de l'esprit.* Paris: Éditions du seuil.

von Neumann J. (1966). *Theory of Self-Reproducing Automata.* Urbana: University Illinois Press.

Waldrop M.M. (1992). *Complexity: The Emerging Science at the Edge of Order and Chaos.* Touchstone Book.

Wolfram S. (1986). *Theory and Application of Cellular Automata.* Singapore: World Scientific.

Zeleny M., Klir G.J. and Hufford K.D. (1989). Precipitation Membranes, Osmotic Growths and Synthetic Biology. In C.G. Langton (ed.), *Artificial Life, SFI Studies in the Sciences of Complexity,* Vol. VI. Addison-Wesley.

PIERRE LIVET

SELF-ORGANIZATION IN SECOND-ORDER CYBERNETICS: DECONSTRUCTION OR RECONSTRUCTION OF COMPLEXITY

So called "Second-Order" cybernetics developed the notions of self-organization in a manner more radical than Wiener, attempting to both give a strong sense to the prefix "auto" and, at the same time, to explain reflexive phenomena, memory, and the relation between the organism and its history, in terms of the emergence of a global effect out of local interactions and the networking of retroactive processes. But the proposed theories didn't always work. Today's connectionist systems are analogous in some ways to the "non-trivial machines" of von Foerster. They lay claim to a common filiate (the formal networks of neurones of McCulloch and Pitts) and, more recently, they attack the problem of the recognition and the representation of recursive, nested structures. This new step is reminiscent of the one which took us from Ashby's homeostat to the application by von Foerster of the concept of recursivity to the problem of non-trivial or memory-equipped machines. We will attempt to show that the analogy is a real one and that von Foerster anticipated many of the current problems. But we will also see that with the development of working models, other more redoubtable difficulties have appeared precisely where Second-Order cybernetics failed to foresee them.

I will limit myself to two examples from the work of von Foerster: his notion of a "non-trivial machine" and his theory of "cognitive tiles". But to understand them, they must be situated in von Foerster's overall project: that of modeling the complexity of the cognitive. He rejected the cybernetics of Ashby, seeing in it nothing but first-order cybernetics, capable of accounting for the phenomenon of auto-regulation, but not of that of cognition and the capacity for reflection and interpretation. In a text which ranges from physics to metaphysics, he suggests that reflection may be obtained by recursion of a function which takes its own values as arguments, thereby

B. Feltz, M. Crommelinck and P. Goujon, (eds.), Self-Organization and Emergence in Life Sciences: 249-263.

progressively leaving behind its input value as it becomes a function of itself. In his project of deconstructing complexity, however, von Foerster adopts Ashby's reductionist view (a point effectively highlighted by Jean Pierre Dupuy[1]). Von Foerster shows, for instance, that memory can be modelled by a simple dependence on the path followed, thereby deconstructing the notion of memory and detaching it from the idea of stored and reactived representations. The problem remains that of being able reconstruct cognitive complexity in its full richness with some combination of the conceptual tools and models obtained by this deconstruction.

Von Foerster was always optimistic in this regard. The deconstruction models preserved an irreducible complexity (recognized by von Neumann when he defined complexity as that which can only be adequately described by the production of the thing itself). Reconstruction could thus launch one into higher levels of complexity, because complexity is added by supplementary loops and by recursion properties. This optimism has to be tempered if the goal of cognitive modeling is not only to obtain complex systems, but ones that conserve cognitive accessibility. These days, connectionist systems (and, in general, theories of dynamic systems) assure a very rich complexity (chaotic systems, for instance) that derives in part from a network structure and from the mutual reapplication of operations. And in studying them one may discover properties neighboring on those sought by von Foerster. But these systems run into the problem of explaining cognitive access to complex phenomena. The limits presented by the intuitions and theories of von Foerster can be better uncovered by a study of the problems encountered by, for example, connectionist systems (and all theories that use complex systems to account for the cognitive).

Furthermore, to illuminate the interest and limits of von Foerster's position, I will make two comparisons: one between von Foerster's notion of a "trivial machine" and the notion of "simply recurrent" networks (used, in particular, by Ellman); the other between von Foerster's cognitive tiles and the adaptive resonance networks of Grossberg. The question in both cases will be whether the modelling tools that permit the deconstruction of complexity also permit their reconstruction in a way which renders them cognitively accessible to other cognitive systems — starting with our own.

1. Dupuy, Jean-Pierre (1994). *Aux origines des sciences cognitives.* La Découverte.

1. "NON-TRIVIAL" MACHINES AND RECURRENT NETWORKS

According to von Foerster, a "trivial" machine is simply a function, *f,* on some input *x,* giving some output *y.* An finite state automaton is a function of two variables: the input, *x,* and the internal state of the automaton, *z.* But x itself may depend on a preceding step in the functioning of the automaton. One should rather, therefore, denote it by *x',* to mark its dependence on the application of f to the preceding *x.* The value of *x'* obviously also depends on the preceding state of the automaton, *z.* And one may reiterate this dependence, working back to the start-up of the automaton.

One may also reformulate the function realized by such an automaton by thinking of it as the single-variable function, *hz*(*x*), which varies with each internal state *z, z', z",* etc., rather than as the two-variable function, *f*(*x,z*). Then it suffices to fix the start-up input value, *x,* because the function h will produce the successive *x's* for each internal state *z,* the latter themselves dependent on the transition table for states of the automaton. The automaton thus defined realizes, therefore, a sequence of functions which depend on changes in the internal states. Finally, one can also realize this sequence of functions by introducing a loop in the system: each output is sent back, after a certain delay D, as input to the automaton, which changes its state in accordance with its function *h* (or *f*). Von Foerster claimed, on this basis, that because the behavior of a "non-trivial machine" at an instant *t* depends on the whole path followed since the start-up of the automaton, such behavior plays the role of a memory trace of the path followed, even though no representation is stored in a separate "memory" location.

One can compare von Foerster's "non-trivial machines" to the "graded states automata" of Servan-Schreiber, Cleermans and McClelland. The goal pursued in proposing these networks is to render a network, as a connectionist system, sensitive to nested contextual structures. For example, determination of the gender of adjectives in embedded sentences requires correctly establishing the co-reference of the relative and main clauses. This is often more difficult in English than French. For instance, in French the sentences "le chien qui *poursuivait* ce chat est très joueur" and "les chiens qui *poursuivaient* ce chat sont très joueurs" are distinguished by the conjugation of the verbs appearing in them, including the embedded one. In English, on the other hand, one of the embedded verbs does not change its form: "the dog that *chased* the cat is very playful" and "the dogs that *chased* the cat are very playful". So all additions in the relative clause have to take account of the fact that the referent is in the plural or singular, as the case may be, even if the verb does not mark this fact. The apparent structure is the same, but the real structure is different. And networks have to learn to be sensitive to this.

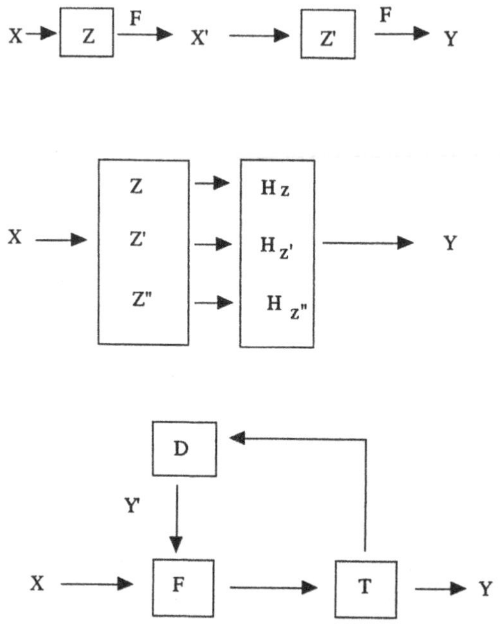

Figure 1. Recursive functions and memory

The type of network to be used for this task has already been proposed by Ellman (1989). It consists of a recurrent network, where the outputs of hidden units are returned at time $t+1$ to a set of input units separated from the others and which play the role of context units[2]. For the rest the network

2. Remember that (according to one of the principal theories, at least) connectionist nets work in the following manner. Entry units transmit their outputs to hidden units, which take the sum of outputs from each entry unit multiplied by a weighting on the connection. The sum is compared to a threshold (determined by a non-linear function *e.g.* the sigmoid function). If the sum surpasses the threshold, it is sent on to exit units (or to other hidden units, if there are several layers of them). The net, thus, transforms the entry vector into the exit vector by means of the weights on the connections. Training of a net consists in modifying the weight of connections. This type of net is called Feed Forward (FFWD) because the activation of units progresses from top to bottom, from the entry units to the exit units. In a recurrent network there exists a way back from the hidden units to the entry ones or from the exit units to the hidden units.

works like a classic FFWD one. In their theory (1994, *Graded States Machines: The Representation of Temporal Contingencies in Simple Recurrent Networks (SRN))*, the authors use this type of architecture, by applying it to sequences of symbols produced by a finite state automaton[3]. A network receives an input vector which codes an element in a sequence of symbols and it must learn to output the following symbol in the sequence. The authors get the network to perform at a satisfactory level and then analyze the activations of each of the hidden units, often considered to be the "representations" that the network has constructed in order to accomplish this task. They conclude that in a network with this type of architecture (a simple recurrent network) the internal representations (the activations of the hidden units) not only encode the symbol that has just been input, but also the data issuing from the input of previous symbols - the data which permits the network to predict the next input on the basis of the preceding one. And they add that "long distance dependencies may be treated by machines more simple than fully recursive ones, at least to the extent that they utilize information with graded states" (1994, 267).

We see thus the retaking up of von Foerster's idea that machines that resemble finite state automata (the authors say that they are imitated by SRN) can show a sensibility to events that occurred long before. Since their treatment of the current symbol depends on the history of their treatment of previous ones, they display the effects of a "memory".

Note firstly that von Foerster's "non-trivial machines" don't use full recursivity. For that, one would need not only that occurrences of h vary as a function of each internal state z, but also that the changes in z' (which satisfy say a transition function, $g(z)$, taking a value for z' in accordance with the preceding z) vary in accordance with the changes in x. Thus the changes in x cause g, or rather g's components, h_1 to h_2, to vary. At time t_0 and in accordance with the state z, for instance, a function h_0 may be applied to x. This will imply, at time $t+1$, a change in the succeeding entry state, from x_1 into x_2 in accordance with a function f_1 (so the inputs are no longer exogenous). Then this change from x_1 to x_2 will change the state z into z' at time $t+1$, that will change the function h, and hence also the state, the type of change in input, the function f_n, etc. Full recursivity implies that one can nest the functions h_n and f_n. A system which permitted such nesting would

3. A finite state automaton can be represented by an oriented graph, each node corresponding to a state of the automaton (and thus to the symbol that it inscribes), and each oriented arc to a transition from state to state. Such an automaton can thus apply a set of rewriting rules to transform one sequence of symbols into another.

dispose of all the resources of recursion, while von Foerster's and those of our authors do not.

If the SRN do instantiated a memory effect by the adaptation of the weights on its connections, or by the bias introduced by the preceding operations, it is only because it disposes of graded "representations". Or, in other words, because in theory its activations are not encoded by natural numbers but by reals (as the network is simulated by a classic machine, the reals reduce, however, to their finite approximations). That evidently gives it a power superior to finite state machines. But it should be noted that finite state automata take into account the structure of the sequence with some sensitivity, because the same symbol may give rise to the inscription of different symbols depending on which of the nodes of the graph the automaton is situated.

The SRN of these authors is thus presented with a symbol and has to predict the following one. In order to be able to say that the network rejects a sequence as ungrammatical, it is sufficient, they claim, that when it is presented with random letters, there comes a moment when it no longer predicts the following letter (but they allow that it may begin by predicting some of the letters in the random sequence). On the other hand it should predict all the "grammatical" letters. The consequence of interest is that, without need of a memory (either in the form of a stack or a register), and although it is only fed symbols one at a time, the network succeeds at a task that depends on the organization of the graph linked to the grammar of the finite state machine, and thus on a structure that is revealed only by the sequence of symbols. And to do this, the network can use only the information provided by the transitions from state to state. By analyzing the clusters of activations from the hidden units in the networks, the authors consider themselves able to show that the activations copied into the 'contextual units' (those that are fed recursively) code the node of the graph related to the treatment in course. With the direct-entry units giving the current symbol, the network can combine the information from the two sources to make a prediction.

It can be claimed, thus, that this network realizes what von Foerster called a "non-trivial machine", because without disposing of a memory in the proper sense of the term, and by using only state transitions, it predicts the symbol to follow in accordance with its treatment of the preceding steps. Moreover, as with von Foerster"s automata, an embryonic memory resides in the loop which forms the recurrent connection with the contextual units, with these latter units sending, at time $t+1$, the time t_0 activations of the hidden units back to them.

Von Foester never specified whether the values taken by these functions had to be discrete or continuous (nor whether some of the functions had to be non-linear, indispensable if the network were to function — but this was probably self-evident to him). Perhaps he can be reproached for this, for it seems that the capacities of these networks derive from the fact that they dispose of graded representations exploiting the full resources of the reals. If we take up the analysis by *clusters,* the network makes finer classifications than those using only the different nodes of the state-transition graph. Around each node is grouped a different set of symbols deriving from the different paths, or sequences of steps, followed before the node was reached.

This is not always the case. The authors claim that the network must contain a greater number of hidden units than would be necessary for simply encoding the node of the graph that has been attained. In other words, it is by exploiting the redundancy with respect to the immediate task that the network encodes differences in the sequence of tasks. But, even though the authors don't mention it, it is undoubtedly difficult to find an equilibrium between enough redundancy to encode additional information, and too much redundancy, and hence too many hidden units, which would lead the network to make overly refined predictions linked to the idiosyncrasies of its learning path (the singularities of the examples given to it) as opposed to those related only to the different possible routes though the graph.

The authors analyze the constraints that force the network to take into account the current symbol, the paths it has followed and even the whole of the preceding sequence of symbols. The first constraint guiding the evolution of the network comes from the process of learning by retro-propagation. This ensures a determinate output from the exit units to the hidden ones, and hence induces the same activation pattern in the hidden units if two entering letters give rise to the same predicted letter at the exit. In the other direction, running from the entry to the hidden units, the letters coded by the entry units influence the hidden units without taking into account the desired outputs. So every context alters the representation in the hidden units. The hidden units, thus, code the association between a given input and the symbol to be predicted. It can be claimed, therefore, that the network learns both to make its states (its internal units) depend on preceding inputs and its outputs depend on its inputs and internal states. Hence it is a realization of a "non-trivial machine" whose states depend on the history of its processing. Note that von Foerster never tells us how the variations in the function h should depend on prior processing (he does not specify whether they depend on the preceding internal states only or on both the internal states and the inputs).

This is also the problem with this type of network. The authors note three phases in the learning process. 1) At the beginning, the activations of the contextual units are unstable, because each new symbol to be processed induces, as a function of the teaching, a new representation in the hidden units and each such representation eliminates the previous one. 2) Once learning has advanced, the contextual units code in a such a way as to activate those hidden units which correspond to the processing of the preceding symbol. The network, thus, takes into account its 'memory' of the previous activation in its ongoing representation activity in the hidden units. 3) Finally, small differences in the coding of preceding activations can indicate differences in the path followed up to that point. This permits, for example, the taking into account of the length of the sequence that preceded the symbol currently being processed.

But in order that all goes well (the authors never say as much, but experiments with networks provide ample evidence), there needs to be sufficient repetition in the symbols being entered to ensure that the first phase culminates with the network having learnt to code the different symbols and to ensure that there are not too many nodes in the graph determining different sequences in response to the same symbol. Otherwise the information held on preceding symbols can be too rich: the network will only learn a couple of letters and the difference between the first and second phase will disappear. Similarly the differences in the length of sequences necessary to change a symbol should not be too great or the three phases will be confused. In fact the network doesn't first learn letters in isolation and then build up to longer sequences. It is confronted by all lengths at every stage of learning. So it is only a careful adjustment of the frequencies of the various factors that enables the network to perform. In particular, when it learns to recognize sub-sequences of identical symbols, nested in other sequences, the network may become insensitive to information that it needs to distinguish two identical sub-sequences in terms of their position in a nested structure. It cannot thus correctly treat problems of relative nesting.

This is to say, somewhat paradoxically, that this "non-trivial machine" does not owe its "memory" properties to the kind of recursivity that von Foerster was thinking about. For if the structure of the sequence to be treated by the network is really recursive (a double nesting with a sub-sequence of symbols nested in a sequence that is exactly *identical* to it), then the network has little chance of recognizing it. This is what makes this type of network so powerful: it is its "recurrent" circuit that makes it sensitive to preceding sequences. But one cannot conclude that the machine is "recursive" and capable of processing nesting of nested structures without difficulty. It recognizes complex structures by using graded representations, but as a consequence it distinguishes complete sets of symbols without being sensitive to the nesting structure proper. It does not seem to be able to both isolate similar sub-sequences and spot their nesting, which is exactly what recursive machines do best (a Lisp machine, for example — it should be noted however that the machine has this structure because of the nature of Lisp and that it is not something that the machine can discover by itself).

Another objection directed at the memory simulation proposed by von Foerster remains valid in this context: a network has no memory, just as a "non-trivial machine" doesn't, except insofar as memory is reduced to "path dependency". But this latter type of memory has some unwelcome properties. It effaces, or mixes up, at much as it preserves. Inversely, one could claim that the "memory" of a classical computer which can always be reactivated (so long as it is not modified) and which preserves without mixing up, has no memory in the human sense either. It is not active, it doesn't modify present perceptions or the meaning that we give to information currently being input. The memory associated with networks has this latter property in surfeit. But it is difficult to see how one might combine the advantages of the two, never mind avoiding their respective inconveniences.

So von Foerster really did anticipate the current situation: his "non trivial machines" share with recurrent networks a sensitivity to the path followed and use this to provide an ersatz memory. But this type of memory seems in complete opposition with what von Foerster saw as the natural extension of "non-trivial machines": their recursivity (at least if by "recursivity" he meant the possibility of self-nesting — which does seem to be the case).

2. COGNITIVE TILES AND ADAPTIVE RESONANCE

Let us see whether we encounter the same anticipation of problems when we consider von Foerster's "cognitive tiles".

In this schema, RSX is a sensor receiver, that interprets inputs from the environment by calculating the relations between the observed activities (external and internal) and itself. *T* is a translator that translates the output into a universal language, accessible to the other cognitive tiles in the system. *D* is a delay loop which feeds back the output of $\Phi(F)$. RSY calculates, from a similar feedback loop, the relations between the system's actions (outputs) and its goals. And at the center we have a finite state automaton ("a non-trivial machine") which calculates the values of various functions in accordance with its own state. Finally *Y* is the exit point from the system seen by an external observer. These states of the system are its eigen states. They correspond to the vectors proper to a network, each vector having particular values *i.e.* those such that multiplication of the proper vector by the matrix constitutive of the network yields the multiplication of this vector by a scalar; the vector shifted in space. The proper values constitute an independent coordinate base for the network.

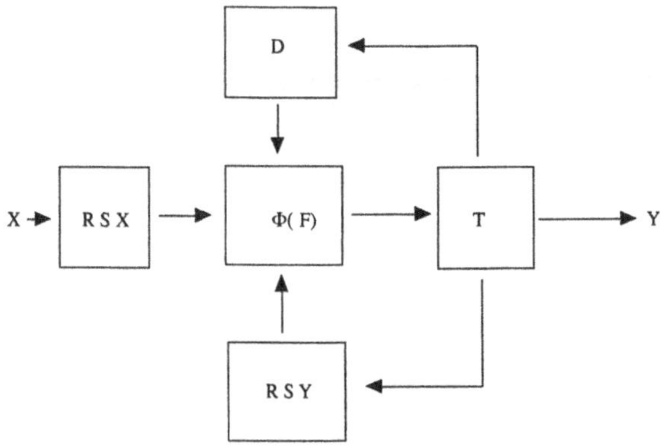

Figure 2. Von Foerster's cognitive tile

This notion of a cognitive tile may be compared to the circuits in Grossberg's theory of adaptive resonance.

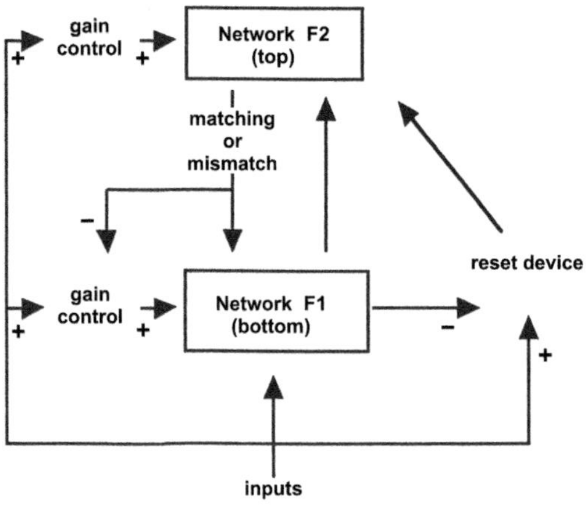

Figure 3. Grossberg Adaptative Resonance Theory

In this schema of Grossberg and Carpenter, *F*1, the first network, encodes in the activation pattern of a short-term memory distributed across a network of trait detectors, a distributed representation of the vector which is to be recognized. The network *F*2 encodes in long-term memory the event to be recognized by way of a more compressed representation of the pattern of *F*1. Learning occurs in the top-down and bottom-up connections between levels *F*1 and *F*2. The path from top to bottom is supposed to make explicit the acquired expectations of patterns, *F*1's prototypes of which are put into correspondence (in resonance) with the pattern of inputs coming from *F*2. Mismatches occurring in response to new events activate the sub-system to the right. This is an orientation system and it resets the recognition codes active in *F*2 and begins a search in memory for a more appropriate recognition code. To the left there is an internal regulation system with "doors" which, above a certain threshold, positively influence either *F*1 or *F*2, thereby accentuating or diminishing the sensitivity of these two networks to the details of their inputs.

At the beginning, *F*1 sends its activation pattern *X* to *F*2, where the corresponding pattern S is transformed by long-term memory traces, thereby activating another pattern *Y*. During this phase, the regulation system on the left is active on both *F*1 and *F*2. Pattern *Y* returns its activation to *F*1 giving rise to pattern *V* (a prototype pattern). This inhibits the activation of the regulatory system on the left on *F*1, but not on *F*2. If *V* does not correspond to the input in *F*1, a new pattern *X** is engendered by *F*1. This inhibits some of the active nodes and diminishes the total activity of *F*1. So the inhibitory signal that *F*1 sends to the orientation system diminishes, permitting this system to send a wave of global updates onto *F*2, inhibiting *Y* and permitting *F*2 to re-launch pattern *X*. The regulatory system then returns to its initial activation (on *F*1 and *F*2). The traces in *F*2 have changed, so the new pattern *Y** is different. Everything starts all over again until the correspondence is correct.

If one looks at the loops being established, one notes that in the first stage there are no loops at all, only activations in parallel (in the link between input, *F*1 and *F*2 and in the parallel link through the doors, from the input on *F*1 and *F*2). Then recurrence occurs between *F*1 and *F*2, blocking the parallel activation of the regulation system on *F*2, but not on *F*2. Once the activation of the reset device (on the right) is sent, this return is inhibited. Then one returns to the first step, and so on.

Let us attempt a comparison with cognitive tiles. In one sense Grossberg's whole system seems to resemble a single chip of von Foerster's mosaic, the RSX chip. For RSX interprets the sensory inputs in accordance with its own activities, and so assures the matching of the activity of the network receiver and network's memory of itself. In another sense, Grossberg's system represents the majority of the chips making up the tiles. We have no need for a universal translator, since it suffices to have a system interpreter at the entrance to each tile. The delay loop *D* is important here, since system *F*2 is a long-term memory, while system *F*1 is a short-term one. The actions of the system, its outputs, can be represented by *F*2's outputs; its relation to the goals of the system, by the matching of the memorized patterns in *F*2 and the activation patterns of *F*1 (once given information about the situation at time t_0, these return information at time $t+1$ on the results of previous actions). Once one has made this comparison it becomes clearer that the function of the chip RSY is to indicate the difference between the effects of actions, the outputs of *F*2, and the "goals" (which are realized when the information of *F*1 and the patterns of *F*2 are matched up). RSY may be represented in part, therefore, by the system that resets *F*2. Finally, the distinctive states of the central chip are realized by the networks *F*1 and *F*2 when they are matched up. This being so, von Foerster's cognitive tile, re-analysed along the lines of Grossberg's model, is seen to be completely reducible to its entry chip (the one interpreting the inputs at the boundary of the system) or to the doubling up of this chip (with a chip of the same style for actions, another one for perceptions).

This comparison shows that von Foerster narrowly imitated communication between human individuals with the dissociation characteristic in individuals between perception, action, re-afference, self and system of communication with others. Remember that a mosaic made up of such chips, or a roof made up of the tiles, cannot, according to von Foerster, function as a collective unless one joins the individuals together at quite specific points: either between the translation and perception modules or between the translation-output and action modules or between the translation-output and delay modules. But we have seen that the translation module is not necessary, because each "tile" can interpret the inputs from the others on its own. If von Foerster had analyzed the perception-interpretation chip, he would have realized this. Furthermore, the heart of the tile, the "Self"-module, is not necessary either: one could just as well consider the "Self" of the network system to be the set of vectors peculiar to the network's matrices. It suffices to construct a "perception-interpretation" chip and to couple it to a chip with an identical architecture, but which functions as an "action interpreter", in order that the system exhibits the "cognitive" properties that von Foerster ascribes to his "tile".

The difficulty is that one can continue to couple the different subs-systems together at their entrance and exit points and that the difference between individual and collective, stipulated in von Foerster's theory, becomes difficult to reconstruct. The problem of "translation" is thus re-encountered. The problem is both internal and external, as von Foerster's schema seems to indicate. He allows messages coming from the translator to arrive at the central tile, which is not a plausible model of communication: we have private information that we cannot communicate, even if it is not in a private language. The problem is also one of categorization in such systems. If we wish to "speak" to ourselves in the way that we talk to others, or if we want to dispose of "representations" that are transmissible not only to others, but in the first case to ourselves, we have to go much further than Grossberg's system. The latter is obliged, like all connectionist systems, to continually update its patterns, thereby running the risk of losing acquired patterns, and is thus condemned to evolve by adaptation to each situation without being able to preserve representational modules potentially transposable from one category to another. From this point of view, the "translation" module of von Foerster seems to be designed to resolve the problem of adjustment between the context-bound classifications of connectionist systems and classifications that can be passed from one context to another.

Von Foerster thus correctly identified certain problems, but his theory was too dependent on a particular philosophical or ideological image of the self to resolve them. To connect the two models, that of memory and that of cognitive tiles, note that von Foerster was faithful to what I call elsewhere the double strategy of inverted reductionism. On the one hand he reduces memory to a dependence on the path followed, and he conceives of reflection as simple recursion (reduced here to a loop that is not properly closed). He practices, thus, a somewhat abusive reductionism. On the other hand, he maintains a notion of self as that center of perception, of re-afference, of the interpretation of action and of communication. It is a notion of self that is far too complex and ought to be reduced if one is constructing a working simulation of perception and its interpretation. On the one hand he reduces complexity, on the other hand he overestimates it. But one must grant that even if his model extends over too large an interval of complexity, it is still within this interval that problems are posed today, and that there is as much continuity between connectionist theories and von Foerster's as there is difference.

By the theory of self-organization we mean, in general, the convergence of two tendencies launched by von Foerster: the tendency to reduce complexity and the tendency to reconstruct it. With it goes the testing and enrichment of the reductionist project by means of attempts to construct "emergent" phenomena (*i.e.* phenomena whose properties are not present at the level of their elements) by the assembly of these elements in such a way as to permit new interactions with the environment and new dynamics for these interactions. In this regard, connectionist systems undoubtedly display the capacity for self-organization. The network in its totality is given to complex behaviors which none of its units, taken in isolation, are capable of.

The problem is then to be able to render this complexity comprehensible and utilizable in interactions with humans. The module interpreting the data produced by each network must be interpretable by cognitive systems like us, or by other networks. And this is to say that the Self-Organizing system must have categorial capacities, must produce canonical categories, so to speak. Only if this is so, can we speak of learning — either by networks or by classical computers. For the moment we call learning in classical computers, the execution of combinations of rule elements which permits the production of new rules, some which may be selected for future use. For networks we call learning the process by which, on the basis of a series of examples, the network generalizes the induced classifications to new data. But in neither case is it possible to determine under what conditions categoricity is assured; that is when the system has the capacity to transfer obtained classifications or rules to neighboring domains or to problems posed by different cognitive systems. Despite the fact that self-organization has become operational, it is not sufficient for defining properly cognitive capacities — even if it is a necessary condition for them.

REFERENCES

Carpenter Gail A. and Grossberg Stephen. Integrating Symbolic and Neural Processing in a Self-Organizing Architecture for Pattern Recognition and Prediction. In Honavar Vasar and Uhr Leonard (eds), *Artificial Intelligence and Neural Networks.* San Diego: Academic Press inc., 387-421.

Hunga G. (ed.) (1970). Molecular Ethology, an Immodest Proposal for Semantic Clarification. *Molecular Mechanisme in Memory and Learning.* New York: Plenum Press, 213-248.

Preiser F.E. (ed.) (1973). On Constructing a Reality. *Environmental Design Research,* Vol. **2**. In Dowden, Hutchinson and Ross, Stroudsburg, 35-46.

Servan-Schreiber, Cleermans and McClelland (1994). Graded State Machines: the Representation of Temporal Contingencies in Simple

Recurrent Networks. In Honavar Vasar and Uhr Leonard (eds). *Artificial Intelligence and Neural Networks.* San Diego: Academic Press, 241-269.
Von Foerster (1969). What is Memory that it May have Hindsight and Foresight as Well? In S. Bogoch (ed.). *The future of Brain Sciences.* New York: Plenum Press, 19-64.

III. EPISTEMOLOGICAL AND CONCEPTUAL APPROACHES

A. TELEOLOGY AND INTENTIONALITY

ROBERT N. BRANDON

TELEOLOGY IN SELF-ORGANIZING SYSTEMS

Teleological language, talk of function and purpose, has long been associated with the appearance of order in the biological world. Indeed, the pre-Darwinian tradition of natural theology (*e.g.,* Paley 1836) gave a clear underpinning for such teleology. The order of nature was a product of God's design and reflected his purposes. In this post-Darwinian era neural selection has taken the place of God's purposes in supporting teleological ascriptions — the ultimate purpose or function of some biological trait, say a wing, is just that effect acted on by natural selection to produce, by evolution, the order of the trait in question. But the recent recognition that order can emerge just from the dynamics of complex systems — no natural selection is needed — leads us to the question of this paper; namely, in what ways, and to what extent, does teleological language properly apply to the self-generated order of complex dynamical systems in biology?

1. TWO ANALYSES OF FUNCTION

Although the philosophical literature on function in biology is crowded and sometimes confusing, two analyses of function have been developed that are both clearly articulated and widely applicable to the biological world. The first bases the function of a biological feature (*e.g.,* a wing) on its causal history, more particularly, its history with respect to natural selection. Generically this sort of analysis is often called the *etiological* model, but a more specific name for it would be the *selected effect* (SE) model. The second analysis is, in contrast, ahistorical. It bases the function of a biological trait on its causal role within some more complex system, thus it is termed the *causal role* (CR) model. In this section I will present these two models, show how they relate to each other, and examine if, and how, either of them support talk of teleology or purpose.

B. Feltz, M. Crommelinck and P. Goujon, (eds.), Self-Organization and Emergence in Life Sciences: 267-281.

1.1 Selected Effect Model

The basic idea behind the SE model of biological function is easy to understand. The function of a trait is that effect (or effects) that caused(s) the trait to have higher fitness than alternative competing versions of the trait, and thus explains the current form and frequency of the trait. Like any interesting analysis of function, the SE model differentiates one (or a few) effect(s) of the trait in question from among all its effects. An analysis of function that failed to mark such a difference would be worthless since function would then simply equal effect, and thus would be a superfluous concept. On the SE model natural selection serves to mark this difference among effects. For instance, if the (SE) function of the red color of a flower is to attract pollinators, then it must be true that: (a) at one time in the history of the lineage in question there was variation in flower color, red being among the variants; (b) this variation was heritable (usually, but not always, this means that the variation has a genetic basis); (c) selection, in the form of pollinator discrimination, acted directly on flower color, not on some correlate of flower color, favoring red over alternative variants; and (d) this selection within the population genetic context of the lineage led to the form and frequency of red flowers we see in the descendent populations today. Elsewhere (Brandon 1990, Chap. 5) I have extensively examined the epistemological difficulties in establishing an SE functional claim (or, alternatively, in establishing that a certain trait is an adaptation for some certain function). Suffice it to say that although these difficulties are formidable, they are not so great as to lead us to complete scepticism with regards to SE-type analyses. Here let me point out the conceptual role (e) plays in differentiating the function from among all effects of the trait. Having red flowers may result in a number of effects compared with having flowers of other colors. For instance, having red flowers may result in a warming of the stem which holds the flower as compared to a white flowered plant. This raised temperature may itself have a number of ramifying effects within the plant' physiology. But none of the effects *are the function* of red flowers if it is exclusively pollinator discrimination that accounts for the superior reproductive success of red flowered plants compared with their differently colored competitors.

Given the sort of selective history sketched above it is easy to see how SE analyses answer (or rather serve as partial answers to) what-for questions, *i.e.* teleological questions. In our case it serves to answer the question of what the red color is for. It is for attracting pollinators. That is its purpose, not increasing the temperature of the stem nor any of a large number of other effects. In other words; the red flower is an adaptation for attracting pollinators.

To put this in abstract form: a what-for question asked of adaptation *A* is answered by citing the effects of past instances of *A* (or precursors of *A*) and showing how these effects increased the relative adaptedness of *A*'s possessors (or possessors of *A*' precursors) and so led to the evolution of *A*. (Brandon 1990, 188).

It would be worthwhile to thoroughly trace the history of the etiological account of function and the more specific SE model. However, in this short paper I cannot do so. Larry Wright (1976) is usually credited with developing the first clear and explicit version of the etiological theory of function. But his account is so general — it is indifferent between a history of natural selection and one of divine creation — that I think it is of little interest to anyone interested in the biological underpinnings of functional ascriptions. In the late 70's and early 80's a number of philosophers of biology offered analyses of function where the selective history of the trait was invoked to ground talk of that trait's function (Wimsatt 1972, Brandon 1981). This sort of analysis has come to dominate recent philosophy of biology (*e.g.,* Lewontin 1978, Gould and Vrba 1982, Sober 1984, Brandon 1990), so that many have seen the SE account as the only legitimate way to justify talk of function or teleology in biology. My own recently expressed views on this are not unrepresentative. Thus, for instance, although I recognized that not all biologists used functional language in the evolutionary sense I argued for, I said, "I believe that ahistorical functional ascriptions only invite confusion, and that biologists ought to restrict the concept [to] its evolutionary meaning..." (Brandon 1990, **24**, 187, quoted in Amundson and Lauder 1994, 449).

1.2 Causal Role Model

A recent article by Amundson and Lauder (1994) has convinced me that I (along with many others) was wrong to insist on a exclusively evolutionary or SE analysis of function in biology. The resurrect Cummins' (1975) causal role (CR) analysis of function. But they explicate and defend its use in biology much more thoroughly than Cummins ever did. Also, and perhaps more importantly, they make clear, for the first time, that the CR model is not a rival alternative to the SE model, but rather that the two are complimentary. We will see how this is so shortly.

> "The basic idea behind the causal role analysis of function is that the function of a trait within a complex system is the effect of the trait that helps to explain

> the behavior of the complex system. Put formally, *K* functions as an *F* in *s* (or : the function of *X* in *s* is to *F*) relative to an analytical account *A* of *s*'s capacity to *G* just in case *X* is capable of *F*-ing in *s* and *A* appropriately and adequately accounts for *s*'s capacity to *G* by, in part, appealing to the capacity of *X* to *F* in *s* (Amundson and Lauder 1994, 448, after Cummings 1975, 762)."

Notice that on this account traits have functions only within a complex system and only relative to some causal account of the workings of that system. To take the above mentioned case of flower color as an example, *X* (red flowers) functions as an *F* (pollinator attractor) in *s* (the system of reproduction in our plants) relative to an analytical account *A* (an account of how flowers receive pollen from other plants) of *s*'s capacity to *G* (reproduce by outcrossing). It does so because *A* does appropriately and adequately account for the system's capacity to reproduce by outcrossing in part by appealing to the red flowers' capacity to attract pollinators. Take away the pollinators, or the other plants, *i.e.,* take away parts of the complex system *s* and red flowers no longer function as pollinator attractors. Similary, relative to another analytical account of these plants' reproductive capacities, *e.g.,* a molecular account that takes for granted the presence of pollen and explains at the molecular level how pollen fertilizes the ovules, red flowers may no longer function as pollinator attractors.

1.3 Two Points Concerning the SE and CR Models

First, it should be clear from the above that the SE and CR analyses of function are quite different. The SE account is explicitly historical — it is true just in case a certain selective history is true. The present functioning of the system in question can only suggest, but can never substantiate, a particular SE account of function. In contrast, the CR model is ahistorical. Present functioning is all that matters to it. This difference makes a difference with respect to the to models' implications for teleology. The SE account does seem to warrant talk of purpose. It does so because if an SE account is true, say of our red flowers, then that account explains why the trait in question is present in the form and frequency we see it. It explains, in a purely mechanistic way[1], why the trait is present in terms of the past effects of past instances of the trait. On the other hand, the CR model neither requires nor supports any attribution of purpose. A particular causal account of a complex system (our plants) may explain an overall capacity of that system (the capacity to reproduce by outcrossing) in terms of the function of

1. See Brandon (1990, Chap. 5) for an account of evolutionary explanations of adaptations that shows how they explain teleological phenomena in a purely mechanistic way.

some part of that system (the capacity to reproduce by outcrossing) in terms of the function of some part of that system (the red flowers' capacity to attract pollinators), without having any implications concerning the origins or maintenance of the trait in question. I will use the felicitous phrase Amundson and Lauder use as a title for their article to describe this feature of the CR model. It is an analysis of "function without purpose".

Second, it should be clear that these two analyses are not mutually exclusive. Both may be true of a particular system, for instance our red flowered plants. Indeed the two are complimentary and fruitful biological research often proceeds via a complex interplay between the two. (Good SE analyses require some CR analyses, good CR analyses can oftentimes suggest plausible avenues, and rule out others, for SE, or evolutionary analyses[2]). Thus, it is a mistake to think of these two analyses as rival philosophical accounts of function.

2. SELF-ORGANIZATION AND GENERIC PROPERTIES

The study of complex dynamical systems has shown that many such systems regularly evolve to certain ordered states (called attractors) without the input of any such order. To take a simple non-living example, water flowing down a drain in a sink will regularly form a spiral (with initial conditions determining whether it is clockwise or counter clockwise). The spiral pattern is a *generic property* of water flowing down a drain, one has to work to produce another pattern. Similarly, developing organisms may have generic properties. One way to think about the study of self-organization in living systems is just to see it as a search for just what are these generic, or typical, properties[3].

2.1 Definition and Example of Generic Property

Generic properties in living systems can be defined simply as forms or morphologies that are highly probable given the dynamics of the developing system. In other words, generic forms are robust in the sense that for a wide range of initial conditions and of parameter values, the generic form develops. In that sense, they are the "natural forms" of the system. Notice the natural selection plays no role in this characterization.

2. See my discussion (Brandon 1990, 180-184) of Kingsolver and Koehl 1985 as an example where this sort of interplay can be seen.
3. For example, this would seem to be a fair characterization of the work of Kauffman 1993 and Goodwin 1994.

Goodwin (1994, Chap. 4) discusses a plausible example of a generic property, the whorls in *Acetabularia acetabulum,* which is a unicellular green alga with a complex life cycle. The part of the life cycle that concerns us is the growth from the fertilized zygote to the mature reproductive stage which consists of a mushroom-like cap on top of a long stalk. In particular, prior to the formation of the reproductive cap, when the stalk is elongating, small branches, called whorls, are formed at the developing tip of the stalk. Goodwin has argued that these whorls are natural forms of this developmental system. He has done so both by modeling this system based on plausible physical and chemical assumptions, and by perturbation experiments on *Acetabularia* during this developmental period. His results indicate that these whorls form quite naturally while the stalk is elongating, *i.e.,* that they are generic in this system.

In *Acetabularia acetabulum* the whorls fall off prior to the formation of the reproductive cap and seem to do the organism no good whatsoever, *i.e.,* they seem to be functionless. (This is not the case in some relatives of *Acetabulum* where the whorls are retained and function in reproduction by housing the gametophores just as the cap does in *Acetabularia.*) If Goodwin is right, these whorls are functionless both from the SE and CR points of view. Selection has not melded these forms, they are the natural by product of the developing system, and they play no role in the overall capacity of the system.

2.2 A More Hypothetical Example and a Mechanical Interlude

To get clearer on the relationship between SE and CR analyses of function in their applications to generic properties, let us consider a more hypothetical example. Consider the characteristic shape of some particular species of tree, *e.g.* the American beech. That shape is a product of the branching pattern of the tree, in particular, the distance between branch points and the angles of the branches. These in turn are properties of the growth dynamics of the meristem. In other words, the overall shape of the tree emerges from the dynamics of meristematic growth. To use a metaphor from computer programming, if you program meristematic growth (branch distance and angle), you get overall tree shape "for free", that is you need not separately try to program overall tree shape.

Let us suppose that natural selection has played no role in molding the dynamics of meristem growth in the American beech[4]. That being the case, then clearly the meristem growth pattern has no SE function, has no evolutionary purpose. But is does have a CR function in determining the overall shape of the tree. That is because à capacity of the large system, the overall shape of the tree, is explained in terms of the growth patterns of the meristems. More fully, the meristems function as shape determiners of the beech tree because our causal account of the emergence of that shape is based on the local dynamics of meristem growth. Thus, in this hypothetical case the meristems have a CR function in determining overall shape, but have no SE function.

To make this point even clearer let us consider an even simpler example involving a non-living mechanism. In a typical piston-driven internal combustion engine there are intake and exhaust valves at the top of each cylinder. They are opened and closed by a camshaft A number of things can go wrong here. In particular when the mechanism linking camshaft to valves is poorly adjusted the valves do not open and close at precisely the right time and a noise, called valve clatter, is produced. The analogue of an SE function in humanly designed devise is, obviously, the intention or goal of the designer. In this case, the designer has no desire for valve clatter, has worked hard to eliminate it, but has failed. Thus there is nothing like an SE function in this case. But let us suppose that the valve clatter creates unusual vibrations in the intake manifold, changing the flow pattern of the gas-air mixture, leading to incomplete detonation of the mixture and ultimately to increased levels of hydrocarbons being released into the atmosphere. We can analyse the complex system, the car, and its exhibited capacity to emit such high levels of hydrocarbons. We account for this capacity, in part, by appealing to the valve's capacity to create vibrations disrupting the efficient flow pattern of the gas-air mixture. In short, we give a CR analysis of valve clatter in which the function of the clatter is to increase hydrocarbon emissions of the car.

Again in this example we have a CR function, but nothing like an SE function. What this example adds to our earlier one is the point that CR-type analyses are in no way embedded in, or committed to, an account of what is good for the system as a whole. The shape of the beck tree may, or may not, have good consequences for the tree. The increased levels of emissions certainly has no good consequences for the owners of, or designers of, the

4. Suppose that only for the sake of making this point, the supposition is itself highly implausible. However selection might be affected by the local consequences of overall shape (*e.g.,* shape affects the effective photosynthetic area of the tree).

car. Yet in both cases we can meaningfully talk of the CR function of parts of those systems.

2.3 Stage 1 Conclusion

Generic properties, as we have characterized them at this point, have no SE functions. They are simply the "default settings" of the developmental systems in question, and so are not the proper objects of teleological language. However, they may or may not have CR functions. In the case of the whorls of *Acetabularia* apparently they do not. Our two hypothetical cases are meant to illustrate the possibility of something having a CR function while having no SE function. Generic properties of developmental systems do potentially have CR functions, *i.e.,* they can certainly play a causal role in a larger more complex system that helps explain the behavior of the system. Therefore, *generic properties can have function without purpose.*

3. TWO SENSES OF GENERIC

The fact that life has evolved on Earth must once presents a methodological problem for biology, namely that we have a sample size of one. If our scientific interests are in the general features of life the sample size of one makes it difficult to distinguish accidental associations from biologically deep generalizations. For instance, on this planet the developers of symbolic languages all happen to be bipedal. Is that a deep biological generalization, *i.e.,* one that we would expect from multiple runs of evolution, or is it a mere accident? Steven Jay Gould offers a vivid analogy to think about this problem (Gould 1989). Think of the evolution of life on Earth as a videotape. If we were to rewind the tape and run it again would we get the same outcome? Gould, and many other evolutionary biologists, think that the evolutionary process is highly contingent, that multiple runs would yield very different results. Creatures like ourselves, for instance, are by no means inevitable. Brian Goodwin represents the opposite pole on this position[5]. He

---

5. It would be better to say *an* opposite pole, since there are at least two important contrasts with Gould's position. Goodwin represents one, where the self-organising properties of life overcome the effects of chance so that evolution of life on Earth in fact represents a highly probable outcome, *i.e.,* other evolutions of life would yield similar outcomes. The other position contrasting with Gould's would be the extreme selectionist position that sees externally imposed selection as largely swamping out chance (*e.g.* Dennett 1995).

thinks that the features we see in the biosphere are largely the generic properties of such Self-Organizing systems and that the contingencies of drift and natural selection play only a minor role in shaping what we see in the biological world. Whatever one's position on this question, the study of generic properties seems to be a, perhaps the, way of addressing it. That is, once we know what are the generic, expectable, features of life, we can see to what extent life as it happened to volve on Earth deviates from this expectation.

3.1 Generic Without Selection

Earlier we characterized a generic property as a highly probable phenotype, given the developmental dynamics of the system. Natural selection plays no role in this sense. Let us term this notion, *generic-without-selection*.

Is this the notion most directly relevant to addressing the contingency of evolution? No, not if our question is: "What are the expectable outcomes of evolution?" That is because natural selection is not just ubiquitous on Earth, but is an all but inevitable feature of life[6]. Thus another notion of generic is suggested.

3.2 Generic-with-selection

The notion of generic-without-selection takes for granted the developmental system in question and asks which traits are highly probable given that system. An evolutionary perspective asks which traits are highly probable given the evolutionary dynamics of the evolving system. These two perspectives will not be equivalent if developmental systems are, to some degree, contingently evolved systems *i.e.,* if developmental dynamics in biology are not simply the playing out of universal laws of physics and chemistry.

Thus we can characterize the notion of *generic-with-selection* as follows: Traits are generic-with-selection if they are highly probable given the evolutionary dynamics of the system. In other words, generic-with-selection traits are robust in the sense that for a wide range of initial conditions and parameter values, the generic traits evolve. Natural selection does play a role

---

6. See Endler 1986 for studies of natural selection in nature. Natural selection is all but inevitable when resources are limited (finite), and so if life has evolved elsewhere in the universe, or if it were to evolve here again, we would expect natural selection to play some role in those evolutionary scenarios.

in this characterisation since it will be included in all (or almost all) of our models of the dynamics of evolution (given its high expectability).

Much of the work in the new field of artificial life can be seen as exploring the domain of generic-with-selection. For instance, Thomas Ray's (1991) virtual computer known as Tierra shows a rich and interesting evolution of an artificial biology. In particular, parasites have evolved in Tierra without in any way being built in from the beginning. Natural selection plays a prominent role in Ray's simulations — here the limiting resource is CPU time. One might take this work as supporting the claim that parasites are an expectable outcome of evolution[7]. That is one might take Ray's work as showing that parasites are generic-with-selection.

3.3 Stage 2 Conclusion

The first notion of generic, generic-without-selection, is best thought of as an ontogenetic or developmental concept. It applies to the expected outcomes of different developmental systems. The second sense, generic-with-selection, is a phylogenetic or evolutionary concept. It applies to the expected outcomes of different evolving systems. Put this way, our Stage 1 conclusion is obvious: SE functions are not applicable to generic-without-selection forms. But, both CR and SE analyses of *selection properties can have an evolutionary specified purpose.*

4. THE MARRIAGE OF SELF-ORGANIZATION AND SELECTION[8]

The point of this paper thus far has been to set out two analyses of function in biology, the SE and CR accounts, and show now they apply to the primary objects of the study of self-organization in biology, namely generic properties. We have seen that there are two different concepts of genericity that can be distinguished and that our two analyses of function apply differently to them. The two conclusions drawn thus far summarize this. The first being that the CR, but not the SE, analysis of function applies to generic-without-selection properties. The second is that both the CR and SE

7. Ray certainly presents his results that way. I must admit to some scepticism here since I am unsure to what extent, if any, my confidence in the claim that parasites are an expectable outcome of evolution would have been shaken by a negative result.

8. I borrow this section title form Kauffman 1993, although, as will become apparent, my conclusions differ somewhat from Kauffman's.

analyses potentially apply to generic-with-selection properties, so that a selection based teleology is applicable to them. Given the distinctions drawn, these conclusions are not at all surprising. But in this, the final, section I want to argue that they are a bit too simplistic to do justice to our topic.

4.1 Selection and Self-Organization are not Mutually Exclusive

It would certainly be a mistake to erect the following mutually exclusive categories for biological traits: those present because of the self-organization of the developmental system and those present because of natural selection. The complete evolutionary explanation of a particular trait's current form and function may well involve both a developmental account of the generic-without-selection aspects of the trait and ecological and genetic account of selection on that trait and the evolutionary response to that selection. For instance, thorns in plants are apparently very easy to make. They have evolved numerous times independently[9]. But, presumably the size, shape and life history characteristics of thorns in particular lineages owe much to selection. Thus if it is true both that (a) from the pint of view of plant development dynamics thorns are easy to make; and (b) in a number of different selective environments thorns are useful, but the optimal size, shape and time of appearance of thorns will differ in different environments, then the complete evolutionary explanation of thorns in a particular plant lineage will involve both facts about the self-organizational properties of thorns and facts about the ecological consequences of the thorns and the population genetic consequences of this.

More generally, selection can maintain and/or modify, and then spread useful generic-without-selection traits. This suggests the following methodological stance: first determine the generic-without-selection properties of the relevant system. Then compare the observed distribution of trait values with the expectation based on genericity. Any deviation from the expectation is then a candidate for explanation in terms of selection[10]. In

9. Not only have thorns evolved independently in different lineages, they have evolved from different structures, some being modified leaves, others modified stems.

10. Or drift. For instance if multiple trait values are equally likely from a developmental point of view, but only one is found in the organisms in question, then two possibilities present themselves: (1) Selection may have eliminated the other equally likely trait values; or (2) The different trait values are selectively neutral, or near neutral, and drift has resulted in the fixation of one. This second possibility should not be ignored, but I will not focus on it in this paper.

other words, the determination of the generic-without-selection properties of the system provides the appropriate null hypothesis against which proffered selectionist hypotheses must be compared[11]. To put the point in still one more way, once we have factored out the generic-without-selection aspects of the trait in question, the residue is then an appropriate candidate for SE functional analysis.

The above is a useful heuristic for evolutionary biologists. It makes non sense to try to explain by selection the presence of traits that are nearly inevitable anyway. Indeed the above heuristic is simply the methodological consequence of the two major conclusions already drawn in this paper. But as I stated earlier, these two conclusions are too simplistic to do justice to the complexities of evolution. Likewise, I will argue, the above methodological maxim, though useful, is itself overly simplistic.

4.2 The Evolutionary Interpenetration of Generic-without-selection and Generic-with-selection

The evolutionary process builds on what is available. The evolutionary potential of a lineage at time *t* depends on the state of that lineage at *t*, not on the state of that lineage at any earlier, or later, time. Developmental systems are the products of evolution and are affected by various evolutionary factors, including natural selection. Thus the traits that are developmentally robust at time *t* for lineage *l* are robust, or generic-without-selection, not in a universal, time-independent, lineage-independent, way, but in a way that is relative to the relevant developmental system, which is itself an evolved system. That means that when we take the developmentally robust trait values as null hypotheses the contrast is not selection versus universal biology, but rather selection with a shallow past versus evolution, including selection, with a deep past[12].

And so, the two conclusions drawn and the methodological heuristic suggested above are too simplistic. They are so because they treat the notion of generic-without-selection as a time-independent, lineage-independent, *i.e.*, universal, concept. But that ignores the quite obvious fact that the developmental systems we encounter today in birds, butterflies, green algae

---

11. This is essentially Kauffman's proposal (1993, 24 and 426).
12. Again, for the purposes of this paper I am focusing on selection as the alternative to genericity, but a fuller treatment would include drift and other evolutionary factors as well.

and green iguanas are products of a long evolutionary past. They are not the predictable products of the workings of the laws of physics and chemistry. In particular, they would not exist as they do were it not for the past workings of natural selection. Space limitations prohibit a detailed defence of this position, but I will offer a brief discussion of reaction-diffusion models to illustrate my point.

Alan Turing (1952) was the first to apply reaction-diffusion equations to pattern formation in biology. This class of models has become the most general explanation of that phenomena and, as such, the example of self-organization with the most extensively confirmed empirical applications[13]. the basics of the model involve two diffusable substances, and *activator* and an *inhibitor*. The activator autocatalyzes the further formation of itself and also catalyzes the formation of the inhibitor. The inhibitor inhibits the formation of both itself and the activator[14]. There are four basic parameters of such a model : the rate of diffusion of the activator, D_a, and of the inhibitor, D_h, and the breakdown rates of both, k_a and k_h. Furthermore, if the region within which the reactions are occurring has boundaries, then the boundaries can either be absorbing or reflecting, the boundary type significantly affecting the resulting pattern. Nijhout (190) has shown how such a model can produce all the wing color patterns of butterflies found in nature. Are the patterns then generic-without-selection since they can be produced by this simple mechanism, a mechanism not tied to life on Earth but applicable anywhere in the Universe one finds such activators and inhibitors? We answer "yes" to this question if we take for granted the values of the relevant parameters and the shape, size and nature of the relevant field boundaries (the wing veins). And with a shallow evolutionary view this is the right answer. But if we take a deeper evolutionary view we see that the values of the relevant parameters (and the nature of the boundaries) are by no means necessary features of wings, but rather are evolved conditions of Lepidoptera. So these patterns have genericity only relative to a contingently evolved background.

Since evolution works on what is available, the shallow perspective is, to personify evolution, oftentimes evolution's view. If a change in the environment of a butterfly species were to select for a new wing pattern,

---

13. For a general discussion of this see Kauffman 1993, my discussion is based on Nijhout 1990.

14. This characterisation is put in terms of diffusable chemical substances, but it is much more general. For instance, it can be applied to host-pathogen co-evolution or any other system that fits the formal structure of reaction-diffusion models (see Comins, Hassell and May 1992).

then that species is stuck, at least for the short term, with its ancestral wing cell boundaries and parameters for diffusion and breakdown. This will constrain, for the short term at least, how it can evolve. Thus in studying this evolution it is appropriate for us to use the methodology suggested above, drawing a rather sharp distinction between generic-without-selection and generic-with-selection. But we must always be mindful that the distinction is itself a contingent reflection of the current state of an evolving lineage. What, from a shallow point of view has function without purpose, may, from a deeper point of view, have an evolutionary purpose.

REFERENCES

Amundson R. and Lauder G.V. (1994). Function Without Purpose: The Uses of Casal Role Function in Evolutionary Biology. *Biology and Philosophy,* **9**, 443-469.

Brandon R.N. (1981). Biologicalteleology: Questions and Explanations. *Studies in History and Philosophy of Science,* **9**,181-206.

Brandon R.N. (1990). *Adaptation and Environment.* Princeton: Princeton University Press.

Comins H.N., Hassell M.P. and May R.M. (1992). The Spatial Dynamics of Host-parasitoid Systems. *Journal of Animal Ecology,* **61**, 735-748.

Cummins R. (1975). Functional Analysis. *Journal of Philosophy,* **72**, 741-765.

Dennett D.C. (1995). *Darwin's Dangerous Idea: Evolution and the Meanings of Life.* New York: Simon and Schuster.

Endler J.A. (1986). *Natural Selection in the Wild.* Princeton: Princeton University Press.

Goodwin B. (1994). *How the Leopard Lost its Spots: The Evolution of Complexity.* New York: Scribners.

Gould S.J. (1989). *Wonderful Life: The Burgess Shale and the Nature of History.* New York: Norton.

Gould S.J. and Vrba E. (1982). Exaptation a Missing Term in the Science of Form. *Paleobiology,* **8**, 4-15.

Kauffman S.A. (1993). *The Origins of Order.* New York: Oxford University Press.

Kingsolver J.G. and Koehl M.A.R. (1985). Aerodynamics, Thermoregulation, and the Evolution of Insect Wings: Differential Scaling and Evolutionary Change. *Evolution,* **39**, 488-504.

Lewontin R.C. (1978). Adaptation. *Scientific American,* **239**, n° 3, 212-230.

Nijhout H.F. (1990). A Comprehensive Model for Colour Pattern Formation in Butterflies. *Proceedings of the Royal Society of London B,* **239**, 81-113.

Paley W. (1836). *Natural Theology, Volume I.* London: Charles Knight.

Ray T.S. (1991). An Approach to the Synthesis of Life. In C.G. Langton, C. Taylor, J.D. Farmer and S. Rasmussen (eds). *Artifical Life II, SFI Studies in the Sciences of Complexity,* Vol. X, Reading (MA): Addison-Wesley.

Sober E. (1984). *The Nature of Selection.* Cambridge (MA): MIT Press.

Turing A.M. (1952). The Chemical Basis of Morphogenesis. *Philosophical Transactions of the Royal Society London B,* **237**, 37.

Wimsatt W.C. (1972). Teleology and the Logical Structure of Function Statements. *Studies in History and Philosophy of Science,* **3**, 1-80.

Wright L. (1976). *Teleological Explanation.* Berkeley: University of California Press.

MARC MAESSCHALCK and VALÉRIE KOKOSZKA

PHENOMENOLOGY AND SELF-ORGANIZATION

1. COGNITIVIST PROJECT AND PHENOMENOLOGICAL PROJECT FOR ATLAN

The current development of cognitivist sciences borrowed to phenomenology the notion of intentionality; with some adjustments, it allows formalizing what is meant by "intelligent behavior" and, therefore, it gives a formal model against which it is possible to determine whether an automaton or an expert computer behaves with intelligence.

The idea of such a borrowing comes from Searle. "As the conditions of satisfaction are interior to the act of speaking, the satisfaction conditions of the intentional state are interior to the intentional state"[1]. In a comment to this quotation, Dreyfus says that "in the light of this theory of intentionality, the phenomenological reduction takes the central importance attributed to it by Husserl. It is a particular act of thought through which we *distract* our attention from the object referred to (and through which we *distract* it from our psychological experience of the object) and *we take it back* on the act, and particularly on its intentional content, and thereby make our object of our representation of the conditions of satisfaction of the intentional state"[2]. Searle had commented already that such a transfer of the attention from the psychic activity of aiming, towards the aim activity itself as an object, implied an implicit criticism of Husserl who, despite the connection established between the operator at the first person and the meaning aims, maintains an abstract conception of intentionality and refuses to consider it as an operating form of causality[3]. Therefore, if the phenomenologist does

1. Cf. Dreyfus H.L. (1991). Husserl et les sciences cognitives. *Les Études Philosophiques*, 1 to 29, 6.
2. *Ibidem.*
3. Cf. Searle J.R. (1983). *Intentionality. An Essay in the Philosophy of Mind.* Cambridge: Cambridge University Press, 65.

B. Feltz, M. Crommelinck and P. Goujon, (eds.), Self-Organization and Emergence in Life Sciences: 283-297.

abandon the standpoint of an observing third person, he does not manage to go beyond a naturalistic conception of causality as a non-intentional relationship. The only way of recovering the asset of the phenomenological displacement towards the intentional activity of the spirit, is by considering such activity as always internal to its own conditions of satisfaction. In reality, the essential points already lie in this single page of Searle published in 1983.

As a matter of fact, in cognitive sciences, intentionality is defined as a quest for or the definition of an aim, *i.e. causality* according to Searle. Intentionality, understood in this restricted sense, is conceived as the particular case of intentionality in general, the latter being defined as an "oriented activity of consciousness... origin of the meanings"[4].

Such a double definition of intentionality implies two remarks. In cognitive sciences, intentionality is physicalized, *i.e.* it is understood, in the case of its general definition as well as in the case of its restricted definition, as an activity in the effective sense. From such a standpoint, we are far away from the notion of act of the consciousness such as conceived in husserlian philosophy, in which the expression "act of consciousness" is never understood purely and simply as an operation of consciousness, an effectuation of consciousness. In order to convince oneself of the semantic shift in cognitive sciences from the notion of act towards that of effectuation, it is enough to remember that, in phenomenology, the act of perceiving is an act of the consciousness, while it is presented as affected of passivity and it depends, essentially, on receptivity.

The second remark concerns the distinction (problematical from a logical point of view) between the particular and the general cases of intentionality. If intentionality in the particular sense is defined as looking for an aim, defining an aim, and if intentionality in the general sense is defined as an oriented activity always origin of meanings, we should conclude that the general case results from the particular case. Consciousness, as an origin-oriented activity, presupposes as its own origin intentionality itself, understood as the definition of an aim, in this case, the aim of signifying. Indeed, anything that appears as particular from the standpoint of consciousness as the origin of meaning, is, from a natural point of view, the origin of consciousness itself, which appears from the operativity of the intentional procedure. Thus the logical contradiction is only apparent, if one considers that it is rather a necessary inversion of perspective from Searle ís standpoint. This inversion does not lead symmetrically to a new relationship

4. Atlan H. (1993). Projet et signification. *Philosophiques*, **20**, 443 to 472, 443.

of constitution, as would be the case in a "reverse phenomenology". On the contrary, it allows to envisage the *reversibility* of the intentional operation itself, as it works sometimes as the definition of a field of meanings and sometimes as the definition of a mode of aiming. As soon as consciousness is no longer the irreversible constituting part, one may free itself from the synthetical *a priori* of transcendental liberty as judging liberty and one envisages an autorevizable procedure at the same able to aim and to be aimed at revision.

The reversal of the intentional order (as a natural indexed relationship and not as a transcendental activity) has the advantage of presenting a reversibility (the aims define the meanings and the meanings simultaneously define the aim) which is necessary in the context of cognitive sciences: the software of a computer defines a possible domain of data, of inputs which may be received in the software. On the other hand, a computer with an intelligent behavior must be able to receive unforeseeable inputs and must therefore be able to adapt his program side (his aim side) to receive them.

If the usage done within the cognitivist sciences of the notion of intentionality involves a transposition in physical terms (effectuation/causality) of intentionality, it also implies, a transposition of its field of emergence and application. As a matter of fact, revealing this eidetic characteristic of consciousness, *i.e.* its being intentional, required the suspension of what Husserl termed the natural attitude, in order to come back to the lived experience of consciousness. The natural attitude is defined by the constant implementation of a presupposition no other than the belief in a being simply given as existing being. For Husserl, the reduction of this attitude which allowed to discover intentionality among other essential features of consciousness aims at coming back to the condition of possibility of this presupposition constantly used by the sciences of nature as well as by human sciences. In other words, as the phenomenological reduction aims at revealing the transcendental basis of the *a priori* constantly used by all natural sciences (intuitive evidence of experience) is able to give a basis to these sciences themselves, in a derived, though not accessory way.

But for the cognitivists, there is no such thing as a reduction or a movement back to the lived experience of consciousness. This means either that the concept of intentionality emerges, in the context of cognitivist sciences, as a fact of consciousness (consciousness gives aims to itself) and not as an essence (eidos-form) of consciousness, and in this case one is limited to the natural attitude, or that the result of the phenomenological enquiry is accepted, that it is recognized that intentionality is a part of the essence of consciousness but that this result has to be translated into the field of natural attitude, has to become a fact in order to become a model testimony of the intelligence of an automate as a concrete effectuation procedure. In either case, far from coming back to the lived experience of consciousness, in order for the concept of intentionality to be operating, one should remain in or come back to the natural attitude.

However, it is also possible to consider the act of staying in the natural attitude in a more decisive way, *i.e.* without interpreting it simply as the necessary condition of the operativity of the notion of intentionality in cognitivist sciences (in which case one would remain in a logic of transposition) but in order to see in it a methodological option as strong as that which gives access, for the phenomenologist, to the eidictic of consciousness, the transcendental reduction. For the cognitivists, the fact that distinguishes the phenomenological attitude from the natural attitude (understood here as the attitude towards sciences) is that the former reduces the world to come back to consciousness while the latter reduces consciousness in order to come back to the facts, to the world.

Both attitude are presented as antagonistic, which was not the case in Husserlís work. For the cognitivists, both of them are characterised by the fact they are not complete. Owing to this negative similarity (being incomplete), the cognitivists assume they can deny one of them to offer the other its conditions of possibility, its basis. It should be noted that from a phenomenological viewpoint, only the natural attitude, the natural rationality aimed at experience can be qualified as incomplete as it may bring out an infinity of results, while the phenomenological rationality is finite, as it deals with essences. Still, this interpretation of the upholding in the natural attitude as a denial of the phenomenological attitude is too weak in comparison with the real cognitivist project. Once both attitudes (phenomenological and natural) have been considered as reverse to one another (one of them reduces the world while the other reduces consciousness), and once it is admitted that they are both incomplete, a progress is still possible if one assigns to the upholding in the natural attitude and, more precisely, to the cognitivism which supports this methodological option, to "reveal the emergence of intentional consciousness or of conscious intention"[5] from the fact, from the world, so that the intentional consciousness can "appear not as an originary, founder phenomenon, but as a secondary, a derived phenomenon"[6]. In other words, cognitivism is assigned a more ambitious project: turning the fact, the

5. Cf. Atlan H., *Projet et signification*, *op. cit*, 469.
6. *Ibidem.*

existing being, the effective into the condition of possibility of consciousness. It is, in fact, the contrary of Husserlís project: consciousness as a condition of possibility of the fact, as the a priori of the existing being. Such a project implies a split between intentionality and consciousness so that intentionality might be the origin of the emergence of the consciousness of itself as an effectuation process.

The relationship between phenomenology and cognitivist sciences for Atlan, lead us to underline the methodological choices involved by the transposition of phenomenological concepts in the context of cognitivist sciences and to reveal a specific project of the theory of consciousness inspired by Searleís indications, in which the active I would be understood internally in the exercise of its intentional causality.

2. THE UNDERLYING CRITICISM OF PHENOMENOLOGY

In Husserl as well as in Frege, according to certain cognitivists[7], the best way of access towards a theory of consciousness would be a propositional reference to a truth on intentional objects[8]. Husserl would have shown the way when he tried to build the identity of meaning on procedures of references to possible objects, *i.e.* merely intentioned.

The mistake of phenomenology is that it maintains an idealism of meaning when it sets the conditions of possibility of those effective cognition procedures. Thus the illusory mission chosen by this philosophy is the clarification of the background knowledge as antepredicative relationships with truth[9], *i.e.* as a pre-science of the world, independent from the procedures of intentionality. From such a standpoint, meaning is the horizon of constitution of validity structures as a pre-conscious relationship with the world. K.O. Apel interprets such a relationship with the world as the experience of a transcendental evidence of the conditions of validity[10], the reference to some thing in common, evoked by Habermas in the *Pensée*

---

7. Especially Mac Intyre and Dreyfus.
8. Cf. Habermas J. (1983). *Le discours philosophique de la modernité*, trad. Chr. Bouchindhomme et R. Rochlitz, Paris: Gallimard, 203 et 369 (Habermas J. (1985). *Der philosophische Diskurs der Moderne*, Frankfurt am Main: Suhrkamp, s. 202-203 u. 363-364).
9. *Ibidem*, 312 (s. 309-310).
10. Cf. Appel K.O. (1986). Le problème de l'Évidence phénoménologique à la lumière d'une sémiotique transcendantale. *Critique*, 89.

postmétaphysique[11], the residue, in a way, of the effective world abandoned by phenomenology.

Among Husserlís intuitions, it would be needed to preserve the way in order to realize his project of constitution of the predicative knowledge of consciousness. He conceives the intuitive presentification of ideal entities[12] to which intuitive acts correspond set up as "autodonation of the intentioned ëobjectí through a linguistic expression"[13]. So that "Husserl *a priori* casts all expressible meanings by language in the mold of cognitive dimension"[14]. "Pure meanings" accomplish the intention of consciousness whose original act corresponds to the pure intuition of an ideality[15], *i.e.* a "sending back structure", a "noem"[16]. Through such a foundation of knowledge on the intuitive relationship with "meanings as such", Husserl defines the relationship between our knowledge and an immediate experience of the world[17], knowable only as a presence[18]. His mistake is that he gives a "metaphysical status" to those pure meanings when they simply correspond to the effectuations of the consciousness intentions.

3. RESISTANCE FROM A PHENOMENOLOGICAL STANDPOINT

Far from being based on the postulate of transparency of consciousness, the aim of phenomenology is to develop a science of the original relationship between knowledge and life by distinguishing the strata of constitution of the consciousness fields as emergences of meaning in the movement of the manifestation of life.

Therefore, the important thing in Husserl is not the ignorance about the effective or the natural world as "pratico-inerte"[19] or as such; the important thing is to know, in any knowledge considered as a cognitive intention, an

---

11. Habermas J. (1993). *La pensée postmétaphysique*, trad. par R. Rochlitz. Paris: A. Colin, 182 (Habermas J. (1988). *Nachmetaphysisches Denken. Philosophische Aufsätze*. Frankfurt am Main: Suhrkamp,s. 182).
12. Habermas J., *Le discours philosophique de la modernité*, *op. cit.*, 172 (s. 173).
13. *Ibidem,* 204 (s. 204).
14. *Ibidem.*
15. *Ibidem,* 203 (s. 202).
16. Cf. Livet P. (1993). Structure noématique et transcendance du Dasein. *Philosophiques*, **20**, 323 à 346, 330 et 331.
17. Habermas J., *op. cit*, 207 (s. 206).
18. *Ibidem,* 205 (s. 205).
19. Cf. Habermas J. (1987). *Théorie de l'agir communicationnel,* t. 2, trad. J.-L. Schelgel, Paris: Fayard (Habermas J. (1987). *Theorie des kommunikativen Handelns*, bd. 2, Frankfurt am Main: Suhrkamp, s. 197-198).

essential relationship with life which stays the most assured guide for the construction of such a knowledge as it depends on an idea about life and testifies, in its own way, to this original perception of an event that affects us. A knowledge which no longer knows its mode of affection through the being-able-to of the world has lost the signification of its own measure and encloses itself in the arbitrary of poetry for poetry that kills the object of "poeticize" itself, life celebrating itself in the art of talking.

From such a standpoint, the phenomenological question refers to the constitution of theoretical and practical attitudes in relation to the original relationship of the subjects with the experience of the world; however, it is by no means a metaphysical dream on the individual experience of a subject who auto-experiences itself[20]. The transcendental reduction leaves the well-known world of consciousness aside, the world objectivised precisely through several intentions which build several distinct horizons of perception. However this operation consists in a movement back to the original world, to the "things themselves", in that they are meaning givers, or in that they fulfil the perceiving activity of consciousness and justify the very idea of it. In Husserl, the transcendental reduction brings us back to the consciousness as a *reflexive unity* constituted by the passive synthesis of its belonging to life. From this passive synthesis, the active synthesis determined by an objectivising intention is determined.

A simple way of explaining this Husserlian conception of consciousness is to give it as an essential characteristic the adjective "*doxic*". The transcendental reduction allows us to understand the doxic essence of our consciousness: it shows the "auto-affection" of life in it through the reflexive unity it constitutes from its original passivity as this "absolute fact" it is for itself as received life. Such an experience of a donation to oneself as living is the origin of consciousness which makes possible the constitution of its reflexive unity in "doxo-theoretical", "doxo-practical", "doxo-esthetic" acts, etc.[21]. Through all these "doxo-logies"í of the world, the original experience of the natural world is shown as life animating the activity of consciousness and differentiating through the synthesis of the living.

---

20. Cf. Habermas J. (1987). *Théorie de l'agir communicationnel,* t. 2, trad. J.-L. Schelgel, Paris: Fayard (Habermas J. (1987). *Theorie des kommunikativen Handelns*, bd. 2, Frankfurt am Main: Suhrkamp, s. 197-198) 143 (S. 205).

21. Cf. Husserl E. (1982). *Recherches phénoménologiques pour la constitution*, trad. par E. Escoubas, Paris: Presses Universitaires de France, 25 (Husserl E. (1952). Ideen zu einer reinen Phänomenologie u. phänomenologischen Philosophie, II. Phänomenologische Untersuchungen zur Konstitution, Husserliana IV, hrsg. M. Biemel, Haag: Martinus Nijhoff, s. 2-3).

Therefore, the husserlian concept of intersubjectivity does not correspond to the concept of a community of knowledge which already presupposes the original experiences linked to the constitution of the body itself and of the psychic egoïty. The community of knowledge is the deployment of a life of free communication characterized by communities of action. Such a life is already ensured it will belong to a spiritual, natural world whose immanent affection determines our language relationship to the community of knowledge.

This affection is a core problem of phenomenology insofar as it founds the access to phenomenalness through the primitive data of sensibility[22]. In comparison with this original passivity of the pure consciousness as a manifestation of life in its auto-affection, intersubjectivity, as it is built by the perception of others, with its intropathic dimension, is a derived experience, a modification of the original relationship with life. Therefore, from a phenomenological point of view, the community cannot be identified with the pragmatic interaction nor with the interpersonal components of the lived worlds: it refers to a more original experience, that of belonging to the flesh of the world and, even more radically, to life as a phenomenologisation of the world being[23]. "That is what the members of the community have in common: the coming in oneself of life, in which anyone of them come in one self as that Self one is. Thus they are at the same time the Same, as the immediation of life and the others as this trial of life is, each time, in themselves one of themselves irreductibly"[24].

This dimension of passivity towards life which introduces to the lived perception of an immanent finality of existence is ignored by cognitivist readers.

Hence, Husserlís problem appears exactly opposed to the problem posed by cognitivist theories of consciousness, insofar as it seeks the immersion of consciousness into life when cognitivists only seek the emergence of consciousness in the dimension of physical effectivity of knowledge as an intentional process. We feel that consciousness as a living fact, or life as the original activity of consciousness are the widest gap between either theories: epistemology as a biology of consciousness or biology of consciousness as an epistemology; natural or phenomenological attitude. In all these reinterpretations of phenomenology, the intention of an object as a central

22. Cf. Henry M. (1963). *L'essence de la manifestation*, Paris: Presses Universitaires de France, (2e Èd. en un volume, 1990), 574.
23. Cf. Henry M. (1990). *Phénoménologie matérielle*. Paris: Presses Universitaires de France, 7.
24. *Ibidem,* 77.

cognitive function, prevails over the original receptivity of consciousness with regard to the phenomenalisation of the natural world.

Such an intention of immersion towards a science of originary experience of life shows that the interest of the phenomenological approach lies in the reflection effort on the constitution of the relationship with life as a being-able-to. From such a standpoint, it is possible to locate the intention of an aim as a possible dimension of the doxic constitution of meaning. Husserlís originality in comparison with the internalist realism (Searle[25]) in the Anglo-Saxon language philosophy is that Husserl always relates the meaning to the effective fulfilment of the cognitive apprehension structures through the manifestation of originary life[26]. This may remind of the accusation of "naive platonism" which was brought against Husserlís standpoint[27]. If Heidegger himself greeted Husserlís perceptiveness about platonism, he also immediately reproached him for too naive a reception, characterized by a non-critical reference to natural reality and, more precisely, by a methodological deficiency in the phenomenological reduction which maintains natural reality itself as an a priori of consciousness[28]. Nevertheless, maintaining natural reality independent from natural attitude is the basis of the transcendental eidetic such as Husserl conceived it. Indeed, the aim is "to show that the idea or the eidé, of which the soul is the collection, are all the less of its own invention that they show, on the contrary, a fundamental connivance with what the manifestation of things have of their ownest, such as they presentify themselves, appear initially from themselves according to the order which, this time, is specific to them (...). Far from being reduced to a simple noetic emanation resulting from a withdrawal of the soul to itself, the eidetic correlatively denotes the emergence principle of all that is manifest; it translates at the same time the life of the present itself"[29].

We are far from the cognitivist interpretation of meaning, which takes back the idea suggested by R. Mac Intyre among others, of a "semantic of the reference". From this standpoint, it is not surprising that cognitivists go on speaking of categories where Husserl would, more rigorously, refer to ideas, as the eidictic is precisely the source through which the original

---

25. Cf. Dreyfus H.L., *Husserl et les sciences cognitives*, *op. cit*., 6.
26. Cf. Solowski R. (1964). The Formation of Husserl's Concept of Constitution. *Phaenomenologica,* **18**, La Haye: Martinus Nijhoff, 131-136.
27. Cf. Husserl E. (1950). *Ideen zu einer reinen Phänomenologie u. phänomenologischen Philosophie, I. Allgemeine Einf,hrung in die reine Phänomenologie*, Husserliana III, hrsg W. Biemel. Haag: Martinus Nijhoff, ß. 22.
28. Cf. Brisart R. (1991). *La phénoménologie de Marbourg*, Bruxelles: Publications des Facultés Universitaires Saint-Louis, 141.
29. *Ibidem,* 139.

phenomenalness is manifested for the consciousness, in the flux of temporal experiences which, being sensations, are not "intentioned" as semantic objects, linguistic signs, but felt as the living Present of time, whose concept seems so repellent to Husserlís cognitivist readers[30].

But it is precisely on this relationship with time of intentionality that the distance is confirmed between phenomenology and the theory of consciousness elaborated by cognitivist sciences. Cognitivists sciences should be compared with Brentano, rather than with Husserl. Husserl tries to build a theory of consciousness temporalisation where memorization itself remains structured by the tensions of the original temporality; so that the flux of life as an auto-affecting living present justifies the relationship to itself of consciousness as the experience of a presence: the psyche[31]. On the contrary, for Brentano, the memory is lead by an associative logic which abolishes the efforts of retention and protension associated with sensible experience. Time is a physical phenomenon associated with perception and not the internal sense which would organise our accumulation of data. It is to such a conception of time that Merleau-Ponty, among others, is opposed in its Phénoménologie de la perception. He writes: "Memories that we evoke in front of someone who has had a limb amputated induce a shadow limb not as an image induces another image in associationism, but because any memory opens again lost time and invites us to take back the situation it refers to. The intellectual memory, in Proustís meaning, has enough with a signal of the past, a past in idea; therefrom it extracts the "characters" or the communicable meaning rather than finding back its structure; but it would not be memory if the object it constructs was not related by some intentional links to the horizon of the lived past and to this past, even such as we would find it back if we went back to these horizons to open time again"[32].

A good example of the conception of time in the phenomenological theory of consciousness is given in the analyses of intropathy.

The intropathy experience belongs to "the pure transcendental experience"[33]. Hence it is possible only according to the transcendental reduction which opens the field of originary experiences of consciousness as constitutive syntheses of its being-in-the-world. Husserl distinguishes

---

30. Cf. Rigal E. (1991). Quelques remarques sur la lecture cognitiviste de Husserl. *Les Etudes Philosophiques*, 101 to 117, 112.

31. Cf. Kokoszka V. (1996). La conception husserlienne de la temporalité entre 1905 et 1910. *Tijdschrift voor Filosofie*, **58** (2), 314 to 341.

32. Merleau-Ponty M. (1945). *Phénoménologie de la perception*, Paris: Gallimard, 101.

33. Husserl E. (1972). *Philosophie première*, t. 2, trad. par A. Kelkel, Paris: Presses Universitaires de France, 249 (Husserl E. (1959). *Erste Philosophie, II. Theorie der phänomenologischen Reduktion*, Husserliana VIII, hrsg. R. Bôhm, Haag, Martinus Nijhoff, s. 181).

immediate intropathy from mediate intropathy[34]. The immediate intropathy reveals the transcendental *life* of the proper self with its "universe of events"[35]. encompassed in the *ego cogito*. The mediate intropathy reveals a "*second transcendental life*"[36], *i.e.* a second form of transcendental life: that of the *alter ego* whose content appears "through organic bodies which may be seized by experience"[37]. Therefore, both the transcendental life of the proper psyche and that of the foreign psyche are revealed through one and the same kind of originary experience. So the pure transcendental perspective encompasses not only the diversity of elements of a proper life (immediately) but also (mediately) a multitude of foreign lifes associated with the diversity of their events. Themselves are also associated indirectly to the diversity of events of foreign psychic lifes.

4. THE NORMATIVE PROJECT OF PHENOMENOLOGY

Through the analysis of the originary relationship with the flux of events in consciousness, phenomenology tries to make possible a *knowledge of the "normative" impulsion of life* which constitutes our moral disposition to the pragmatic interaction in order to institute a concrete ethical order. The knowledge of it leads to an ethos, or a wisdom[38]. Therefore, phenomenology is not an eidetic description of deeper experiences which structure our relationship with the world, as a theory of generative grammar would search the structural descriptions which allow the implementation of the linguistic competence in a determined linguistic field. When it tries to locate itself within the dynamic synthesis (the flux) of the life which constitutes our

---

34. Husserl E. (1972). *Philosophie première*, t. 2, trad. par A. Kelkel, Paris: Presses Universitaires de France, 249 (Husserl E. (1959). *Erste Philosophie, II. Theorie der phänomenologischen Reduktion*, Husserliana VIII, hrsg. R. Bôhm, Haag, Martinus Nijhoff, s. 191 (s. 137).
35. *Ibidem*, 249 (s. 181).
36. *Ibidem*.
37. *Ibidem*.
38. Cf. Husserl E. (1987). *Fichtes Menschheitsideal, Drei Vorlesungen,in Aufsätze und Vorträge* (1911-21), Husserliana XXV, hrsg Th. Nenon u. H.R. Sepp, Dordrecht/Boston/ Lancaster, 267 à 293.

presence, phenomenology builds a normative knowledge of the living[39]. It is able to denounce all the perspectives which are opposed to this regional knowledge of the being on the basis of this fundamental ontology of the lived world.

As a knowledge which avoids the question of the originary origins as far as the finality of life is concerned, phenomenology does state the conditions of a concrete involvement of the subject for life, contrary to what Duméry feared, for instance. However, Duméryís[40] critics towards the phenomenological description are not only of historical interest. They show there is a persistent misunderstanding towards the phenomenological knowledge, when it is considered as the description of the universal structures of human experiences perceived in their nude immediacy, as Ricoeur writes[41]. Phenomenology tries to reach the originary disposition of the subject towards life, as an answer to the injunction of life, the homo capax, the being-able-to[42].

Such a restoring (normative) reduction to the being-able-to of life is possible for phenomenology as it takes into account the immediacy of the relationship of all intentional horizons to the originary origins. For phenomenology, immediacy does not mean a residue or a trace or a confuse report, without any relational clarity, or simply effective. It is the clear and distinct idea for any form of life that it is effectively, fully and authentically the realization of the essential part of originary life[43]. Such an apodictic evidence of the immediacy of the relationship to the deeper origins brings back to the being-able-to of life which makes possible the "dimensionalisation of existence" according to a unified (synthetic) process

39. Cf. Levinas E. (1929). Sur les Ideen de M. E. Husserl. *Revue Philosophique de la France et de l'Étranger*, 230 à 265.
40. Cf. Dumery H. (1957). *Critique et religion, Problèmes de méthode en philosophie de la religion*. Paris: SEDES, 171.
41. Cf. Ricoeur P. (1994). Phénoménologie de la religion. *Lectures 3, Aux frontières de la philosophie*. Paris: Seuil, 263 à 271, 266.
42. The *fiat* of the consciousness according to the ß. 122 of *Ideen I, op. cit.*
43. Already in the ß. 78 of *Ideen I, op. cit.*
44. Cf. Ladrière J. (1982). Philosophie et langage. *Annales de l'Institut de Philosophie et de Sciences morales*. Bruxelles: Éd. de l'Université Libre de Bruxelles, 21 à 38, 33 et 34: "Indeed, the 'theion' is not only one more dimension, next to the others: as it is the representation, under the form of dimensionality, of the deeper moment of the universal deployment, as it represents the form of any deployment, it is, as a dimension, the immanent structuration of all the others."

which institutes its diversity in totality, in cosmos. And this source of dimensionality is the absolute life or the *theion*[44], the conviction of which (perceived from the apodictic evidence) forms the disposition to answer to the concrete injunction by "historialising" the auto-affection of originary life.

REFERENCES

Atlan H. (1993). Projet et signification. *Philosophiques*, **20**, 443 to 472.

Appel K.O. (1986). Le problème de l'Évidence phénoménologique à la lumière d'une sémiotique transcendantale. *Critique*, 89.

Brisart R. (1991). *La phénoménologie de Marbourg*, Bruxelles: Publications des Facultés Universitaires Saint-Louis, 141.

Dreyfus H.L. (1991). Husserl et les sciences cognitives. *Les Études Philosophiques*, 1 to 29, 6.

Dumery H. (1957). *Critique et religion, Problèmes de méthode en philosophie de la religion.* Paris: SEDES.

Habermas J. (1993). *La pensée postmétaphysique*, trad. par R. Rochlitz. Paris: A. Colin, 182 (Habermas J. (1988). *Nachmetaphysisches Denken. Philosophische Aufsätze.* Frankfurt am Main: Suhrkamp, s. 182).

Habermas J. (1987). *Théorie de l'agir communicationnel,* t. 2, trad. J.-L. Schelgel, Paris: Fayard (Habermas J. (1987). *Theorie des kommunikativen Handelns*, bd. 2, Frankfurt am Main: Suhrkamp, s. 197-198).

Habermas J. (1983). *Le discours philosophique de la modernité*, trad. Chr. Bouchindhomme et R. Rochlitz, Paris: Gallimard, 203 et 369 (Habermas J. (1985). *Der philosophische Diskurs der Moderne*, Frankfurt am Main: Suhrkamp, s. 202-203 u. 363-364).

Husserl E. (1987). *Fichtes Menschheitsideal, Drei Vorlesungen,in Aufsätze und Vorträge* (1911-21), Husserliana XXV, hrsg Th. Nenon u. H.R. Sepp, Dordrecht/Boston / Lancaster, 267 à 293.

Husserl E. (1982). *Recherches phénoménologiques pour la constitution*, trad. par E. Escoubas, Paris: Presses Universitaires de France, 25 (Husserl E. (1952). Ideen zu einer reinen Phänomenologie u. phänomenologischen Philosophie, II. Phänomenologische Untersuchungen zur Konstitution, Husserliana IV, hrsg. M. Biemel, Haag: Martinus Nijhoff, s. 2-3).

Husserl E. (1972). *Philosophie première*, t. 2, trad. par A. Kelkel, Paris: Presses Universitaires de France, 249 (Husserl E. (1959). *Erste Philosophie, II. Theorie der phänomenologischen Reduktion*, Husserliana VIII, hrsg. R. Bôhm, Haag, Martinus Nijhoff, s. 181).

Husserl E. (1950). *Ideen zu einer reinen Phänomenologie u. phänomenologischen Philosophie, I. Allgemeine Einf hrung in die reine Phänomenologie*, Husserliana III, hrsg W. Biemel. Haag: Martinus Nijhoff, ß. 22.

Henry M. (1990). *Phénoménologie matérielle.* Paris: Presses Universitaires de France, 7.

Henry M. (1963). *L'essence de la manifestation*, Paris: Presses Universitaires de France, (2e Èd. en un volume, 1990), 574.

Kokoszka V. (1996). La conception husserlienne de la temporalité entre 1905 et 1910. *Tijdschrift voor Filosofie*, **58** (2), 314 to 341.

Ladrière J. (1982). Philosophie et langage. *Annales de l'Institut de Philosophie et de Sciences morales.* Bruxelles: Éd. de l'Université Libre de Bruxelles.

Levinas E. (1929). Sur les Ideen de M. E. Husserl. *Revue Philosophique de la France et de l'Étranger*, 230 à 265.

Livet P. (1993). Structure noématique et transcendance du Dasein. *Philosophiques*, **20**, 323 à 346, 330 et 331.

Merleau-Ponty M. (1945). *Phénoménologie de la perception*, Paris: Gallimard.

Ricoeur P. (1994). Phénoménologie de la religion. *Lectures 3, Aux frontières de la philosophie.* Paris: Seuil, 263 à 271, 266.

Rigal E. (1991). Quelques remarques sur la lecture cognitiviste de Husserl. *Les Études Philosophiques*, 101 to 117, 112.

Solowski R. (1964). The Formation of Husserl's Concept of Constitution. *Phaenomenologica,* **18**, La Haye: Martinus Nijhoff, 131-136.

Searle J.R. (1983). *Intentionality. An Essay in the Philosophy of Mind.* Cambridge: Cambridge University Press, 65.

III. EPISTEMOLOGICAL AND CONCEPTUAL APPROACHES

B. Explanation

PAUL THOMPSON

A ROLE FOR MATHEMATICAL MODELS IN FORMALIZING SELF-ORGANIZING SYSTEMS

In what follows, I argue that, in the context of theorizing about non-linear systems, a model-theoretic account of the structure of scientific theories is superior to the widely accepted axiomatic-deductive and linguistic formalization account. Self-Organizing systems are largely non-linear systems and, hence, for Self-Organizing systems, a model-theoretic account is a superior account of theory structure as well as of explanation and theory confirmation.

1. A SKETCH OF THE STANDARD VIEW OF THEORY FORMALIZATION

For much of this century, theories have been construed as axiomatic-deductive structures. On this conception, a theory consists of a set of statements, all of which are laws and a small subset of which are axioms. Axioms are statements that are self-evident and highly general. In, principle, all the other statements (laws) of the theory can be deduced from the axioms. The laws are statements about the causal structure of the world. This conception is linguistic and the syntax of the language of a theory is first-order predicate logic with identity. The semantics is provided by correspondence rules which are part of the theory and directly link the formal system to the phenomenal world. In effect, the correspondence rules define an empirical model of the formal system. That empirical model is understood as logically equivalent to the phenomenal system to which the theory applies.

Two common examples of axiomatic-deductive theories are Euclidean Geometry and Newtonian Mechanics : the former mathematical, the latter scientific. Once one accepts Euclid's axioms or Newton's axioms, all the

B. Feltz, M. Crommelinck and P. Goujon, (eds.), Self-Organization and Emergence in Life Sciences: 301-313.

other propositions of Euclidean Geometry or Newtonian Mechanics can, in principle, be deduced.

Euclidean Geometry exemplifies the power of an axiomatic-deductive structure. This power is in part due to the availability of the tools of deductive logic such as indirect proof. Indirect proof rests on the fact that the derivation of a contradiction from the axioms of a theory is clear evidence that the axiom set is inconsistent and the theory flawed. A classic use of indirect proof is found in investigation of the parallel line axiom in Euclidean Geometry. Attempts to demonstrate that the parallel line axiom was a genuine axiom of the geometry have a long history. In the nineteenth century, several mathematicians attempted to provide a demonstration using an indirect proof (*a reductio ad absurdum* technique). In this technique one assumes the opposite of what one is attempting to establish, demonstrates that the opposite assumption leads to a contradiction and concludes that the original assumption was correct. The parallel line axiom states that through a point outside of a given line one and only one line can be drawn parallel to the given line. This axioms allows of two negations. First, one can assume that no lines parallel to the given line can be drawn through the point. Second, one can assume that more than one line parallel to the given line can be drawn through the point. Replacing the parallel line axiom with either of these negations of it was found not to lead to a contradiction. Instead, two new consistent geometries were discovered. The first is known as Reimannian Geometry (spherical or curved geometry). The second is called Hyperbolic geometry. This clearly demonstrates the power of an axiomatic-deductive formal structure.

Newtonian Mechanics is also held to be an example of an axiomatic deductive structure. Newtonian Mechanics is held to have four axioms:

1. all bodies tend to remain in a state of rest or uniform motion unless acted upon by an external unbalanced force;
2. force equals mass times acceleration;
3. for every action there is an equal an opposite reaction;
4. for any two bodies the force of gravitational attraction between them is equal to the product of their masses divided by the square of the distance between them.

Newtonian Mechanics, however, is a less impressive example of an axiomatic-deductive structure than Euclidean Geometry because the deductions of most of the laws from the axioms require numerous subsidiary and simplifying assumptions. This dramatically affects the deductive integrity of the structure and, for instance, makes the use of indirect proof and many other tools of deductive logic impossible. This, in turn, reduces the power f the axiomatic-deductive formalisation when applied to Newtonian Mechanics. Given that Newtonian Mechanics is one of the best examples of an axiomatic-deductive formalisation of a scientific theory, it appears that scientific theories, in practice, are not pure axiomatic-deductive structures. Hence, their deductive integrity is compromised and, as a result, much of the strength of axiomatic-deductive formalization is compromised in the case of scientific theories. This need for subsidiary and simplifying assumptions does not appear to be solely a function of our current state of knowledge but even if it is, the requirement of complete or almost complete knowledge of the empirical world in order to provide an axiomatic-deductive formalization is an extremely demanding and unachievable restriction on actual scientific theorizing.

In the view of most philosophers of science, what this view of theories loses in deductive integrity, it more than gains in codifying explanation, prediction, unification and confirmation — the elegance of the logic of these aspects of this conception is beguiling.

Explanation, in this conception, consists of deduction using laws (theorems of the theory) and, in its idealized form, has the following pattern:

Logic deduction

$$L1, L2, L3, \ldots L_n$$
$$C1, C2, C3, \ldots C_m$$
$$\longrightarrow E$$

where E is the explanandum (the thing or event to be explained), L_i is a law drawn form the theory, and C_i is a statement of an initial condition. The conjunction of L_i's and C_i's is called the explanans. If the conjunctio of L_i's and C_i's is used to deduce an event or thing that has not yet occurred or is not yet known to have occurred, one is predicting what will be the case or will be discovered to have been the case rather than explaining what is known to be the case. This explanation scheme has three central virtues. First, the relevance of the statements in the explanans to the occurrence of the explanandum is guaranteed by the requirement of deduction. Second, the invocation of laws from a theory connects the theory directly and intuitively to phenomena. Third, the explanation of a thing or event, although only citing a limited number of laws from the theory brings to bear the entire force of the theory. This is so because the theory is an integrated deductive whole.

Unfortunately, the deductive integrity of the formalization is compromised in another way. Many of the laws appealed to in scientific theories are probabilistic or statistical. Deduction from these laws is not possible and explanation becomes probabilistic.

Confirmation has a similar structure.

$$T \rightarrow (E \rightarrow R)$$

where T is a theory, E is an experiment, and R is the result of the experiment (the statement is read as: if the theory T is true then if experiment E is undertaken then result R will be found). As has been pointed out by numerous writers, but most notably by Sir Karl Popper, this logic of "confirmation" actually only allows disconfirmation with deductive certainty. That is, one can reject a theory if $(E \rightarrow R)$ can be deduced from it but E fails to yield result R. Confirmation using this logical scheme is, at best, a probability function. That is, the more instances of experiments leading to the results expected on the basis of the theory, the higher the probability that the theory is true. This is not a deductive procedure but an inductive one.

As can be appreciated, this view of theories and the ways in which they connect to phenomena is elegant and simple. It still plays an important role in aspects of scientific theorizing and reasoning. Unfortunately, it fails to capture many important aspects of theorizing in the physical and biological sciences. Many of these shortcomings are now well known (see: Suppe, Suppes, Van Fraassen, Beatty, Thompson, Lloyd). In what follows, I point to another shortcoming of this view: its inability to deal with Self-Organizing systems. In order to deal with Self-Organizing systems, I argue in the final section of the paper, one need to employ a richer mathematical framework and this richness is found in a alternative conception of the structure of scientific theories called the semantic conception.

2. ARTIFICIAL LIFE AND NON-LINEARITY

One of the prime features of the received view is the deductive nature of the formal structure of a theory. One of the hallmarks of a non-linear system is the absence of a unique solution to the equations describing the system. As a result deducing future states of the system is impossible in principle. More importantly, deducing other equations of the system from those given in its definition is impossible in principle.

Artificial life and Self-Organizing systems in general are non-linear systems. Hence, although they can be modeled using mathematical equations to describe the dynamics of the system, they, in principle, cannot be given a deductive formalization in first-order predicate logic.

In what sense are non-linear systems, in principle, not capable of being rendered as deductive structures? A property of non-linear systems is that small differences in the initial conditions of the system result in very large differences after a short period of time. Or, put another more precise way, small differences between two systems at time t result in significant divergences in the trajectories of the systems in a phase space after only a few temporal sequences. These small differences can be extremely small and beyond our capacities with the most refined measurement devices to detect or describe. Differences in the measurement of a variable at the 10th decimal place or beyond are significant enough to cause these major divergences. Accepting the most optimistic prognosis for accuracy of measurement in the future as technology increases, the limitations of accuracy will still result in these divergences. And in systems that have a large number of variables, the divergences will be more dramatic. As a result predictability will be impossible "in principle" for any meaningful sense of in principle[1].

A key feature of a deductive structure is that elements of a system being described can be discovered by deduction from the axioms. In addition, in principle, the axioms and additional generalizations deduced from them allow the prediction of future states from a given present state of the system. The set of equations that define a non-linear system do not enable, as a fundamental feature of (differential and integral calculus) mathematics, a rigorous deduction of other equations governing the behavior of the system. Whereas in axiomatic-deductive structures, all the behaviors of the system and all the state transitions of the system are deducible in principle from the axioms. A theory is not the description of a state a system but a description of its behavior. Non-linear systems, in principle, are such that the set of equations defining the system do not permit a deduction of the future states of the system. The most common method of dealing with this fact is the use of computer simulation in what are termed "numerical experiments." In this way, the behavior of an abstract system is simulated by numerically integrating the transformation equations ("Numerical integration" is a

1. The concept of "possible in principle" is not easy to capture. On the surface, it seems straight forward: "possible in principle" means "possible without logical contradiction." Understood in this strict sense only conjunctions of contradictory statements describe in principle impossible states of affairs. However, if a state of affairs is so complex that only the invocation of the capacities of an omnipotent God operating outside of the material realm can describe it, the line between impossible in principle and merely impossible in fact becomes thin to the point of vanishing. Under such circumstances one would be justified in asking what work the distinction between "in principle" and "in fact" is doing.

method of finding approximate values for definite integrals. It is used frequently in cases where no analytic methods are available).

An example will make the above points more forcefully.

A number of aspects of model construction and organization and self-organization come together in a field of inquiry known as artificial life. artificial life is the study of simulations of carbon-based living organisms. Those simulations can be mechanical devices, computer-based models, conceptual mathematical models, and carbon-based entities. The only significant distinction between artificial life and "natural" life is that humans rather than nature are responsible for the existence and characteristics of the "organisms." The formal models of these structures are almost always non-linear.

One of the achievements of artificial life is the demonstration that complex behaviors can be simulated on a computer screen by means of a few local rules of organization. One clear example of this is the computer simulation of flocking. Birds often move in flocks in which the pattern of the flock — a result of the flying behavior of each bird — and the dispersion and reformation in the face of an obstacle is seen as a complex co-ordinated activity. The same is true of the behavior of schools of fish and the herd movements of cattle. Craig Reynolds has simulated flocking behavior on a computer screen. His entities (called Boids which are devoid of material relevance) behave in accordance with three rules of behavioral tendencies:

1. to maintain a minimum distance from other objects in the environment including other Boids;
2. to match velocities with Boids in its neighborhood; and
3. to move towards the perceived center of mass of the Boids in its neighborhood.

These are rules governing individual Boids. They are rules of local control. There are no further rules for the aggregate: the flock of birds. Aggregate behavior emerges from the behavior of individuals governed by these rules. The result on the computer screen is that when a number of individual Boids is given a random starting position, they will come together as flock and will "fly" with grace and naturalness around obstacles by breaking into sub-flocks and then regrouping into a full flock once around the object. The flock's actual behavior when confronted with an object emerged from rules that only determined the behavior of individuals. To watch the Boids on the screen is to watch a complex co-ordinated aggregate behavior.

This example illustrates all of the above outlined assumptions of artificial life. It illustrates the primacy of the organization of entities over the properties of the matter of which they consist. It illustrates that there are no rules governing the aggregate behavior; only rules which govern the behavior of all entities (rules that are local and distributed over the local domain of entities). The aggregate behavior emerges from the individual (uncoordinated) behavior of the entities. In some cases, several independent (from the point of view of potential organizational independence) systems may interact to produce a higher order complex behavior. One could view this higher order system as a single larger system with a slightly larger set of rules or as the interaction of several individually organized systems. Ultimately, the distinction is irrelevant as long as none of the rules under either description exercises global control.

The Boid example also illustrates the assumption that control is not global but local. There are no rules of co-ordination for the aggregate. Co-ordination is a function of the rules of behavior of the individual entities. An important feature of local, distributed control is the importance of a neighborhood. The behavior of interacting entities is specified in terms of neighboring entities: their positions or states. This system of entities interacting according to local rules based on a neighborhood, in effect, is the heart of the concept of organization. And such systems can be described using precise mathematical models in terms of state spaces as described above in the context of the semantic conception of a scientific theory. The emphasis in artificial life and in the semantic conception of theories is on the dynamics of systems. In both cases those dynamics are specified in terms of organization.

Finally, the Boids example illustrates the assumption that complex behavior is the outcome of a few local rules. The essential point of this assumption is that simple behaviors of interacting elements are the basis for high level organizational complexity and that the attempt to formulate rules at higher levels (globally) to describe high level complex behaviors is wrongheaded. Chris Langton (a leading exponent of artificial life has claimed that the quest for global rather than local control mechanisms (rules) is the source of the failure of the entire program of modeling of complex behaviors up to the present including, especially, much of the work on artificial intelligence.

3. MATHEMATICAL MODELS AND THEORY FORMALIZATION

The semantic conception of theories has a relatively short history the beginnings of which Frederick Suppe has traced to von Neuman (Suppe

1988). Two other early initiators and advocates were Evert Beth in 1948-49 (Beth 1948, 1949; see also 1961) and Patrick Suppes in 1957 in his Introduction to Logic (Suppes 1957). Beth advanced what has become known as a state space approach while Suppes advanced a set-theoretical predicate approach.

During the late 1960's and the 1970's the semantic conception was consolidated and extended by a number of philosophers from a variety of perspectives — most notably, Bas van Fraassen, Frederick Suppe, and Patrick Suppes. Over the last 15 years, John Beatty, Elisabeth Lloyd and I have been extending and applying the semantic conception in the context of biology and, in particular, in context of evolutionary theory and genetics (see Beatty 1980a, 1980b; Lloyd 1983, 1984, 1986, 1987; Thompson 1983b, 1985, 1986, 1987, 1988a, 1988b, 1989).

The semantic conception is so called because scientific theories are formalised in terms of models (*i.e.,* semantic structures) and, hence, an adequate formal approach to the structure of scientific theories consists in the direct specification of the models (*i.e.,* the semantics) and not in the specification of a linguistic axiomatic-deductive system (*i.e.,* a syntax). The significant differences, therefore, between syntactic and semantic accounts are the nature of an adequate semantics of a scientific theory and the nature of an adequate (logically and heuristically) formalization of a scientific theory. On the syntactic conception, the semantics of a theory are provided by correspondence rules. On a semantic conception the semantics of a theory are provided directly by defining a class of models. For Patrick Suppes, the class of models is directly defined by defining à set-theoretical predicate. For Bas van Fraassen and Frederick Suppe, the class of models is defined in terms of a phase space or state space (*i.e,* a topological structure).

In a semantic conception, a theory is defined directly by specifying in mathematical English the behavior of a system. Most importantly, laws do not describe the behavior of objects in the world, they specify the nature and behavior of an abstract system. This abstract system is, independently of its specification, claimed to be isomorphic to a particular empirical system. Establishing this isomorphism, as I shall argue below, requires the employment of a range of other scientific theories and the adoption of theories of methodology (*e.g.,* theories of experimental design, goodness of fit, etc.).

4. THEORIES AND PHENOMENA

The fundamental concept in terms of which the relation between a theory and phenomena is articulated in the semantic conception a mapping function. The mapping function is best described as isomorphic (in algebraic contexts such as groups and rings) and homeomorphic (in topological contexts). An isomorphism or homeomorphism is a one-to-one correspondence between the elements of one or more sets resulting from abijective mapping. A bijective mapping is a "one-t-one" and "onto" mapping (that is, it is both injective and surjective). Abstractly, two groups *AG, IB* and *AG*', *MB* where l and *m* are operators on *G* and *G*', respectively are isomorphic if for any *x* and any *y* in *G*, *f*(*alb*) =*f*(*a*) *m f*(*b*) where *f* is function that maps *G* into *G*'.

The essential feature of an isomorphism is the preservation of structure and behavior. The elements of *G* and *G*' may be entirely different but the structure of the groups is preserved and hence the behavior of the elements with respect to the operators is the same in both groups. When applied to models, models are isomorphic if there is a one-t-one correspondence which preserves relations, functions and constants. Confirmation of a theory consists in establishing the required correspondence between a mathematical model and a physical system. Explanation and prediction consists in using a theory for which an acceptable level of confirmation has been achieved.

Confirmation is complex involving, among other, theories of measurement, experiment and experimental design. A theory of measurement provides an agreed upon standard in terms of which observed phenomena are compared as well as a set of principles governing the conditions under which measurements are made. These principles ensure that the measurements are in accord with the theory of the experiment. A theory of the experiment specifies a broad conceptual framework within which experiments can take place. It specifies such things as what assumptions based on other scientific theories can be employed (for example electromagnetic theory and quantum theory when using an electron microscope in a biological experiment), the possibility and role of simplifying assumptions, correct patterns of inference, etc. A theory of experimental design specifies the exact nature of the technique of experimentation. The appropriate methods for controlling extraneous variables is an important component specified by a theory of experimental design.

Under the strictures of the above theories, experimental observation of physical systems results in date. Confirmation consists in comparing the structure and behavior of the physical system that emerges from this data with the structure and behavior of the theory (mathematical model) within the scope of which this physical system falls. One of the most straightforward ways of performing this comparison is to represent the data in a "observation space" analogous to the "phase space" in which the theory is represented. In these spaces, states of both physical system, the comparison of the observation space and the phase space of the theory is uncomplicated and based on an identity relation — the observation space and the phase space will have the same dimensionality and the points representing states of the system will be identically located within the space. Unfortunately few physical systems for which we wish to confirm a theory are this uncomplicated. Two common physical systems which introduce significant complications are statistical ones and non-linear ones.

In the case of non-linear systems the main complicating feature is the fact that there is soldom a unique analytic solution to the differential equations defining the system. The comparison between the theory and the observations for the purpose of confirmation is then unidirectional. The actual structure and course of state transitions of the physical system as observed must correspond to at least one of the possible solutions to the differential equations specifying the theoretical system.

In the context of scientific explanation, a theory (mathematical model) explains a set of phenomena if:

- the system defined by the theory is isomorphic to the phenomenal system in which the set of phenomena to be explained occurs, and
- the set of elements of the mathematical model which are mapped onto the set of relevant phenomenal objects within the phenomenal system can be shown, within the mathematical model, to be a consequence of the structure or behavior of the model.

5. CONCLUSION

Self-organization systems are in most cases non-linear systems. Theories about non-linear systems are best understood in terms of the semantic conception of theory structure — a model-theoretic account of scientific theories — rather than a syntactic conception — an axiomatic-deductive structure account. In the semantic conception, the connection between a theory and phenomena is understood in terms of a relation of isomorphism. This relationship enables models of non-linear systems to explain phenomena and to be confirmed or disconfirmed by phenomena. The account of explanation and confirmation in the syntactic conception is not at all well suited to theories about non-linear systems.

REFERENCES

Beatty J. (1980a). Optimal-Design Models and the Strategy of Model Building in Evolutionary Biology. *Philosophy of Science,* **47**, 532-561.

Beatty J. (1980b). What's Wrong with the Received View of Evolutionary Theory? In P.D. Asquith and R.N. Giere (eds). *PSA 1980,* t. 2, East Lansing: Philosophy of Science Association.

Beatty J. (1987). On Behalf of the Semantic View. *Biology and Philosophy,* **2**, 17-23.

Beth S. (1948). *Natuurphilosophie.* Gorinchem: Noorduyn.

Beth E. (1949). Towards an Up-to-Date Philosophy of the Natural Sciences. *Methodos 1,* 178-185.

Beth E. (1961). Semantics of Physical Theories. In H. Freudenthal (ed.), *The Concept and the Role of the Model in Mathematics and Natural and Social Sciences.* Dordrecht: Reidel, 48-51.

Drazin P.C. (1992). *Nonlinear Systems.* Cambridge: Cambridge University Press.

Kellert S.H. (1993). *In the Wake of Chaos.* Chicago: The University of Chicago Press.

Langton C.G. (ed.) (1989). *Artificial Life.* New York: Addison-Wesley.

Lloyd E. (1984). A Semantic Approach to the Structure of Population Genetics. *Philosophy of Science,* **51**, 242-264.

Lloyd E. (1986). Thinking about Models in Evolutionary Theory. *Philosophica,* **37**, 87-100.

Lloyd E. (1987). Confirmation of Ecological and Evolutionary Models. *Biology and Philosophy,* **2**, 277-293.

Rosenberg A. (1985). *The Structure of Biological Science.* Cambridge: Cambridge University Press.

Rosenberg A. and Williams M. (1986). Discussion of Fitness as Primitive and Propensity. *Philosophy of Science,* **53**, 412-418.

Ruse M. (1973). The Philosophy of Biology. London: Hutchinson & Co. Ltd.

Ruse N. (1977). Is Biology Different from Physics? In R. Colodny (ed.). *Logic, Laws, and Life.* Pittsburgh: Pittsburgh University Press.

Schaffner K.F. (1969). Correspondence Rules. *Philosophy of Science,* **36**, 280-290.

Suppe F. (1967). *On the Meaning and Use of Models in Mathematics and the Exact Sciences.* Ann Arbor: University Microfilms International (Thèse de doctorat).

Suppe F. (1972a). Theories, their Formulations, and the Operational Imperative. *Synthèse,* **25**, 129-164.

Suppe F. (1972b). What's Wrong with the Received View on the Structure of Scientific Theories? *Philosophy of Science,* **39**, 1-19.
Suppe F. (1974). Theories and Phenomena. In W. Leinfellner and E. Kohler, (eds). *Developments in the Methodology of Social Science.* Dordrecht: Reidel, 45-91.
Suppe F. (1976). Theoretical Laws. In M. Prezlecki, K. Szaniawski and R. Wojcicki, *Formal Method in the Methodology of Empirical Science.* Wroclaw: Ossolineum.
Suppe F. (1979a). *The Structure of Scientific Theories,* 2[e] ed. Urbana: University of Illinois Press.
Suppe F. (1979b). Theory Structure. In P.O. Asquith and H.E. Kyburg Jr. (eds). *Current Research in the Philosophy of Science.* East Lansing: Philosophy of Science Association.
Suppe F. (1988*). The Semantic Conception of Theories and Scientific Realism.* Urbana: University of Illinois Press.
Suppes P. (1957). *Introduction to Logic.* Princeton: Van Nostrand.
Suppes P. (1962). Models of Data. In E. Nagel, P. Suppesand, A. Tarski (eds). *Logic, Methodology and Philosophy of Science: Proceedings of the 1960 International Congress.* Stanford: Stanford University Press, 252-261.
Suppes P. (1968). The Desirability of Formalisation in Science. *Journal of Philosophy,* **65**, 651-664.
Thompson P. (1983). The Structure of Evolutionary Theory: A Semantic Approach to Studies. *History and Philosophy of Science,* **14**, 215-229.
Thompson P. (1985). Sociobiological Explanation and the Testability of Sociobiological Theory. In J.H Fetzer (ed*.) Sociobiology and Epistemology.* Dordrecht: D. Reidel, 201-215.
Thompson P. (1986). The Interaction of Theories and the Semantic Conception of Evolutionary Theory. *Philosophica,* **37**, 73-86.
Thompson P. (1987). A Defence of the Semantic Conception of Evolutionary Theory. *Biology and Philosophy,* **2**, 26-32.
Thompson P. (1988a). The Conceptual Role of Intelligence in Human Sociobiology. In H.J. Jerison et I.L. Jerison (eds). *Intelligence and Evolutionary Biology.* New-York: Springer-Verlag.
Thompson P. (1988b). Logical and Epistemological Aspects of the "New" Evolutionary Epistemology. *Canadian Journal of Philosophy,* **14** (supplementary), 235-253.
Thompson P. (1989). *The Structure of Biological Theories.* Albany: State University of New York Press.
Thompson P. (1993). *The Structure of Biological Theories.* Albany: State University of New York Press.

Van Fraassen B.C. (1970). On the Extension of Beth's Semantics of Physical Theories. *Philosophy of Science,* **37**, 325.

Van Fraassen B.C. (1972). A Formal Approach to Philosophy of Science. In R.E. Colodny (ed.). *Paradigms and Paradoxes.* Pittsburgh: The University of Pittsburgh Press.

Van Fraassen B.C. (1980). *The Scientific Image.* New-York: Oxford University Press.

Van Fraassen B.C. (1981). Theory Construction and Experiment: An Empiricist View. In P.D. Asquith et R.N. Giere (eds). *PSA 1980,* Vol. 2, East Lansing: Philosophy of Science Association, 663-677.

ROBERT C. RICHARDSON

EXPLANATION AND CAUSALITY IN SELF-ORGANIZING SYSTEMS

ABSTRACT

There are two broadly different approaches to understanding scientific explanation. The first approach encompasses what is sometimes called a "causal", or "mechanical", approach to explanation (*e.g.,* Railton 1978, 1989 ; Salmon 1984, 1989). The goal is to reveal the causes, or mechanisms, responsible for the phenomena we observe. Causal realism becomes a requirement for scientific explanation. The second approach emphasizes the virtues of unification (*e.g.,* Friedman 1974; Kitcher 1981, 1989, 1993). The goal is to develop a system of laws capable of describing observed phenomena in the most economical way. Organization and systematic unification become the central goals of scientific explanation.

Explanations in terms of self-organization promise to give us explanations of observed order. Thus, Stuart Kauffman (1993) claims that the problem for twenty-first century science is to explain "organized complexity", including ecosystems, communities, organisms, genetic regulatory systems, and neural systems. His exploration of the "origins of order" emphasizes that across disparate domains simple general principles suggest that there is a natural and spontaneous order in complex systems, that it is systems which are at the "edge of chaos" that are most evolvable, and that selection maximizes this evolvability. His "statistical mechanics" for complex systems needs to be understood in terms of the explanatory unification it afords, and fits poorly with the aspirations of a causal/mechanical model of explanation.

B. Feltz, M. Crommelinck and P. Goujon, (eds.), Self-Organization and Emergence in Life Sciences: 315-340.

1. CAUSAL MODELS OF EXPLANATION

According to "causal" or "mechanical" models of explanation, the central problems in understanding explanation depend on distinguishing genuine explanatory relations from spurious ones, and on developing principled reasons for maintaining asymmetry and relevance in explanations. The natural suggestion, given this conception of the problem, is that explanatory relations are causal relations. Wesley Salmon says "... explanation involves revealing the mechanisms at work in the world. Mere subsumption of phenomena under generalizations does not constitute explanation. Explanation involves understanding *how the world works*" (Salmon 1989, 156). Accordingly, the focus is on explaining the individual case (see Cartwright 1989). We want to explain the origin of life, the fall of the Roman Empire, or the Cambrian explosion. The results and the explanations are unique. Some of these explanations may be irreducibly probabilistic, as when we explain the distribution of phenotypes in Mendel's peas, the fixation of Sickle cell alleles in west African populations, or, perhaps, the decline in the American dollar.

In even the simplest deterministic cases, it is clear that simple deductive relations are not sufficient to delimit good explanations. To take but one of many classical problems (see Salmon 1989 for more detail), one can explain a lunar eclipse using Newtonian mechanics given the relative positions and motions of the sun, moon, and earth ; and it is possible to deduce the prior positions given the later positions of the bodies and their motions. To take a probabilistic example, one can predict the distribution of genotypes or phenotypes under selection, given prior distributions for them and knowledge of linkage patterns; and one can equally well project the prior distributions from later ones given suitable additional information. Though the former count as explanations, the latter do not. The subsequent lunar position does not explain the eclipse any more than the subsequent genotypic or phenotypic distribution explains prior frequencies. Explanatory relations are asymmetric. Deductive relations are not sufficient to distinguish spurious from actual explanations. Something more is needed, and the obvious suggestion, given a realistic perspective, is that explanatory relations must be causal relations. We explain an eclipse by citing its causes, and in particular by showing how it was brought about by having the sun, earth, and moon in the previous positions given the laws of celestial mechanics. We explain the distribution of genotypes by citing their prior distributions and the selection pressures. We cannot explain an eclipse by citing the subsequent positions, or the distributions of genotypes by citing their consequences. Salmon once thought that statistical relevance — and in particular screening off-would suffice to discriminate genuine from spurious explanations (1971) by distinguishing causal relations among statistically significant ones, but he has since abandoned that view. The key, in his view, nonetheless lies with causality. In a later work, *Scientific Explanation and the Causal Structure of the World* (1984), he begins with statistically relevant relations, and supplements them with a causal model. He accepts the view that statistically relevant properties may not be causally relevant, and causally relevant properties may not be statistically relevant (cf. Fetzer 1981, 92). Nonetheless, explanatory relations must be statistically relevant and causal. So thought Salmon has given up the attempt to glean causal relations from statistical relations, nevertheless, on his view explanatory relations are also causal relations: to explain an occurrence is to show how it fits into the causal structure of the world (Salmon 1971, 276).

The view has its attractions. It also has its problems. One of the most serious problems is its inability to deal with explanations cast in more abstract terms, such as the explanation of the trajectory of a photon as following a geodesic, relying on the structure of space-time rather than some more immediate causal processes. These do not seem to fit the causal/mechanical model very well, as was noticed by Clark Glymour (1984). Salmon claims such explanations are *consistent with* causal/mechanical explanations even if these are not themselves structural explanations (see Salmon 1989, 181 ff.), and in some cases they even seem to be complementary (especially, see Brandon 1990). However, if these are good explanations, and not causal/mechanical in character, then we must somehow supplement the account of explanation to incorporate explanations which are not causal/mechanical. Another problem concerns the epistemological character of causal claims. This includes questions concerning whether our knowledge of more general causal claims depends on our knowledge of more specific claims, how we can know that a specific causal claim is correct, and whether there are the epistemic resources to ground the "ontic" claims of the causal theorist[1]. The point remains, though, that an exclusives focus on singular causal relations is myopic. Perhaps the causal/mechanical view is inadequate. Perhaps it is only incomplete. Let's look at the alternative.

1. For an insightful and able discussion of difficulties with Salmon's view, see Hitchcock (1995).

2. UNIFICATION AND SCIENTIFIC EXPLANATION

In a landmark paper, "Explanation and Scientific Understanding", Michael Friedman defends the view that the essence of scientific explanation derives from the fact that "science increases our understanding of the world by reducing the total number of independent phenomena that we have to accept as ultimate or given" (1974, 15). A good explanation should enhance understanding, and unification does just this. Systematic unification is gauged by whether it minimizes the number of premises and maximizes the number of conclusions. The goal is to explain the most with the least. Thus, Friedman explains, the kinetic theory of gases serves to unify a variety of otherwise disparate phenomena including not only the Boyle-Charles Law, but also Graham's law and a theory of specific heats. It does this by embedding an explanation of macroscopic phenomena within a Newtonian context, recognizing the deep similarities between terrestrial mechanics, celestial mechanics, and statistical mechanics. Friedman's attempt to explicate this fundamental intuition was flawed. The application of the idea depended on understanding theories as sets of sentences, and using the number of fundamental laws as a measure of simplicity. His strategy depended on finding an index of what counts as *an independently acceptable sentence,* and defining a measure of the complexity of a theory in terms of the cardinality of the set of sentences comprising it. A good explanation then would reduce the cardinality of the set of independently acceptable sentences, thus effecting a unification. The specific measure of complexity Friedman used depended on the number of K-atomic sentences; that is, the number for which there is no partition possible, either because there is no set of equivalent sentences or because any partition involves sentences which are not independently acceptable (cf. Friedman 1974, 17). Philip Kitcher (1976) showed that this approach has the unfortunate consequence of excluding a variety of explanations which draw on diverse domains; in any case, the syntactic criterion of complexity does not appear to be well-defined when we are confronted with theories having distinct vocabularis. There is, nonetheless, something fundamentally attractive about the basic intuition that unification is somehow fundamental to scientific explanations, or at least is part of what we sometimes want from explanations.

There are alternative ways to capture this intuition. Kitcher takes an entirely different approach, emphasizing common *patterns of explanation* as the key to understanding unification. He says,

> "Understanding the phenomena is not simply a matter of reducing the 'fundamental incomprehensibilities' but of seeing connections, common patterns, in what initially appeared to be different situations. ... Science advances our understanding of nature by showing us how to derive descriptions of many phenomena, using the same patterns of derivation again and again, and, in demonstrating this, it teaches us how to reduce the number of types of facts we have to accept as ultimate..." (1989, 432).

Instead of minimizing the number of fundamental principles, Kitcher seeks unification in general argument patterns, which apply in diverse domains. The idea, he says, "behind unification is the generation of as many conclusions as possible using as few patterns [of explanation as possible]. It is also important that the instantiations of the patterns should be genuinely similar" (1989, 434). Kitcher illustrates the idea with a variety of case studies, including those found in classical genetics, evolutionary biology, and chemistry. He explains that as he understands it, unification consists in deploying similar explanations for a variety of phenomena. Instead of relying on relatively few principles, Kitcher sees unification flowing from a reliance on relatively few patterns of explanation. It is not clear how having common patterns of explanation constitutes substantial unification ; but the main point I want to make is that there is more than one way to elaborate the demand for unification[2]. The idea is still to explain more with less.

I propose to set aside, for the moment, questions of how to explicate unification. Instead, I will turn to two examples where the differences between causal accounts and unification are particularly salient. Both involve probabilistic explanations. Both are suggestive. Neither fits comfortably with a causal/mechanical approach. Someone committed to a causal/mechanical model might be inclined to regard this as showing that unification leads us in the wrong direction, embracing spurious explanations. Alternatively, one possibility would be to recast them to fit a causal/mechanical approach better. To some extent, this can be done. I am content to show that there is a natural understanding of the cases which does not fit the causal/mechanical paradigm. Someone committed to unification would conclude that the causal approach sometimes mislocates the explanation of an accepted phenomenon.

R.A. Fisher's (1930) explanation of the prevalence of a 1:1 sex ratio of males to females is a useful case, in part because of its elegance and in part

2. Friedman says that unification is a reductionist triumph. Kitcher's form of unification is not reductionist, as Todd Jones (1995) shows. In spite of Friedman's suggestion to the contrary, even the theory of statistical mechanics is Newtonian in a broad sense, and in the systems it treats-in particular, gases-the only explanatory models are statistical and probabilistic.

because it has come to be so broadly accepted in the biological community. Fisher argued on general grounds that the optimal reproductive strategy is to invest equally in males and females. Darwin evidently came very close to this explanation in *The Descent of Man*:

> "Let us now take the case of a species producing... an excess of one sex-we will say the males-these being superfluous and useless, or nearly useless. Could the sexes be equalized through natural selection? We may feel sure, from all characters being variable, that certain pairs would produce a somewhat less excess of males over females than other pairs. The former, supposing the actual number of the offspring to remain costant, would necessarily produce more females, and would therefore be more productive. On the doctrine of chances a greater number of the offspring of the more productive pairs would survive; and these would inherit a tendency to procreate fewer males and more females. Thus a tendency towards the equalization of the sexes would be brought about" (1871, 316).

It is not clear that Darwin saw how to generalize this argument to show how a population producing an excess of females would be subject to similar correction. An excess of females would, after all, be neither superfluous nor useless. I expect this simple fact explains why Darwin could not extend the point. Fisher did see how the generalize the point, and the central feature of his analysis requires that we look at more than one generation. Fisher showed that a 1:1 ratio is a stable equilibrium, and that under natural assumptions no other ratio is stable[3]. Offspring can be viewed as making contributions to parental fitness, depending on the number of offspring they sire. If on average, a population invests more in females than males, then there will be a differential advantage to individuals that invest more heavily in males ; and if on average a population invests more in males than females, then there will be a differential advantage to individuals that invest more heavily in females. That is, individuals deviating from the norm will have offspring that are at a reproductive advantage. Given that the tendencies are heritable, selection should favor a 1:1 ratio. If males and females require an equal reproductive investment to bring them to maturity, the ratio of the two sexes defines the relative fitness of the two reproductive types. Fisher gave this an elegant formal characterization. If the cost to produce a son is c_m and the cost to produce a daughter is c_f, and if the

---

3. Those assumptions are not always met, and deviations from the 1:1 ratio are explained accordingly. The central case is the skewed ratio of females to males in the social insects, but there are dramatically divergent ratios in other species as well. These can generally be explained by the way they defy the assumptions which drive Fisher's argument. It is, in fact, one of the attractive features of Fisher's argument that it explains the prevalence of 1:1 ratios while also making the deviations intelligible.

percentage of sons is p, then the total benefit accruing to a parent will be a function of the benefits due to sons, b_m, and to daughters, b_f:

$$b_m pT/c_m + b_f(1\text{-}p)T/c_f$$

An alternative investment in sons of p^* will result in a differential advantage insofar as

$$b_m p^* T/c_m + b_f(1\text{-}p^*)\ T/c_f$$

is greater than the benefit to any alternative strategy. If sons and daughters are equally costly, then sons will provide a greater return when females are more common, and daughters will provide a greater return when males are more common. The equilibrium point is one in which the average numbers of sons and daughters is the same.

Elliott Sober (1984) calls such explanations "equilibrium explanations", contrasting them explicitly with causal explanations. He says,

> "Equilibrium explanation shows why the actual cause of an event is, in a sense, explanatorily irrelevant. It shows that the identity of the actual cause doesn't matter, as long as it is one of a set of possibilities of a certain kind" (140).

Whereas identifying the cause of some state of the population depends on specifying the actual cause, and knowing the actual history, equilibrium explanations such as Fisher's do not. The explanation is an explanation of why the equilibrium state-in this case, a 1:1 ratio-is stable and others are not. Sober says that "the details of a population's past often do not matter to its present configuration" (1984, 141). This might seem incomprehensible. Neglecting the history would be neglecting the cause. Equilibrium explanations do not depend on any particular causal history, though that does not imply that there are no causes for what we observe. Whether we are explaining the generality of a 1:1 sex ratio, or its occurrence in some given species of altricial birds, Fisher's explanation does not cite specific causes. Indeed, there likely is no single cause[4]. Kitcher adopts a similar position. Kitcher says that, confronted with a sex ratio very near 1:1, the right explanation of this is that there are selection pressures which favor the evolutionary equilibrium. Though we might be able to detail a "complete causal history" based on sperm and egg production, mating patterns, and the like, this information is, in a way, irrelevant to what we want to explain.

---

4. Explanations which depend on showing something is an "evolutionarily stable strategy" provide other clear cases of equilibrium explanations (see Maynard Smith 1975, 1976, 1982). This has been applied broadly, including sexual behavior and animal aggression. The emphasis on stable equilibria has the consequence that such explanations do not offer us any immediate insight into evolutionary history ; that provides some reason to be skeptical about them as explanations (see Richardson 1984).

He says "... the causal approach seems to err by overlooking the fact that the particular phenomenon to be explained is one example of a class, all of whose members instantiate a general regularity" (1989, 426).

Genetic drift provides an analogous case. Evolution is clearly a stochastic process in the sense that, given an initial distribution of genes, or genotypes, or phenotypes, with realistic parameter values we are at best able to project a probability distribution of the relevant states. This is some times associated with genetic drift. Genetic drift is simply the "error" in transmission of types from generation to generation, arising from finite population size. Drift is standardly treated using models incorporating infinite, or effectively infinite, ensembles of finite populations. Given a single gene with two alleles with frequencies p_i and p_j, in the absence of selection ensembles of populations initially polymorphic at that locus will tend to disperse across a wide range, from populations fixed for one allele to populations fixed for the alternative allele (see Falconer 1989 or Roghgarden 1979). These changes in sub-populations are random if there is no selection operating, and so though different sub-populations become differentiated there will be no change in the allelic frequencies in the overall population. Since the extremes are absorbing states in which the only source of change is mutation, as the ensemble disperses over the space each population will eventually become monomorphic. In the limit, the ensemble of populations will bifurcate into a bimodal distribution at the two extremes. The frequencies of populations fixed for the alternative alleles should be the same as the initial frequencies of the alleles, namely p_i and p_j. This is the neutral case, and should be understood as defining the probability with which a neutral allele will go to fixation over a given time as a function of the effective population size. (This process is illustrated in Figure 1).

The neutral case is an ideal, assuming that there are no selective differences. If we focus on selection alone, in the absence of drift and mutation, we have a deterministic process: given a frequency distribution at one time, and selection coefficients, then there should be a unique distribution of frequencies in the next generation. Mathematically, this is modeled using infinite population sizes in which sampling error, and therefore drift, could not occur. With finite populations, drift has the effect of exploring adaptive zones. In the deterministic case, the change in z_t, the mean value of a trait at t, will be

$$\Delta z_t = z_{t+1} - z_t - h^2\sigma^2(sz_t - z_t)$$

where s is the selection coefficient and $h^2\sigma^2$ is the heritable variance. In an infinite population, these changes would be a deterministic function of fitness, and, short of equilibrium, would result in an increase in the frequency of more fit individuals, and in average fitness (see Figure 2). In finite populations (cf. Lande 1976; Sewall Wright 1931, 1932), drift captures the extent to which changes tend not to be correlated with fitness differences ; they are random with respect to fitness. Fitness values determine the strength and location of the central tendency within an ensemble of populations, and drift becomes the amount of dispersal around the mean value (cf. Richardson and Burian 1992 this is illustrated in Figure 3).

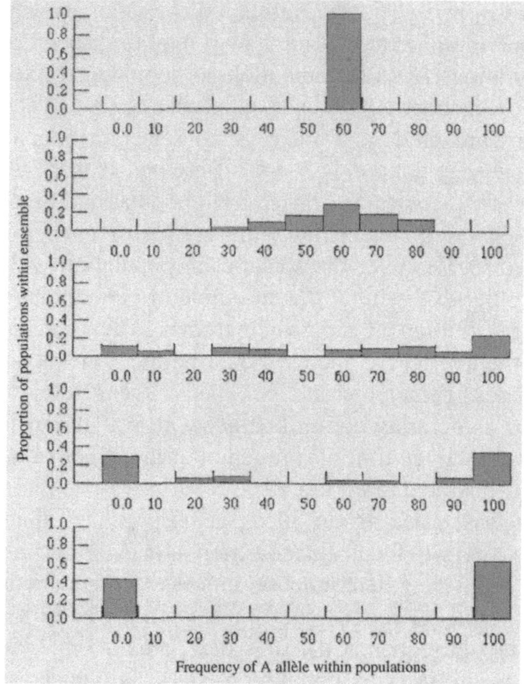

Figure 1. Drift in a population without selection.
An illustration of the effect of drift in a population in the absence of selection. The x axis represents the frequency of the *A* allele within populations, and the *y* axis the proportion of populations in the ensemble. Initially, all the populations in the ensemble have 60 %. A alleles (and therefore 40 % of the alternatives). Over time, the populations disperse, with some reaching fixation for *A* and some becoming fixed for alternatives to *A*. In the absence of mutation or outcrossing, populations which become fixed remain fixed. At the limit, all populations become fixed for A or an alternative. Given that initially the frequency of *A* is 60 %, the expected outcome is that 60 % of the populations become fixed for *A*.

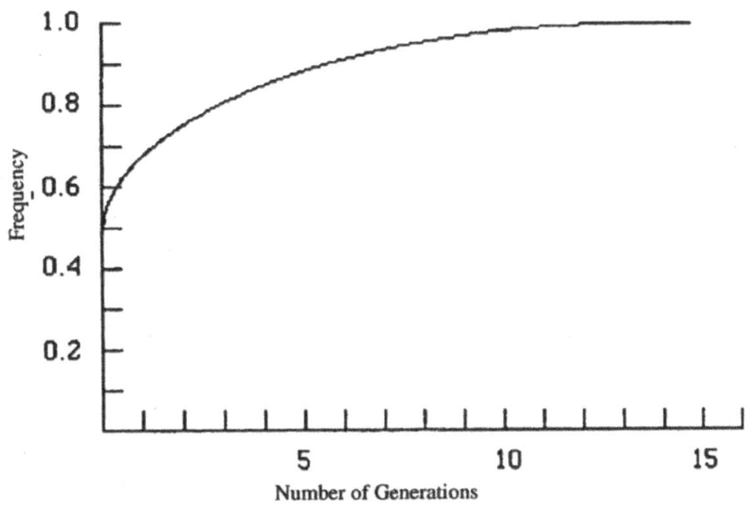

Figure 2. Changes in frequency due to selection.

Change in a trait, or the frequency of a gene, under selection in the absence of drift. The change from one generation to the next is a function of the heritable variance, and the differences of fitness (or the selection differential). If selection is the only factor operating, then fitness at $t + 1$ is a deterministic function of the fitness at t.

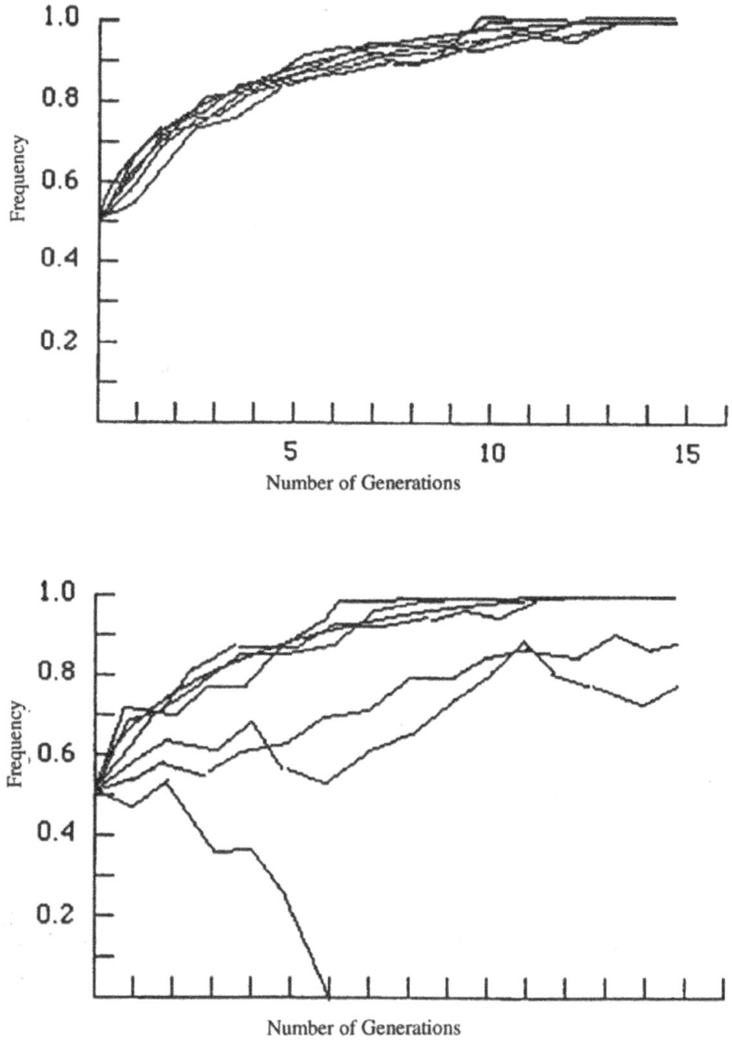

Figure 3. Changes in frequency under selection with drift.
Change in a trait, or frequency of a gene, with both selection an drift. The mean expected change is a function of the heritable variance, and the selection differential as above. The variance in the frequency changes are a measure of the significance of drift, and is in turn a function of the population size. The top figure illustrates a smaller population size.

What this provides is an abstract framework in which to deal with changes in the frequency of genes, genotypes, or phenotypes. It describes the more general patterns of change, or expected patterns of change. Once again, the actual causal history does not matter. We may know, for example, that a population has undergone some change in the distribution of genes, perhaps bringing one allele to fixation. Given parameters on the population, we can determine the probability which this would happen as a result of chance alone (cf. Lande 1976; Falconer 1989). This may be the right explanation from an evolutionary perspective, even though it does not invoke the specific causes in a given case. Again, this does not imply that there are no causes but hat the details of the causal history are not necessary to explain the result. The adequacy of the explanation depends entirely on whether it adequately captures patterns of change. This is not a equilibrium explanation, but like equilibrium explanations, the significance of drift depends on its ability to explain changes because they conform to a more general pattern. Again, no causal history features in the explanation.

It is possible for a causal theorist to retain or recast some of the cases here. Fisher's explanation, for example, is suggestive of mechanisms even though it abstracts from any particular causal history. It is consistent with any number of historical scenarios and any number of selective regime : so long as an evolutionary trajectory results in a 1:1 ratio because a 1:1 ratio was favored, then whatever the specific cause, it will conform to Fisher's model. In fact, the range of causes might be heterogeneous in the extreme. This much supports the view that even if these are not causal/mechanical explanations, they are compatible with them. If we insist on recasting the explanation within a causal/mechanical explanation, Fisher's model is not only abstract but incomplete. One key question is whether, were we to discover, say, that a 1:1 sex ratio is ancestral rather than derived, we would retain Fisher's explanation as an explanation of 1:1 sex ratios. One committed to a causal theory of explanation would not[5]. Fisher would disagree.

---

5. This is the position which Robert Brandon has embraced in conversation. It is a natural extension of his work in Brandon (1990) and in Brandon (forthcoming), where he shows how well explanations in terms of natural selection fit a causal/mechanical model. I think that a causal theorist has substantial latitude here. The prevalence of a 1:1 ratio, or its presence in a given species, might be explained causally as the result of stabilizing selection, even if a 1:1 ratio is ancestral.

3. DISPLACEMENT IN FAVOR OF CAUSAL FACTORS

Salmon accepted, at one point, that an explanation of the pressure of a gas in terms of classical themodynamics, rather than statistical mechanics, does not fit the causal/mechanical model of explanation (1984). Neither do equilibrium explanations such as Fisher's or explanations of populational changes in terms of drift. On a causal/mechanical view, a satisfactory explanation requires some appeal to the underlying mechanisms[6]. These explanations do not depend on appeal to these mechanisms, and as a result these explanations might simply be rejected. This is not an eccentric position, though it is not one I think is defensible.

Alexander Rosenberg, for example, takes up the view that evolutionary processes are stochastic, urging that, despite appearances, evolutionary phenomena are deterministic, and that drift does not make "the theory of natural selection" probabilistic (1988, 198). Rosenberg seems convinced that even small measurement error will compromise predictability, and that measures of gene frequencies are subject to this sort of error. Apparently, the point he has in mind is that, for example, in attempting to assess the frequencies of various genes in a population, we inevitably resort to sampling; and this is an error prone procedure. This sort of issue is important methodologically, and there certainly is uncertainty in such sampling; but is not a source of anything but epistemic uncertainties. Rosenberg goes on to suggest that even drift is fundamentally an epistemic matter, contingent on our lack of knowledge of initial conditions. He poses this question:

> "Could drift actually be a way of referring to those unknown non-evolutionary force that interfere and deflect evolution from the outcomes which deterministic forces like selection, mutation, and migration, would otherwise secure?" (1988, 195).

Rosenberg imagines a case in which there is a small population of giraffes, which is affected by poachers with a preference for longer necked individuals. As a result of their activity, the population is shifted toward shorter-necked variants. In the absence of knowledge of the poacher's activities, a biologist would be inclined to think that any shift was a matter of drift rather than selection. He says that if he can generalize from the example, the conclusion is that "from a position of omniscience, there is no

---

6. Brandon requires that what he calls a "complete adaptation explanation" include not only evidence of selection, but an ecological grounding for differences in adaptedness (1990, ch. 5, esp., 165). Brandon recognizes that most evolutionary explanations fall far short of the ideal of completeness he defends.

need for the notion of drift; that evolution simply moves faster among small populations, when their gene frequencies change at all; and that the phenomena and the theory of natural selection are thorough-goingly deterministic ones" (1984, 197). The only significance for drift becomes that of a "place holder" for a variety of "unknown evolutionary forces", and does not support the conclusion that evolutionary processes are indeterministic.

It is certainly true that there are many cases in which the evidence leaves us unable to determine whether some evolutionary change is due to selection, or to drift. However, Rosenberg is simply wrong to think that this warrants the conclusion that an observed change is due to drift, simply because there are no known selective agents. Epistemic probabilities do come into play in estimating objective probabilities. That does not make the objective probabilities less objective. It only makes them less certain. An assessment of the significance of drift or of selection, specifically, depends on estimates of variation and frequency. Frequency distributions of amino acid variants, as revealed in gel electrophoresis, were used for some time in attempting to distinguish the forces controlling genetic variation, though it is now generally accepted that they do not provide sufficient resolution (Lewontin 1974). The use of DNA sequency, or screening for nucleotide variation, does allow us to distinguish between the effects of drift and selection (cf. Kreitman 1983; Riley, Hallas and Lewontin 1989). Given an estimate of population size, it is possible to determine how much change — that is, how many gene substitutions — could result from drift alone in the absence of selection. Population size is subject to the problems of estimation, as are estimates of genetic variation, but those are not the focus of the analysis. Given an estimate of population size, and an estimate of the degree of variation between populations, it is possible to ask how likely some observed change is to be due to drift (cf. Lande and Arnold 1983a and 1983b). With a small population, any change is possible. Any could, theoretically, be due to drift, though some are so unlikely that they can be disregarded. With a larger population, drift becomes less significant. However population size and variation are assessed, though, showing that selection is responsible for some difference between populations depends on knowing the likely effects of drift. Even in the absence of any known selective agents, we can, in some cases, determine whether a change could be due to drift, or whether there must be selection acting on a trait (see Kane and Richardson 1990; and for a definitive treatment, Culver, Kane and Fong 1995). This does not require knowing the selective agents[7]. Drift becomes a

7. This does not imply that a complete adaptation explanation in Brandon's sense does not include an ecological explanation of adaptive advantage; the point is merely that it is possible to know that there is selection operating without knowing the cause of the selective advantage.

null hypothesis against which selection must be tested and discerned. Rosenberg's treatment of the role of drift has the relationship of drift and selection inverted.

Even with this corrected, the prospect still remains that drift is only "a way of referring to congeries of ... non-evolutionary forces, ones that are responsible for changes in gene frequencies, but not for their *evolution*" (Rosenberg 1988, 199).Barbara Horan (1994) adopts a similar line, claiming that, however useful statistical models of evolution might be, they are "theoretically unnecessary, given the underlying deterministic character of evolutionary processes" (1994, 79). The interesting question, she thinks, is not whether there are statistical generalizations employed (because these can be useful with deterministic and indeterministic systems), but whether the fundamental principles are irreducibly probabilistic[8]. She motivates the general point by appeal to statistical mechanics, saying this:

> "The molecules of an ideal gas and particles in Brownian motion are two examples. Their behavior is governed by deterministic laws. In these cases, as is well known, a Laplacian supercalculator in possession of a complete state description and knowledge of the relevant laws could predict and explain the behavior of the ensemble in terms of the physically nonrandom behavior of its individual members" (1994, 81).

The conclusion is supposedly that statistical properties are "theoretically unnecessary". Genetic drift is given a parallel treatment. She recognizes that drift is described as being chance fluctuations in gene frequencies. The idea that the fluctuations are genuinely random, she says, is a mistake. Many processes that affect populations are indiscriminate, in the sense that the effect on individuals, on genotypes, or on genes, is uncorrelated with any features which would affect the level of adaptedness. Nonetheless, she claims, the underlying *process* may be quite deterministic[9]. Horan concludes that evolutionary theory is deterministic (1994, 93-94). The most that follows is that evolutionary processes are, at some level of description, deterministic. I do not know whether this is true. Even if it were true, it would not be enough to yield the conclusion that evolutionary theory is deterministic.

---

8. Robert Brandon and Scott Carson (in preparation) provide an aggressive and interesting defense of the view that evolutionary processes are fundamentally probabilistic, and that appeals to "deterministic hidden variables" to explain evolutionary phenomena is counterproductive.

9. This view is reinforced by the way the propensity interpretation is often defended. Mills and Beatty (1979 suggest that we should think of a case in which there are two dogs on an island, and one is hit by lightning. This would hardly give us reason to think the survivor was more fit. Lightning strikes might appear to be random, but be quite deterministic at root.

Let's look at this argument more carefully. It is true that the thermodynamics of the 19th century was deeply deterministic, embedded in a Newtonian framework. It is also true that the idealized molecules of the gas laws were understood as Newtonian particles: elastic, solid, and particulate. So when we compare the explanation of, say, some change in pressure using Boyle's law, with an explanation couched in statistical thermodynamics, it might be tempting to think that the latter is a better mechanical explanation and that the former is somehow incomplete. This is an illusion, as James Woodward (1989) has shown us. Though gases are conceived within statistical thermodynamics as aggregates of molecules, the prospect of tracking the trajectories and interactions of each molecule — or of any molecule — is clearly unreal. Appealing to the underlying molecules and their collisions in explaining, say, a change in pressure is "a trivial, non serious explanation of the behaviour of the gas" (Woodward 1989, 363). It also is not what is done in statistical thermodynamics. Instead of appealing to the mechanical processes which underlie the behavior of gases, we abstract from such causal processes and focus on aggregate behavior. The explanation is irreducibly statistical, and shows that the apparently deterministic laws at the macroscopic level are probabilistic at a deeper level. Holding out for a more complete, deterministic, causal/mechanical explanation is holding out for something we do not have and should not expect. These are akin to equilibrium explanations, insofar as the explanation offered abstracts from the underlying cause, making it "explanatorily relevant". Moreover, even if we were in a position to produce a serious explanation at the level of the underlying mechanical processes for some specific change, we should be unsatisfied with it. We might be able to explain some transaction, but we would miss the pattern into which the pieces fit. Ultimately, the pattern is more important here in explanation than the pieces. The pattern needs to be understood probabilistically. Thus, there is no causal /mechanical explanation to supplant the statistical one, and if there were, it would explain particular effects but neglect the phenomena which the statistical model explains.

This is even clearer in the biological cases. Once we shift to a focus on finite populations, all evolutionary explanations become probabilistic. Drift provides a measure of the variance around the expected value, as a function of population size. Again we are presented with statistical patterns, without depending on the specific causes of evolutionary change. It might be tempting to think that we could do better by looking to the specific causes;

but that would be to fall into the same mistake. Though there might be a constellation of causes for any evolutionary change, in light of the complexity of ecological interactions, actually tracking or understanding those effects will be possible only in the simplest cases. The appeal to ecological factors is, often, not a serious explanation; or, more carefully, it is not a serious explanation in the absence of the more general statistical case, which it depends on. Again, actual explanations do not follow the recommendations a causal model would indicate, but abstract from these processes and focus on aggregate behavior. And again, the explanations are irreducibly probabilistic. Were an explanation in terms of causal features actually available, again we should be dissatisfied with it. As in the case of statistical mechanics, we might be able to deal with any particular case but we would miss the pattern. As Fisher wrote:

> "The investigation of natural selection may be compared to the analytic treatment of the Theory of Gases, in which it is possible to make the most varied assumptions as to the accidental circumstances, and even the essential nature of the individual molecules, and yet to develop the general laws as to the behavior of gases, leaving but a few fundamental constants to be determined by experiment" (1922, 321-322).

What an analysis in terms of mechanisms or causes is bound to miss are the patterns. Missing the patterns, we miss the explanation. It may be that no event will be unpredictable by a Laplacean demon, but one restricted to the movement of atoms in the void will be oblivious to the patterns which animate the world, and which define what counts as being of the same kind. Thus, though two populations might evolve in different directions, without the influence of selection, and though two other populations mightevolve in the same direction, differing nonetheless in the particulars, knowing the particulars will not displace an understanding of the patterns. In many ways, this is the heart of the appeal to unification.

4. SELF-ORGANIZATION AND THE ORIGINS OF ORDER

Stuart Kauffman says that what we need "is a new kind of statistical mechanics, one which analyzes the properties of complex systems with very many coupled elements. By understanding the characteristic structure and behaviors of the members of such ensembles, we may be able to understand both the emergence of order in organisms and its adaptive evolution" (1993, 182). Kauffman offers a formal framework which allows him to pose problems about the constraints that self-organization imposes on the evolution of complex systems, and the relation of self-organization and selection. The central theme running through *The Origin of Order* is that "the order in organisms may largely reflect spontaneous order in complex systems" (1993, 173). The idea is the elegant one that in systems which are composed of an array of simple elements, with modest numbers of connections between them, over time the system will spontaneously assume a regularly ordered form. The order in these systems is often described as "emergent"[10]. This includes a wide variety of what are otherwise apparently very different systems: genetic regulatory networks, protein metabolism, neural networks, ecological systems, and economic systems[11].

Let's look more carefully at this "statistical mechanics" for Self-Organizing systems. The statistical mechanics of the 19th century focused on systems which were inherently *dis*organized. In the case of (perfect) gases, we are confronted with systems of particles, each with six variables describing position and momentum. Since the system is disorganized — that is, a ideal gas is a simple aggregate of its component particles — a system at one point in the phase space simply wanders over the entire phase space. Statistically, we can provide a probability that the system will occupy some volume of the phase space which is described by the macroscopic data. We describe the macroscopic phenomena as statistical averages of unknown microscopic aggregates. In more organized systems, the situation is dramatically different. Biological systems are thermodynamically open systems, and thus subject to different principles (cf. Prigogine 1984; Wiley and Brooks 1988); moreover, their organization insures that not every region of the phase space is equally open to exploration. Kauffman attacks the problems these more highly organized systems present by turning to Boolean networks, systems of binary variables coupled according to switching functions of arbitrary complexity.

The strategy is important because it displays one important sense in which what he provides is a "statistical mechanics" of organized systems. Boolean networks offer large numbers of coupled elements which Kauffman says provide analytically tractable and reasonable approximations to real

10. Appeals to emergence often have a nearly mystical character, offering nothing of substance. For discussion of the concept of emergence as it applies here, see Bechtel and Richardson (1992), and Bechtel and Richardson (1993), chapter 9.

11. This and an array of related issues are ably discussed in David Depew and Bruce Weber in Darwinism Evolving (1995). Depew and Weber are sympathetic with Kauffman's approach, and emphasize Kauffman's view that natural selection maintains systems at the edge of chaos. This is an important aspect of Kauffman's views which I do not emphasize here. See Burian and Richardson 1992 for a perspective that differs from that of Depew and Weber, and their discussion in chapter 16, especially 454 ff.

12. This may turn out not to be an innocent idealization. Boolean networks are discontinuous, and with continuous activation functions some attractors may turn out to be unstable.

systems[12]. They are governed by three fundamental parameters: the number of nodes (N), the number of connections per node (K), and the number of Boolean functions (P) governing the nodes. The state of any Boolean node is either 1 (on) or 0 (off), depending on the states of the input nodes and the Boolean switching function. Figure 4 depicts a very simple Boolean network with three units, leaving eight possible states and sixteen possible Boolean functions for the system.

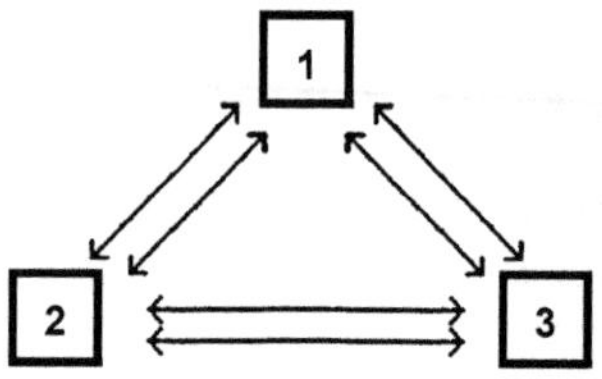

1	2	3
+	+	+
+	-	+
-	+	+
-	-	-

2	3	1
+	+	+
+	-	-
-	+	-
-	-	-

1	3	2
+	+	+
+	-	+
-	+	+
-	-	-

Figure 4. A simple Boolean network with three modes. Boolean networks are sets of two valued switches, which are either on (+) or off (-) as a function of the inputs to the system. In this case, each node takes inputs from two other nodes (K=2), there are eight possible states, and sixteen possible Boolean functions. Unit 1 in this case is an AND gate, assuming a + value if both units 2 and 3 are + in the previous cycle. Units 2 and 3 are OR gates, assuming a + value if any other unit assumed a + value in the previous cycle, otherwise assuming a - value.

Unit 1 is an *and* unit, and units 2 and 3 are both *or* units. Kauffman sets about to find what he calls the "typical" behavior of Boolean networks given these parameters. He considers networks with varying values for K, varying from $K = N$ (so that every element is connected to every other element) to $K = 1$ (so that every element modulates only one other). With N nodes, there are 2^N possible configurations of nodes. A unit can be either on or off in response to any of these configurations, and as a result for K inputs there will be 2^{2^*K} possible Boolean functions. For each K and N, Kauffman assigns both connections and Boolean functions at random across the N units. He explains, "we generate networks with random wiring diagrames and random logic, and ask whether orderly behavior emerges nonetheless" (1993, 192). In general, what Kauffman found is that with a K value near, the system is maximally disordered and chaotic — that is, highly sensitive to initial conditions —. As one would expect on probabilistic grounds alone, cycles lengths are on average $\sqrt{2}^{N/2}$, where there are 2^N possible states. (This is simply the number of steps at which the probability of finding one of the possible states by random search is 0.5.) With large value of N, this is clearly an enormous number. Moreover, since the average cycle length is rather long, the number of state cycles which will form is small. Moreover, change is chaotic in the sense that the state of the system changes unpredictably with small changes in components: the "landscape", Kauffman says, is uncorrelated. Even with values of K much below N, Kauffman tells us that these general features are retained. As K decreases, though, the spontaneous order is more marked. When $K = 2$, so that a unit receives input from only two other units, Kauffman tells us that there is a large amount of spontaneous order. The average cycle length is $\sqrt{N}$, which means that a Boolean network with 10 000 elements would, on average, limit itself to 100 states; accordingly, the number of attractors is also $\sqrt{N}$. The explanation for this kind of spontaneous order, Kauffman tells us, is simple. Some Boolean functions are canalyzing functions: they are such that a particular input from a unit necessitates some state of the regulated unit, independently of the state of other input units. An and unit is a canalyzing function in this sense, since if one input unit is off, the regulated unit will be off; similarly, or units are canalyzing, since if one input unit is on the regulated unit will be on. This has the effect, with reasonably short state cycles, of "freezing" some units on and some off. The result is that the only units changing are functionally isolated from other. Attractors are small. Change is localized, and as result the state of the system changes minimally with changes in components : the "landscape" is highly correlated. Kauffman says "Random NK Boolean networks with $K = 2$ inputs to each of 100 000 binary elements yield systems which typically localize behavior to attractors with about 317 states among the $2^{100\,000}$ possible alternative states of activity. ... note and remember that *our intuitions about the requirements for order in very complex systems have been wrong*" (1993, 235).

Notice that Kauffman's strategy is one that samples statistically from the range of possible networks with given parameter values for *N*, *K*, and *P*. He does not attempt to explain the behavior of any of these randomly generated networks, and there would be little interest in doing so. He asks, instead, what the expected behavior of such complex systems would be, apart from any preferred order. This is a statistical question. Again, the prospect for an explanation based on the wiring diagrams of these Boolean networks is not a serious one, any more than we might explain human behavior from a wiring diagram of the brain if we had one. This is no objection to Kauffman's approach. It is important to know whether the patterns he has discerned apply equally well to real systems; and to know this we would need to ask the question whether the parameter values he uses are realistic. This will bring us closer to a causal/mechanical explanation, but the actual explanations offered still will be irreducibly statistical, and will not cite causes. If explanation consists in coming to understand the common patterns in the phenomena, then these are good explanations. Notice that we could not attempt to describe in detail the behavior of any of these randomly constructed Boolean networks, and our explanation does not at all depend on being able to do so. This does not mean that the behavior of such systems is necessarily well understood. For example, it would be possible, with higher values of *K*, for spontaneous order to emerge, with a restriction on the number of Boolean functions which are used. Again we would end up with forcing structures and frozen components. Determining what count as reasonable parameter values is an important project. This does not carry us, though, to the project of understanding the detailed organization of individual systems. We still remain at the more abstract level, focused on the statistical behavior and how it varies with changes in parameter values.

The generic properties of random *NK* systems thus do provide a kind of "statistical mechanics" for complex systems (Figure 5 provides an overview). They are properties which would be expected independently of selection, as a consequence of organization alone. If Kauffman is right, they provide robust properties across a variety of systems — from immune systems and genetic regulatory circuits to neural networks and economies — that are statistically typical, but not universal. In any given case, the task of detailing the mechanisms might elude us, but the overall pattern gets an explanation event if particular cases do not. Kauffman highlights several such general results.

Control Parameters:

N: the number of nodes; *K:* the number of connections per node; *P:* the number of Boolean functions

Varying values for *K*, $0 < K < N$

2^N possible configurations
2^{2*N} possible Boolean functions

$K = N\text{-}1$		$K = 1$
$\frac{\sqrt{2^N}}{2}$	mean cycle lenght	$\sqrt{N}$
$\frac{N}{e}$	number of attractors	$\sqrt{N}$
uncorrelated disordered chaotic	landscape	correlated ordered stable

Figure 5. Complex systems and their dynamics.
With Boolean functions, *NK* random networks are governed by three control parameters. Kauffman systematically varies *K* from 1 to *N*-1, and finds there are systematic dynamic differences. The characteristics of networks with *K* = *N*-1 and *K* = 1 are as indicated. The critical values lie at around *K* = 2, where systems are neither chaotic nor frozen, and intermediate in the amount of order.

He says, at one point:

> "Boolean networks, among the most general class of massively parallel-processing systems, exhibit three broad regimes of behavior. Systems may be in the ordered regime with frozen components, in the chaotic regime with no frozen components, or in the boundary region between order and chaos where frozen components just melt. ... The central idea is that, if a network is deep in the frozen phase, then little computation can occur within it. At best, each small unfrozen, isolated island engages in its own internal dynamics functionally uncoupled from the rest of the system by the frozen component. In the chaotic phase, dynamics [are] too disordered to be useful. Small changes at any point propagate damage to most other elements in the system. Coordination of ordered change is excessively difficult. At the boundary between order and chaos, the frozen regime is melting and the functionally isolated unfrozen islands are in tenuous shifting contact with one another. It seems plausible that the most complex, most integrated, and most evolvable behavior might occur in this boundary region. It is not yet unambiguously clear that this hypotheses is correct" (199, 219).

The idea that there is adaptation to the edge of chaos is central to Kauffman's vision. It is at the edge of chaos that evolvability is at its greatest ; and the role of natural selection is to favor parameter values which produce systems in this region[13].

It is natural to wonder whether Kauffman's simulations offer adequate explanations of the biological phenomena we see, in the absence of the contingent and variable causal factors which the history provides (cf. Burian and Richardson 1991). Gabriel Dover put the point briskly:

> "The central issue is at what point do Kauffman's statistical structures bear upon evolving, historically processed genomes and ontogenies as we know and love them? There are times when the bracing walk through hyperspace seems unfazed by the nagging demand of reality" (1993, 705).

The point can be easily illustrated in terms one of Kauffman's central results. Consider, in particular, the proposition that it is at the boundary between order and chaos at which evolvability is maximized. The highly ordered regime is one in which perturbations have unpredictable effects: there is no correlation between the initial and perturbed states. There is no heritability. The "edge of chaos" is simply the region in which there is heritable variation. There are reasons to doubt how significant this result is. It is true that heritable variation is necessary for evolution. It is then true that systems at the "edge of chaos" would be "evolvable", because it is only here that there would be heritable variation. This is a consequence which is well known among evolutionary biologists: evolution requires heritable variation. However, the fact that the boundary at the "edge of chaos" gives us heritable variation does not guarantee that Kauffman's *NK* Networks tell us how heritable variation is realized in actual, historically evolved, biological systems[14]. One tempted to reject Kauffman's explanation on these grounds, as "unfazed by the nagging demands of reality", is drawn by the attraction of causal explanations, responsive to the details of history. Kauffman's "statistical mechanics for complex systems" is inspired by an alternative vision in which the goal is to find the abstract patterns independently of the details of history. It fits poorly with causal/mechanical models of explanation, and much more comfortably with an emphasis on explanatory unification.

---

13. The expression is evidently due to Christopher Langton. The frozen regime are his Class I and II rules, and the chaotic regime involve his Class iii rules. The regime which provides for interesting "order" are the intermediate Class IV rules. See Langton (1989) and Waldrop (1992) for an accessible discussion.

14. This point was forcefully made to me in discussion by Henri Atlan. Atlan apparently takes this to undermine the credibility of Kauffman's explanations.

REFERENCES

Bechtel W. and Richardson R.C. (1992). Emergent Phenomena and Complex Systems. In A. Beckermann, H. Flohr and J. Kim (eds), *Emergence or Reduction?* Berlin: Walter de Gruyter, 257-288.

Bechtel W. and Richardons R.C. (1993). *Discovering Complexity.* Princeton: Princeton University Press.

Brandon R. (1990). *Adaptation and Environment.* Princeton: Princeton University Press.

Brandon R. (1996). Testing Adaptationism: A Comment on Orzack and Sober. *The American Naturalist,* **148**, 189-201.

Brandon R. and Carson S. The Indeterministic Character of Evolutionary Theory: No 'No Hidden Variables Proff but No Room for Determinism Either. *Philosophy of Science,* **63**, 315-337.

Burian R. and Richardson R.C. (1991). Form and Order in Evolutionary Biology. In A. Fine, M. Forbes and L. Wessels (eds). *PSA 1990* (Philosophy of Science Association) Vol. 2, 267-87.

Cartwright N. (1989). Capacities and Abstractions. In P. Kitcher and W.C. Salmon (eds). *Scientific Explanation.* Minneapolis: University of Minnesota Press, 349-356.

Culver D., Kane T.C. and Fong D.W. (1995*). Adaptation and Natural Selection in Caves.* Cambridge: Harvard University Press.

Darwin C. (1871). *The Descent of Man, and Selection in Relation to Sex.* Reprinted with an introduction by J.T. Bonner and R.M. May. Princeton: Princeton University Press, 1981.

Depew D.J. and Weber B.H. (1995). *Darwinism Evolving.* Cambridge: Bradford Books/MIT Press.

Dover G. (1993). On the Edge. *Nature,* **365**, 704-706.

Falconer D.S. (1989). *Introduction to Quantitative Genetics.* Third Edition, Essex: Longman.

Fetzer J. (1981). *Scientific Knowledge.* Dordrecht: D. Reidel.

Fisher R.A. (1930). *The Genetical Theory of Natural Selection.* Oxford: Oxford University Press.

Friedman M. (1974). Explanation and Scientific Understanding. *Journal of Philosophy,* **71**, 5-19.

Glymour C. (1982). Causal Inference and Causal Explanation. In R. McLaughlin (ed.). *What? Where? When? Why?* Dordrecht: D. Reidel.

Hitchcok C.R. (1995). Discussion: Salmon on Explanatory Relevance. *Philosophy of Science,* **62**, 302-320.

Horan B.L. (1994). The Statistical Character of Evolutionary Theory. *Philosophy of Science,* **61**, 76-95.

Jones T. (1995). Reductionism and the Unification Theory of Explanation. *Philosophy of Science,* **62**, 21-30.

Kane T.C. and Richardson R.C. (1990). The Phenotype as the Level of Selection: Cave Organisms as Model Systems. In A. Fine, M. Forbes and L. Wessels (eds). *PSA 1990,* Philosophy of Science Association, Vol. 1, 151-164.

Kauffman S. (1993). *The Origins of Order.* Oxford: Oxford University Press.

Kitcher P. (1976). Explanation, Conjunction and Unification. *Journal of Philosophy,* **73**, 207-212.

Kitcher P. (1981). Explanatory Unification. *Philosophy of Science,* **48**, 507-531.

Kitcher P. (1989). Explanatory Unification and the Causal Structure of the World. In P. Kitcher and W.C. Salmon (eds). *Scientific Explanation,* Minneapolis: University of Minnesota Press, 410-506.

Kitcher P. (1993). *The Advancement of Science.* Oxford: Oxford University Press.

Kreitman M. (1983). Nucleotide Polymorphism at the Alcohol Dehydrogenase Locus of Drosophila melanogaster. *Nature,* **304**, 412-417.

Lande R. (1976). Natural Selection and Random Genetic Drift in Phenotypic Evolution. *Evolution,* **30**, 314-334.

Lande R. and Arnold S. (1983a). The Measurement of Selection on Correlated Characters. *Evolution,* **37**, 1210-1226.

Lande R. and Arnold S.J. (1983b). The Measurement of Selection on Correlated Characters. *Evolution,* **39**, 502-522.

Langton C.G. (ed.) (1989). *Artificial Life: Santa Fe Institute Studies in the Sciences of Complexity 6.* Redwood City CA: Addison-Wesley.

Lewontin R.C. (1974). *The Genetic Basis of Evolutionary Change.* New York: Columbia University Press.

Maynard Smith J. (1975). *The Theory of Evolution.* London: Penguin.

Maynard Smith J. (1976). Evolution and the Theory of Games. *American Scientist,* **64**, 41-45.

Maynard Smith J. (1982). *Evolution and the Theory of Games.* Cambridge: Cambridge University Press.

Mills S. and Beatty J. (1979). The Propensity Interpration of Fitness. *Philosophy of Science,* **46**, 263-288.

Mitton J.B. (1994). Molecular Approaches to Population Biology. *Annual Review of Ecology and Systematics,* **25**, 45-70.
Railton P. (1978). A Deductive-Nomological Model of Probabilistic Explanation. *Philosophy of Science,* **45**, 206-226.
Railton P. (1989). Explanation and Metaphysical Controversy. In P. Kitcher and W.W. Salmon (eds). *Scientific Explanation.* Minneapolis: University of Minnesota Press, 220-252.
Richardson R.C. (1984). Biology and Ideology: The Interpenetration of Science and Values. *Philosophy of Science,* **51**, 396-420.
Richardson R.C. and Burian R.M. (1992). A Defense of Propensity Interpretation of Fitness. In A. Fine, M. Forbes and K. Okruhlik (eds). *PSA 1992,* Philosophy of Science Association, Vol. 1, 349-362.
Riley M.A., Hallas M.E. and Lewontin R.C. (1989). Distinguishing the Forces Controlling Genetic Variation at the Xdh Locus in Drosophila pseudoobscura. *Genetics,* **123**, 359-369.
Rosenberg A. (1988). Is the Theory of Natural Selection a Statistical Theory? *Canadian Journal of Philosophy,* **14**, 187-207.
Rosenzweig M.L. (1966). Community Structure in Sympatric Carnivora. *Journal of Mammology,* **47**, 602-612.
Roughgarden J. (1979). *Theory of Population Genetics and Evolutionary Ecology: An Introduction.* New York: MacMillan.
Salmon W.C. (1971). *Statistical Explanation and Statistical Relevance.* Pittsburgh: University of Pittsburgh Press.
Salmon W.C. (1984). *Scientific Explanation and the Causal Structure of the World.* Princeton: Princeton University Press.
Salmon W.C. (1989). Four Decades of Scientific Explanation. In P. Kitcher and W.C. Salmon (eds). *Scientific Explanation.* Minneapolis: University of Minnesota Press, 3-252.
Sober E. (1984). *The Nature of Selection.* Cambridge: Bradford Books/MIT Press.
Waldrop M.M. (1992). *Complexity: The Emerging Science at the Edge of Order and Chaos.* New York: Simon and Schuster.
Woodward J. (1989). The Causal Mechanical Model of Explanation. In P. Kitcher and W.C. Salmon (eds). *Scientific Explanation.* Minneapolis: University of Minnesota Press, 357-383.
Wright S. (1931). Evolution in Mendelian Populations, *Genetics,* **16**, 97-159.
Wright S. (1932). The Roles of Mutation, Inbreeding, Crossbreeding, and Selection. *Proceedings of the Sixth International Congress of Genetics,* **1**, 356-366.

BERNARD FELTZ

SELF-ORGANIZATION, SELECTION AND EMERGENCE IN THE THEORIES OF EVOLUTION

1. INTRODUCTION

Since Darwin, the hypothesis of natural selection has gained such a considerable weight in the explanation of biological evolution that, in the context of the synthetic theory, the pressure of the selection is now seen as the single source of order in the evolution of the living. However this thesis has been undermined by recent research on self-organization. In particular, S. Kauffman, in several books, tries to demonstrate that an organization can arise independently from any pressure of selection, and pleads for a combination of self-organization and selection inside of one theory explaining biological evolution. I would like to try to clarify this debate by presenting a summary and an epistemological analysis of the ideas of S. Kauffman. Therefore I will first refer to the research work on the explanation at stake in the synthetic theory, in order to investigate the compatibility of selection and self-organization regarding the synthetic theory. Then, I will refer to other areas of biological sciences using the selection-oriented explanation: immunology and neuroscience. Indeed in each of these fields, recent theories make use of a combination of the principle of selection together with a preexisting organization. An analysis of the possible combinations between selection and self-organization will be most helpful.

The concept of emergence is present all through this book. Self-organization is often associated with emergence, which classically means the appearance of a level of complexity more advanced than the existing components of a system. Furthermore current epistemological research on emergence opposes emergence and reduction; in this context some studies indicate that the concept of selection can play a central role in favor of the non-reductionist assertion of the autonomy of the biological sciences, in

Bernard Feltz, Marc Crommelinck and Philippe Goujon, Self-Organization and Emergence in Life Sciences: 341-360.

relation with physical and chemical fields. With the concepts of self-organization and selection associated to the logic of explanation, we now have the main elements to clarify the concept of emergence in connection with theories related to self-organization. This will be the object of the last section of this contribution.

2. S. KAUFFMAN AND THE RESEARCH ON THE LAWS OF COMPLEXITY

2.1 Boolean Networks and Laws of Complexity

S. Kauffman refers to the theoretical analysis of the behavior of networks of automatons to tentatively describe what he calls the "laws of complexity". He compares then these behaviors with several biological processes, in order to support the idea that the organization of the living is not only the result of natural selection but is also produced by some complex networks spontaneously aiming to order. I would like first to make a general presentation of the problem, with the practical example of the analysis of the genome. Afterwards I will deal with the compatibility of this view with the hypothesis of the natural selection.

A network is defined by its elements and the connections between these ones. In a Boolean network, each element is a Boolean function which associates a binary response — 1 or 0 — to all inputs reaching this element at time t. If one element receives K inputs, there are $2K$ possible combinations of the inputs he can receive and the Boolean function associates exactly one response "1 or 0" to each combination. The calculation of the state of the network at time $t+1$ implies consequently, for each component, the integration of several inputs at time t depending on the Boolean operator in place.

Generally such a network does not lead to a stable situation where each component would have a fixed value, but well to a cycle of states, in which the network goes through a repetitive sequence of different combinations of the values 1 and 0, the same sequence of combinations being endlessly repeated. The length and the quantity of possible cycles are important characteristics of networks behavior. These can be studied generally in function of three network characteristics: the number N of components, the number K of connections affecting each component and the Boolean function characterizing each element. Such networks can present three forms: chaotic, ordered and complex.

In the chaotic form, active subnetworks populate the entire network and isolate some "frozen" areas, where the automatons keep a fixed value: 0 or 1. This means that any minor local modification is spread to the rest of the network. In other words, such networks are hypersensitive to initial conditions. Furthermore the length of state cycle attractor increases exponentially with the number of automatons. Such a form can be observed in highly connected networks.

For example, if $K = N$, which means that all components are connected with each others, the length of the state cycle is the square root of the number of states. For $N = 200$, with 2 possible states for each automaton — 0 or 1 —, the network presents 2^{200} or 10^{60} different possible states. Consequently the length of the state cycle to which the network tends is 10^{30}, which is basically unobservable. Kauffman has indeed calculated that, if it takes a millionth of a second to the network to switch from state to state, 10^{30} millionths of a second correspond to billions of times the 15 billions of years of age of our universe. We have therefore a network of which the stabilized state cycle is materially unpredictable[1].

However, even in this context, there are indications tending to show a reduction of the disorder. Indeed for the networks where $K = N$, the number of attractors is equal to N/e, where e is the logarithmic constant, equal to 2,71828. The number of attractor pools is thus very small compared with the length of the state cycle.

The chaotic form can be observed in highly connected networks, but also in the networks where K is much smaller, in particular where $K = 4$ or $K = 5$. In the ordered form, starting from $K = 2$, the networks have a completely different behavior: both the number and the length of the cycles of states become the square root of the number of components. If $N = 100\ 000$, for instance, the network having $2^{100\ 000}$ different states, the average length of the cycles of states is 317. So an order appears spontaneously since, if it takes one millionth of a second to the network to switch from state to state, the cycles of states are fully processed in 317 millionths of a second. The difference with the networks where $K = N$ is significant, and it is on this particular property of $K = 2$ networks that Kauffman has focused his work.

---

1. Kauffman 1995, 82. The chaos according to Kauffman is slightly different from the concept of determinist chaos. Indeed, in the determinist chaos, the hypersensitivity to initial conditions makes the evolution of a determinist system unpredictable. In the networks of Kauffman, for a large K, this hypersensitivity to initial conditions is also a characteristic of the networks of which the unpredictability is due as well to the size of the state cycle which materially cannot be processed until it becomes stable. For more details, cfr Kauffman 1993, 471.

The ordered form is characterized by the fact that "frozen" subnetworks, the elements of which keep a constant value — 0 or 1 —, isolate areas where the automatons have varying values, with complex evolution, but do not interfere with the other subareas of the network. The modification of an automaton does not lead to a chain reaction through the "frozen" subnetworks, but keeps confined to a particular activity island. In other words, an alteration of the network structure does not imply a fundamental change of behavior.

Finally, the complex form refers to the analysis of the behaviors of the networks where $K > 2$, which shows that, in certain conditions, one may observe the transition between the chaotic and non chaotic networks. Kauffman refers to "systems at the edge of chaos" to characterize the transition between the chaotic and the ordered form.

2.2 Boolean Networks, Ontogenesis and Theory of Evolution

One attractive way of interpreting these networks of Boolean automatons is to consider the genome of a living cell as a network of automatons where a gene corresponds to a Boolean automaton and where each state cycle attractor corresponds to a cellular type in the given organism.

There are several valid reasons to justify such an interpretation. First of all, from a genetic point of view, most of the genes seem to be regulated by a small number of molecular inputs. Kauffman refers to the works of Monod, Jacob and Lwof on the regulation of the lactose operator in E. coli which show that this regulation is bound to two inputs: the allolactose and the repressive protein. Furthermore, the regulation modes of the genes are diverse and the Boolean functions are able to reflect these various modes in a logical way. For example, the Boolean operators NOT IF, OR and AND allow to express many instances of genes regulation. Finally, in the hypothesis of the representation of the genes as Boolean functions where $K = 2$, these functions are active or inactive as regards the transcription, and on the other hand they are only controlled by two inputs. Such a model brings into play a set of sixteen Boolean functions, fourteen of which are "canalyzing Boolean functions", *i.e.* they show characteristics of stabilization of the network, which duly correspond to the observations made concerning the regulation of the genome[2].

2. Kauffman 1995, 105.

There are thus some valid logical reasons to state the interpretative hypothesis of the genome as a network of Boolean automatons with $K = 2$. However, the strength of that hypothesis mainly comes from the conclusions that may be drawn. Kauffman develops four characteristics of the ontogenesis which are related to that hypothesis.

First of all, each cellular type is confined to an infinitesimal fraction of the possible patterns of the genetic activity, which is precisely described in the reduced dimension of the state cycle attractors in the model of automatons networks. Considering the example of the human race, the human genome includes a figure of about 100 000 genes - as it was estimated at the time of Kauffman's publication. According to the adopted formalism, the state cycle attractor include an average of 317 genes, which means that each cellular type of the human race only brings into play a tiny part of the human genome, which is reflected at the experimental level indeed.

Then, if each cellular type corresponds to the implementation of about 317 genes, the time taken by the differentiation process must correspond to such an implementation; the time of a state cycle in the Boolean formalism must correspond to the time of a cellular cycle in experimental biology. For instance, if the expression of a gene of the human being takes about one to ten minutes, a cellular cycle should take about 317 to 3 170 minutes, or 50 hours, which corresponds to the margin actually observed. Moreover, according to this hypothesis, the average time of a cellular cycle should be a function of the square root of the number of genes. Kauffman deals with it by referring to various organisms throughout the phylogenesis and shows clearly how this data correspond to the relation in question[3].

Third consequence, such a model allows one to predict the number of cellular types according to the number of genes of the organism. When $K = 2$, the average number of attractive state cycles is about the square root of the number of genes. As regards the human being whose total number of genes is about 100 000, we must thus expect to observe a figure in the region of 317 different cellular types. But at present we only know 256 human cellular types. In this case as well, there is a relation to be set between the number of cellular types expected according to the Boolean model and the number of cellular types observed within various species in the course of evolution. A parallel evolution of the number of attractors and the number of cellular types according to the quantity of ADN may be thus highlighted[4].

3. Kauffman 1993, Figure 12.11, 485; Kauffman 1995, Figure 5.5, 108.
4. Kauffman 1991, 84; Kauffman 1993, Figure 12.7, 462; Kauffman 1995, Figure 5.6, 109.

Fourth characteristic of the ontogenesis conveyed by the model: the homeostasis and the branching pathway of differentiation. According to the Boolean model indeed, the modification of one element does not lead to deep changes in the network state. On the contrary, slight modifications generally lead back to the preceding state. The concept of homeostasis precisely describes such a behavior: a tiny modification brings a return to the previous situation. However there are exceptions. Indeed, when the network is on the limit between two attractive pools, a slight modification may drag the network along towards a neighboring attractive pool: such a branching off towards another cellular type stands for the phenomenon of cellular differentiation. Consequently, a cellular type may change to another cellular type corresponding to the next attractive pool. It is not any cellular type that may transform into any other cellular type. The Boolean model implies indeed a branching pathway of differentiation. And this is exactly what we may observe in cellular biology: an ectoderm cell is likeable to change into a retinal cell but not into an intestinal cell! Regarding cellular differentiation as well, the Boolean model offers particularly enlightening interpretations.

Finally, one last implication relates to the fraction of the genes which determines the difference between the cells. According to the Boolean model, in ordered form, two networks in a state of equilibrium include a great number of "frozen" elements in constant state, presenting consequently the same behavior in varying state cycles, in diverse cellular types. As regards the mammals, 70 % of the genes are considered to be part of that common core which is active simultaneously in all cellular types of a mammalian organism. Furthermore, in one single organism, the part of the genome leading to cellular differentiation is itself very tiny. Let's take the case of a plant including about 20 000 genes: the difference between the cellular types is related to the activity of about 1 000 genes, which means 5 %. It is precisely the proportion expected in the Boolean model.

On the basis of these arguments and several analysis relating to other organization levels of the living, Kauffman defends the general idea that "selection is not the only source of order in ontogenesis"[5]. On the one hand, he tends to prove that order emerges spontaneously in certain conditions. That is the reason why one chapter of his work (1995) is entitled Order for free, *i.e.* order which is non profitable, which is not determined by the logic of selection. The concept of self-organization refers precisely to the spontaneous tendency towards the organization of complex systems "at the

5. Kauffman 1995, 111.

edge of chaos". But on the other hand, Kauffman does not want to break with the darwinian tradition. So he keeps considering selection as a major key in the evolution processes of the species. "Rather, the task must be to include Self-Organizing properties in a broadened framework, asking what the effects of selection and drift will be when operating on systems which have their own rich and robust self-ordered properties. For in such cases, it seems preemently likely that what we observe reflects the interactions of selection processes and the underlying properties of the systems acted upon"[6]. The main thesis of Kauffman defends a conception of evolution which links up self-organization and selection. "...if we ever are to attain a final theory in biology, we will surely, surely have to understand the commingling of self-organization and selection"[7].

I would precisely like to contribute to that conjunction. Therefore I will first refer to an epistemological analysis of the selectionist explanation.

3. SELECTION EXPLANATION AND SELF-ORGANIZATION

3.1 Natural Selection and Explanation

At first I will show that the selectionist explanation is not incompatible with other types of explanations. I will therefore refer to the work of E. Sober. In his epistemological analysis of natural selection, Sober introduces a double distinction, first between the equilibrium explanation and the causal explanation, then between the variational explanation and the developmental explanation[8]. These distinctions are very enlightening in that context.

Sober considers that, in the equilibrium explanations, the causes are not a necessary condition for the effects. The equilibrium explanations offer a statement of the possible causes, but none of them is necessary for its effect. Indeed, in the theory of evolution, the forces — mutation, migration, selection, drift — represent causes that lead the population through a sequence of gene frequences. For a given population, a causal explanation will account for the impact of a particular cause, brought to the fore by a given experimental procedure or by specific observations — which is actually rarely possible. At a more general level, the synthetic theory proposes more often so-called equilibrium explanations that show how the

6. Kauffman 1993, 23.
7. Kauffman 1995, 112.
8. Sober 1993, 139 and foll.

event that we want to explain may result from various ways. The causal explanation insists on the present trajectory of a given population, whereas the equilibrium explanation refers to the possible causes.

Furthermore, from a very similar viewpoint, Sober distinguishes between the variational explanation and the developmental explanation. To make it clear, he takes the example of a class of children who are all able to read. How can we explain that all these children are able to read at the same level? There are two explanatory strategies. In the developmental strategy, the story of each individual is described. These stories are then aggregated and incorporated into one single story, relating to the actual group. In the selective strategy, a selection criterion is defined at the admission to the class, which is sufficient to explain the composition of the population. These perspectives represent two different explanations. The developmental story tells us why each individual as such presents such given reading level, whereas the selective story shows why the class is made up of individuals at a certain level instead of others.

Sober insists on the originality of the darwinian viewpoint which sets at once the explanation of evolution at the level of the population. The population is not considered as made up of aggregates of individuals but as an irreducible reality which has its own logic of explanation. To a certain extent, in such a context, the frequency of a feature in a given population may be explained by natural selection, even though the possession of that feature by the individuals within the population may not. Sober goes further, "the idea of endogenous constraints on the changes a species may undergo is hardly unknown in evolutionary theory. My claim is that natural selection stand in opposition to this sort of mechanism"[9]. Sober considers his position as a kind of antireductionism. "Change in a set of objects is not accounted for in terms of changes in those objects"[10].

Sober's ideas are quite stimulating and help to precise the contribution of the theories on self-organization to natural selection. The concept of natural selection is specifically relevant as far as the population as a whole is concerned; the explanations given are related to that specific level. However, I think that such a viewpoint is not incompatible with the developmental explanation. On the contrary, it seems to me that the synthetic theory conveys some connection with the mechanisms of change within the objects. The population genetics refers to the concept of mutation related to the morganian genetics. It is a clear example of a mechanism of change within the unitary organism. More, if we refer to the works of

9. Sober 1993, 154.
10. Ibidem, 155.

J. Monod, we see that the pressure of mutation is subjected to a double constraint level: there is selection at the organism level and selection by the external environment. "Let's say that the 'initial conditions' of selection encountered by a new mutation include at the same time and indissoluble, the external environment and all the structures and performances of the telenomic system"[11]. In other words, Monod takes into account the capacities of change at the individual level: the concept of necessity refers to two levels of selection, one related to the organism and one to the environment.

However, Monod, who is a molecular biologist, does not study the nature of the organismic constraints. There is indeed no developmental explanation; his system is based on an analysis of the possibilities of change of an isolated organism. Elsewhere, I developed the idea that the synthetic theory of evolution has the same characteristics as the systemic approach that links directly the population level to the genetic level by short-circuiting the hierarchical sublevels such as the "cell" and the "organis"[12]. This is how I interpret the self-organization perspectives: the theory links together the genetic, cellular and organismic levels, and it is complementary to the explanation in terms of natural selection — which refers to the population as a whole. The variational and the developmental explanations are not mutually exclusive, but are complementary. If not connected with the dynamics of individual evolution, the population explanation remains incomplete and inadequate. This viewpoint is closely akin to the ideas of Kauffman who tries to link together the self-organization issue and natural selection.

If the compatibility of both explanations is claimed, we should nevertheless show why they are not mutually exclusive. Indeed, a strictly deterministic developmental explanation, in the style of Lamarck, may lead to calling into question the theory of natural selection. Even if we postulate the compatibility of both types of explanations, we still have to propose a cohesive view showing the relevance of each perspective.

Actually, since Darwin, the selectionist explanation has been integrated into other areas of biology. In particular, since the fifthties in immunology, and more recently in neurosciences, the concept of selection has moved to a key position in several very fundamental theories of these fields. From a prospective point of view, it seems relevant to me to refer to these ones in order to analyze the modes of combination of a selectionist principle and a developmental explanation.

---

11. Monod 1970, 141.
12. Feltz 1992.

3.2 Instruction and Selection: Selectionist and Developmental Explanation in Immunology and Neurosciences

The explanation by selection has been recently integrated into important theories in immunology and neurosciences. G. Edelman proposes an epistemological reading of such explanatory systems, based on the distinction between explanation by instruction and by selection, in conjunction with an approach using the "recognition system". I would like to refer to these analyses in order to tentatively define the articulation between the selectionist explanatory frame and a developmental explanation.

"Recognition" is meant as "the connection, adaptative and continuous, of the elements of a physical field, to new events arising in elements belonging to another physical field, more or less independant of the first one"[13]. This recognition can be performed in two ways. In an instructive process, the recognizing structure is built of information coming from the structure to be recognized. While the selective process sees as a prerequisite an existing variability on which operates a selection with a specific mechanism.

Recognition following a selective process is the basis of the Theory of Clonal Selection by MacFarlane Burnet in immunology. Indeed, for Burnet, the organism is able to synthesize a great variety of lymphocytes. Each lymphocyte has on its surface a site of antibody linking specific to one molecule, or one type of molecule. When an outside molecule, the antigen, appears, it connects itself to the antibody with the most complementary configuration to its own. From a certain level of complementarity antigen-antibody, the lymphocyte bearing the antibody starts dividing intensively: that is the amplification. The outcome is that the composition of the population of lymphocytes is drastically modified. The new population is made of a larger amount of lymphocytes able to be connected to this specific antigen. The organism is therefore better immunized against this antigen. We have then a situation of differential reproduction by clonal selection linked to the post-multiplication of a certain type of lymphocyte.

This theory definitely involves a selectionist logic. It presents a set of characteristics that will also be found in neurosciences and in theories of evolution. The explanation works at the population level, and involves dynamics of cellular populations. It implicates also a generator of diversity, being in this case a mechanism producing antibodies using lymphocytary

13. Edelman 1992, 100.

DNA. It includes also a type of heredity enabling the perpetuation of the induced modifications, being here the division of the lymphocytes into clones. Eventually it contains a selection device; the amplification of the clonal division leads to a varying importance of the clones in function of the encountered antigens.

This same scheme of a recognition system based on a selective process is present in the Neuronal Group Selection Theory by Gerald Edelman in neurosciences. Indeed, the set up of the general structures of the central nervous system of superior mammals, and in particular of the human species, occurs primarily through mechanisms similar to those controlling the development of other organs. Even if these mechanisms remain for a large portion hypothetical, all current models have as a prerequisite a strict genetical determinism, which fits precisely with an explanatory model by instruction.

The set up of the detailed structure of the central nervous system, the detail of synaptic linking at the level of the cerebral cortex, follows another logic which leads to a considerable variability of the final structures. Still regarding explanatory principles, let me just mention that, in the context of the theory of selection of neuronal groups, the learning or memorizing processes for instance are linked to a first phase of redundant and mutual connections between different cards associated to the organs of sense and to the sensorimotor behavior of the animal. This phase corresponds to the generator of diversity of the general scheme. Moreover a mechanism of selective stabilization leads to the strengthening of neuronal circuits, the neuronal groups of G. Edelman, which are activated, to the detriment of the non-active circuits. One can understand that such mechanisms enable the adjustment of the animal's behavior, gestures and postures, to the signals coming from different sensorial cards.

The behavior adaptation does not obey to the application of a preexisting program. It is the sensorimotor activity of the entire cartography which selects the neuronal groups with the adequate output or behavior. It is a clear reference to the explanatory scheme by selection which brings Edelman to define his theory as Neural Darwinism.

Indeed this explanatory framework is an opportunity to illustrate the darwinian theory of evolution, as interpreted by J. Monod, among others. Actually, we have a correspondence between two physical systems, the population and the eco-system. The mechanisms of recognition include a generator of diversity, the random mutations. The mechanisms of heredity ensure the continuity in time of induced mutations. The natural selection is precisely the mechanism of amplification of the frequence of genes bearing the favorable mutations, as a consequence of a double necessity: the functioning of the organism and the interaction with the other individuals and species in the ecosystem.

In the perspective of a contribution of the self-organization concept to the theories of the biological evolution, what do we learn from these preliminary analyses? First of all, the explanatory strength of the selectionist scheme in fields as varied and experimentally controlled as immunology and neurosciences noticeably reinforces the credibility of the selectionist scheme. We have there a dynamics of explanation which opens the way to highly motivating perspectives. The selectionist scheme enables to consider the diversity and the multiple modes of adaptation of the living to the environment where it has appeared.

On the other hand, these references to several fields prove that the selectionist scheme is compatible with a multiplicity of specific mechanisms. Whatever it relates to — generators of diversity, mechanisms of continuity in time or the selection itself —, mechanisms referring to completely different structures may be involved. In a way we find back the distinction in the empirical philosophy between general theory and model, as a specific interpretation of the theory. Such a distinction clarifies the possible articulation to the issue of self-organization: these various models confirm the position of Sober who sees the concept of selection as meaningful at the populational level. But, in each of these fields, the selective dynamics is bound to different mechanisms of developments, and these mechanisms are fully part of the explanatory scheme. On one hand, we definitely have an articulation between populational selectionist explanation and developmental explanation. And on the other hand, the developmental explanation relates to a mechanism producing diversity. This very condition seems to me as a basic element without which the selectionist scheme is not founded.

A same articulation between developmental and selectionist explanation is in my view applicable to the theories of evolution providing that the self-organizational perspectives take the form of developmental explanations which can be interpreted in terms of production of diversity.

And I think that it is well the case. Self-organization relating to networks "on the edge of chaos" as described by Kauffman can be interpreted as a way to incorporate the organismic necessity basically recalled by Monod, organismic necessity still to be exposed to the external necessity. The pressure for mutation named by Monod is part of an organismic pressure that self-organizational models enable to take into account. Moreover several models have already demonstrated their richness regarding the possible articulation between the genetic, organismic and populational levels. I think in particular to the work of G. Weisbuch and H. Atlan who have modelized the evolution of a population of Boolean networks with 6 genes exposed to a constant pressure for mutation and have highlighted situations of "ponctuated equilibrium"[14].

Far from being mutually exclusive, the variational and developmental explanations match each other for a true synthetic theory articulating organismic and populational constraints. The self-organizational models at the same time find here their whole relevance and can develop their explanatory strength.

4. SELF-ORGANIZATION AND EMERGENCE IN LIFE SCIENCES

4.1 The Concept of Emergence

The concepts of emergence and self-organization are often associated since self-organizational models tend to account for the emergence of an organization from a situation of disorder. This association presents us with some serious problems. This is the reason why I would like to explain what is at stake when we consider things from that point of view.

In a historical study on the concept of emergence, A. Stephan specifies different ways of approaching that concept. I would like to refer to that study, from a non-critical point of view at first — I will develop a more critical analysis with the self-organization issue[15].

The concept of emergence refers to the relations of an organized entity with its constituting elements. The "naïve" formulation of the concept of emergence is something like "the whole is more than the sum of the parts". The varying conceptions of emergence relate to the varying ways of considering that "more than" idea.

In a set of conceptions, "more than" means the impossibility to account for the characteristics of the whole according to the characteristics of the parts. Four historical conceptions noted by Stephan seem to match that view. In the 19th century, J. St. Mill presents the emergence in terms of non-additiviness. For this writer, "a law (of transition) is homopathic if the effects of a complex cause equal to the sum of the effects of a partial

14. For a more detailed analysis, see Weisbuch 1989 ; Atlan 1987 ; Feltz 1992.
15. Stephan 1992.

cause"[16], while a law is said heteropathic if not homopathic. A typical homopathic example is the first law of Newtonian dynamics : $F = m.a.$

M. Bunge tries to formalize this notion of "more than" in terms of novelty. A property P of a system is hereditary if P is the property of a component of this system. In the opposite case, P is qualified as emergent, collective, systemic. Bunge proposes a formalism which leaves open the explanation of emergent properties in function of the properties of the parts. This approach of novelty matches with an intuitive conception of emergence and the spontaneous reference to this concept in the area of self-organization.

A third view of emergence refers to the notion of non-predictibility. Since it is relevant in the context of self-organization, I would suggest to analyze in more details the formalism proposed by Stephan.

"Imagine the following situation;

(a) If the conditions $B_1, \ldots, B_k$ are met simultaneously, then the microstructure $(C_1, \ldots, C_n; O)$ will come into being out of some particles $C_1, \ldots, C_n$ according to a transition law T_L.

(b) For each system x having the microstructure $(C_1, \ldots, C_n;O)$: x instantiates property P according to a property law P_L"[17].

We have then a transition law which expresses the constitution of a microstructure based on its components in function of conditions B_i and a property law expressing a global property of this microstructure in function of its components C_i. The emergence defined in terms of unpredictability may consequently include two separate levels: either it is not possible to predict the transition to the constitution of the microstruture; or it is not possible to predict the property of this microstructure in function of its components.

Stephan goes through several alternatives of which certain can more particularly clarify the issue of self-organization. A first alternative refers to what the current litterature calls determinist chaos. In non-lineary systems, a small divergence at the level of the initial conditions can lead to significant differences in the evolution of the entire system. If one cannot reach an acceptable degree of accuracy in the knowledge of the initial conditions, it becomes impossible to predict the evolution of the system.

Another approach of unpredictability is linked to the concept of "ultimate law", through which Broad wants to express the fact that some events are not predictable as long as they did not occur. This does not mean that the devil of Laplace could not predict them; it practically means that their

16. Mill 1843, quoted by Stephan 1992, 28.
17. Stephan 1992, 33.

occurrence only can bring the elements of explanation of the occurrence itself.

A forth view of emergence, which is close to the preceding one, accounts for impredictibility in terms of non-deductibility. Originally the idea dates back to the views of Hempel and Nagel on the reductions between theories. The law concerning a structure's macro-level is said to be emergent if there is no theory on the micro-level which is able to account for the law in question. Nagel, on that basis, proposes two conditions: connectibility of the concepts and deductibility of the laws for a reduction of the macrolevel theory by the micro-level theory.

Such a view, based on the notion of non-deductibility, may lead to several interpretations. Hempel and Nagel for example consider that emergence refers to the state of knowledge. That no current theory may account for the properties of the macro-level according to those of the micro-level is not a reason to say that such a reduction is impossible. Simply, it has not been realized yet.

Stephan takes up the position of Broad who goes much further and claims that, for some phenomenon's, we may speak of absolute emergence in the sense that such a deduction would be impossible.

A last view of emergence which is proposed by Stephan goes further than the mere impossibility to explain the behavior of the whole according to the parts. Stephan refers to the notion of Downward Causation, relating to Sperry who tries to account for the impact of the mental on the body regarding human beings, avoiding any dualistic view. Stephan summaries the ideas of Sperry in four points. (1) The microstructure of a system completely determines the emerging macro-properties. Which implies that two physical systems presenting the same microstructures will have the same macro-properties. (2) Neither the macro-properties of the systems, nor the relational properties of the parts may be reduced to the non-relational properties of the parts. (3) The properties of the system are holistic properties different from the properties of the parts. (4) The properties of the system have a causal impact on the parts of the systems. Besides the micro-determination by the parts, you have to consider the macro-determination by the system as well.

Avoiding any dualistic representation, Sperry tries to account for the impact of mental activity on behavior. Stephan underlines the difficulty of Sperry's position, *i.e.* among others the divergence between point (1) and point (4). He underlines as well how difficult it would be to imagine a process accounting for such a downward causation.

4.2 Emergence and Self-Organization

This historical survey on the concept of emergence will allow us to clarify its relation to self-organization. It seems to me that the relation suggested between self-organization and emergence is related to the notion of novelty. The surprise, the wonder indeed, of S. Kauffman when discovering the "ordered" behavior of the networks where $K = 2$ is significant. The analysis of the Boolean networks makes it clear that a spontaneous organization, unpredictable before the experiment, may result from an apparent disorder. The concept of emergence refers historically to the various connotations of novelty, unpredictability, emergence of order from disorder, emergence of an organization from chaos.

Nevertheless, referring to the notion of unpredictability leads to a more cautious approach of the concept. Indeed, the networks of Boolean automatons are strictly determinist systems and tend to account for order according to the properties of the elements — *i.e.* the Boolean operators. The basic logic adopted by the searchers in the field of self-organization is thus strictly reductionist.

Even Broad's argument, which claims that the properties of the macro-structures can only be explained a posteriori, after the occurrence of the phenomenon, is not very convincing since it has always been the case in scientific explanation. Chemists did not deduce the properties of H_2O from a separate analysis of H_2 and O_2. On the contrary, the properties of each single element were not only thoroughly analyzed, but the properties of water were completely and independently analyzed too. The reductionist theory tends to account — not always successfully — for the properties of the compounds on basis of the properties of the parts, long after the phenomenon of composition has occurred. In other words, even in that matter, the research program on self-organization must deal with strictly reductionist criteria.

Going deeper into the issue of the predictability and deductibility criteria will allow us to make some progress. Linking up emergence and unpredictability may lead to some ambiguity. Indeed, even in the cases of networks where $K = N$, we are in a situation of unpredictability, but this situation is not in conflict with the presupposition of a determinist world. In that respect, speaking about emergence, like in the case of "determinist chaos", can be ambiguous since it considers as equal the uncertainties linked to the weather forecasts and the uncertainties linked to the structure of the central nervous system. But it seems to me that these two things have such different characteristics that, even if some similarities exist, a concept of emergence takes into account these differences.

The notion of deductibility is in my opinion richer since it incorporates the multiplicity of hierarchical levels characterizing the biological events. And in philosophy of biology, the emergence is analyzed as the inverse of the reduction. The issue of reduction between theories is the subject of an abundant literature in which it appears that it is very difficult to narrow a theory of macro-level down to a theory of micro-level. And, paradoxically, it is the condition of connectability between concepts which is right away the most difficult to implement. Another approach reducing biologics down to physico-chemicals tends to restrict the concept of biological function. There too, serious problems arise because contemporary biologists describe the function in connection with natural selection, and it is not clear to which physico-chemical law natural selection could be reduced to. These conclusions lead E. Mayr, for instance, to distinguish between constitutive and explanatory reductionism. At a distance from vitalism, Mayr postulates that living matter has the same properties as inert matter, which is constitutive reductionism. But the difficulties to establish the deductibility between the theories of several hierarchical levels of the living bring Mayr to define an explanatory non-reductionism. The various hierarchical levels characterizing the living can be the subject of theories specific by their concepts and methods. These theories are compatible with the lower levels, but non-reducible to the theories of these lower levels. In this context, the thesis of the unity of science breaks up with the aim of a unitary theory explaining the entire nature and takes the form of a set of theories being compatible and articulated with each other, but the theories of the micro-level would not reduce the theories of the macro-level.

In this context, the concept of emergence has a very obvious epistemological connotation. It can be distinguished from the intuitive perception of emergence associated to the concept of novelty. It stands aloof from the concept of self-organization, in the sense that self-organization tends to modelize the appearance of organization in a strictly determinist context, in function of the elements. The epistemological concept of emergence would tend consequently to set the self-organizational approach into reductionist dynamics.

Eventually arises the question of downward causation. A major objection to this theory is related to the mechanisms of such a causality. On this particular point, the theory of selection of neuronal groups by G. Edelman is in my opinion very exciting. The mechanisms of selective stabilization lead exactly to the reinforcement of activated neuronal circuits, to the detriment of non-activated circuits which tend to degenerate. Edelman proposes there a mechanism by which the behavior can influence the structure itself from which it has been originated. We have definitely there an example of downward causation[18]. On the top of that, such a causality complies strictly with the position of Mayr regarding the constitutive reductionism, since, in Edelman's work, there is not a single word about dualism or vitalism. The approach of emergence expressed in terms of downward causation is then in my view more relevant in the case of neurosciences and more specific of the living, since it integrates the levels of hierarchization without ignoring constitutive reductionism.

Interestingly, the same concept of selection leads Mayr to postulate his explanatory non-reductionism and Edelman to build the theory he names neuronal darwinism. The theory of natural selection enables Mayr to distinguish himself from a strict physico-chemical reductionism. Through the theory of selection of neuronal groups, Edelman proposes a vision of the evolution of the central nervous system which eludes the strict genetical determinism. In the analysis of the theories of evolution, I have supported the idea that self-organization and selection are two complementary approaches which refer to developmental and variational explanations, by insisting on the necessity to produce a variety on which a selection can be made.

In the context of emergence, self-organization and selection are not only complementary, they cannot be disconnected since self-organization without selection does not lead to emergence. It is only when self-organization is producing variety, that selection can play its specific emergentist role. Paradoxically, emergence, in the epistemological meaning, does not find its origin in self-organization but well in the undetermination that it leaves open for the sake of selection. Selection appears then more than ever as the implementation of emergence.

REFERENCES

Atlan H. (1986). Emergence of Classification Procedures in Automata Networks as a Model for Functional Self-Organization. *J. Theor. Biol.,* **120**, 371-380.

Artlan H. (1987). Self-Creation of Meaning. *Physica Scripta,* **36**, 563-576.

18. The notion of behavior is specific of the level of the entire organism or the individual; it does not imply anyhow a dualistic approach from the anthropological angle. The mechanism of selective stabilization permits then to explain the impact of a higher level of organization on the underlying structure itself.

Bechtel W. and Richardson R. (1993). *Discovering Complexity. Decomposition and Localization as Strategies in Scientific Research.* Princeton: Princeton University Press.

Beckerman A., Flohr H. and Kim J. (1992). *Emergence or Reduction? Essays on the Prospects of Non-reductive Physicalism.* Berlin, New York: Walter de Gruyter.

Bunge M. (1967). *Scientific Research.* Berlin, London: Springer Verlag.

Burnet F.M. (1959). *The Clonal Selection Theory of Acquired Immunity.* Nashville: Vanderbilt Univ. Press.

Canguilhem G. (1970). Le tout et la partie dans l'histoire de la pensée biologique. In Canguilhem. *Études d'histoire et de philosophie des sciences.* Paris: Vrin, 319-333.

Changeux J.P. (1983). *L'homme neuronal.* Paris: Fayard.

Delsol M. (1991). *L'évolution biologique en vingt propositions. Essai d'analyse épistémologique de la théorie synthétique de l'évolution.* Paris: Vrin.

Edelman G. (1987). *Neural Darwinism. The Theory of Neuronal Group Selection.* New York: Basic Books.

Edelman G. (1988). *Topobiology.* New York: Basic Books.

Edelman G. (1990). *The Remembred Present: A Biological Theory of Consciousness.* New York: Basic Books.

Edelman G. (1992). *Biologie de la conscience.* Paris: Odile Jacob.

Feltz B. (1991). *Croisées biologiques. Systémique et analytique. Écologie et biologie moléculaire en dialogue.* Namurs: Erasme.

Feltz B. (1992). Auto-organisation, développement et théories de l'évolution. *Uroboros,* II, 105-130.

Feltz B. (1994). Neurosciences et réductionnisme. In Feltz B. et Lambert D., (eds). *Entre le corps et l'esprit. Approche interdisciplinaire du Mind-Body Problem.* Liège: Mardaga, 181-215.

Feltz B. (1995). Le réductionnisme en biologie. Approches historique et épistémologique. In Feltz B. (ed.). Le réductionnisme dans les sciences de la vie. *Revue philosophique de Louvain,* **93**, 9-32.

Feltz B. (1997). Temps et nouveauté dans les sciences de la vie. In Florival and Greisch (eds). *Création et événement.* Louvain: Peeters.

Feltz B. (sous presse). Pertinence et limites de l'explication sélectionniste. In Exbrayat J.M. et Flatin J. (eds). *Quelques problèmes de l'évolution biologique et leur philosophie.* Paris: Vrin.

Kauffman S.A. (1991). Antichaos and Adaptation. *Scientific American,* August 1991, 78-84.

Kauffman S.A. (1993). *The Origins of Order. Self-Organization and Selection in Evolution.* Oxford: Oxford University Press.

Kauffman S.A. (1995*). At Home in the Universe. The Search for Laws of Complexity.* London, New York: Penguin Books.

Kim J. (1992). "Downward Causation". In Emergentism and Non-reductive Physicalism. In A. Beckerman, H. Flohr and J. Kim (eds*), Emergence and Reduction?* Berlin, New York: Walter de Gruyter, 119-139.

Mayr E., (1989). *Histoire de la biologie. Diversité, évolution et hérédité.* Paris: Fayard.

Monod J. (1970). *Le hasard et la nécessité. Essai sur la philosophie naturelle de la biologie moderne.* Paris: Seuil.

Nagel E. (1974). *The Structure of Science. Problems in the Logic of Scientific Explanation.* London: Routledge and Kegan.

Rosenberg A. (1985). *The Structure of Biological Science.* Cambridge. Cambridge University Press.

Sarkar S. (1991). Reductionism and Functional Explanation in Molecular Biology. *Uroboros,* I, 67-74.

Sober E. (1993). *The Nature of Selection. Evolutionary Theory in Philosophical Focus.* Chicago: Chicago University Press.

Stephan A. (1992). Emergence. A Systematic View on its Historical Facets. In A. Beckerman, H. Flohr and J. Kim (eds), *Emergence or Reduction?* Berlin, New York: Walter de Gruyter, 25-48.

Weisbuch G. (1989). *Dynamique des systèmes complexes. Une introduction aux réseaux d'automates.* Paris: InterEdition.

Wimsatt W.C. (1979). Reduction and Reductionism. In P.D. Asquith and H.E. Kyburg (eds). *Current Research in Philosophy of Science. PSA,* 352-377.

GPSR Compliance

The European Union's (EU) General Product Safety Regulation (GPSR) is a set of rules that requires consumer products to be safe and our obligations to ensure this.

If you have any concerns about our products, you can contact us on ProductSafety@springernature.com

In case Publisher is established outside the EU, the EU authorized representative is:

Springer Nature Customer Service Center GmbH
Europaplatz 3
69115 Heidelberg, Germany

Batch number: 10762627

Printed by Printforce, the Netherlands